Optimizing Current Strategies and Applications in Industrial Engineering

Prasanta Sahoo
Jadavpur University, India

A volume in the Advances in Civil and Industrial Engineering (ACIE) Book Series

IGI Global
DISSEMINATOR OF KNOWLEDGE

Published in the United States of America by
IGI Global
Engineering Science Reference (an imprint of IGI Global)
701 E. Chocolate Avenue
Hershey PA, USA 17033
Tel: 717-533-8845
Fax: 717-533-8661
E-mail: cust@igi-global.com
Web site: http://www.igi-global.com

Library of Congress Cataloging-in-Publication Data

Names: Sahoo, Prasanta, editor.
Title: Optimizing current strategies and applications in industrial engineering / Prasanta Sahoo, editor.
Description: Hershey, PA : Engineering Science Reference, [2019] | Includes bibliographical references.
Identifiers: LCCN 2018049694| ISBN 9781522582236 (h/c) | ISBN 9781522582243 (eISBN)
Subjects: LCSH: Industrial engineering--Research.
Classification: LCC T56.24 .O68 2019 | DDC 620.0068/5--dc23 LC record available at https://lccn.loc.gov/2018049694

This book is published in the IGI Global book series Advances in Civil and Industrial Engineering (ACIE) (ISSN: 2326-6139; eISSN: 2326-6155)

British Cataloguing in Publication Data
A Cataloguing in Publication record for this book is available from the British Library.

The views expressed in this book are those of the authors, but not necessarily of the publisher.

For electronic access to this publication, please contact: eresources@igi-global.com.

Advances in Civil and Industrial Engineering (ACIE) Book Series

Ioan Constantin Dima
University Valahia of Târgovişte, Romania

ISSN:2326-6139
EISSN:2326-6155

Mission

Private and public sector infrastructures begin to age, or require change in the face of developing technologies, the fields of civil and industrial engineering have become increasingly important as a method to mitigate and manage these changes. As governments and the public at large begin to grapple with climate change and growing populations, civil engineering has become more interdisciplinary and the need for publications that discuss the rapid changes and advancements in the field have become more in-demand. Additionally, private corporations and companies are facing similar changes and challenges, with the pressure for new and innovative methods being placed on those involved in industrial engineering.

The **Advances in Civil and Industrial Engineering (ACIE) Book Series** aims to present research and methodology that will provide solutions and discussions to meet such needs. The latest methodologies, applications, tools, and analysis will be published through the books included in **ACIE** in order to keep the available research in civil and industrial engineering as current and timely as possible.

Coverage

- Productivity
- Urban Engineering
- Structural Engineering
- Engineering Economics
- Coastal Engineering
- Quality Engineering
- Earthquake engineering
- Ergonomics
- Materials Management
- Transportation Engineering

The Advances in Civil and Industrial Engineering (ACIE) Book Series (ISSN 2326-6139) is published by IGI Global, 701 E. Chocolate Avenue, Hershey, PA 17033-1240, USA, www.igi-global.com. This series is composed of titles available for purchase individually; each title is edited to be contextually exclusive from any other title within the series. For pricing and ordering information please visit http://www.igi-global.com/book-series/advances-civil-industrial-engineering/73673. Postmaster: Send all address changes to above address.

Titles in this Series

For a list of additional titles in this series, please visit: www.igi-global.com/book-series

Optimization of Design for Better Structural Capacity
Mourad Belgasmia (Setif 1 University, Algeria)
Engineering Science Reference • copyright 2019 • 283pp • H/C (ISBN: 9781522570592) • US $215.00 (our price)

Reusable and Sustainable Building Materials in Modern Architecture
Gülşah Koç (Yildiz Technical University, Turkey) and Bryan Christiansen (Global Research Society, LLC, USA)
Engineering Science Reference • copyright 2019 • 302pp • H/C (ISBN: 9781522569954) • US $195.00 (our price)

Measuring Maturity in Complex Engineering Projects
João Carlos Araújo da Silva Neto (Magnesita SA, Brazil) Ítalo Coutinho (Saletto Engenharia de Serviços, Brazil) Gustavo Teixeira (PM BASIS, Brazil) and Alexandro Avila de Moura (Paranapanema SA, Brazil)
Engineering Science Reference • copyright 2019 • 277pp • H/C (ISBN: 9781522558644) • US $245.00 (our price)

Big Data Analytics for Smart and Connected Cities
Nilanjan Dey (Techno India College of Technology, India) and Sharvari Tamane (Jawaharlal Nehru Engineering College, India)
Engineering Science Reference • copyright 2019 • 348pp • H/C (ISBN: 9781522562078) • US $225.00 (our price)

Contemporary Strategies and Approaches in 3-D Information Modeling
Bimal Kumar (Glasgow Caledonian University, UK)
Engineering Science Reference • copyright 2018 • 313pp • H/C (ISBN: 9781522556251) • US $205.00 (our price)

New Approaches, Methods, and Tools in Urban E-Planning
Carlos Nunes Silva (University of Lisbon, Portugal)
Engineering Science Reference • copyright 2018 • 407pp • H/C (ISBN: 9781522559993) • US $205.00 (our price)

Designing Grid Cities for Optimized Urban Development and Planning
Guiseppe Carlone (Independent Researcher, Italy) Nicola Martinelli (Politecnico di Bari, Italy) and Francesco Rotondo (Politecnico di Bari, Italy)
Engineering Science Reference • copyright 2018 • 305pp • H/C (ISBN: 9781522536130) • US $210.00 (our price)

Design Solutions for nZEB Retrofit Buildings
Elzbieta Rynska (Warsaw University of Technology, Poland) Urszula Kozminska (Warsaw University of Technology, Poland) Kinga Zinowiec-Cieplik (Warsaw University of Technology, Poland) Joanna Rucinska (Warsaw University of Technology, Poland) and Barbara Szybinska-Matusiak (Norwegian University of Science and Technology, Norway)
Engineering Science Reference • copyright 2018 • 362pp • H/C (ISBN: 9781522541059) • US $225.00 (our price)

IGI Global
DISSEMINATOR OF KNOWLEDGE

701 East Chocolate Avenue, Hershey, PA 17033, USA
Tel: 717-533-8845 x100 • Fax: 717-533-8661
E-Mail: cust@igi-global.com • www.igi-global.com

Table of Contents

Detailed Table of Contents

The widespread deployment of inexpensive sensors, processors, embedded systems, etc., as well as the latest advances in data storage, analytics, cloud, etc. triggered significant changes in industrial engineering. This chapter aims to examine the adoption of the internet of things (IoT) in industry and the challenges for higher education. In this respect, the authors tried to explain the relationship between major concepts: IoT and industrial internet of things (IIoT). At the same time, they focused on the presentation of some IIoT enablers that could be viewed as building blocks for IIoT higher education curricula. In order to ensure the required qualifications and to develop the necessary skills for current employees, management staff, and students, academic engineering programs must undergo important changes. Some of these changes are also discussed in this chapter. Nevertheless, since there are so many uncertainties that lie between today and the future, higher education programs need to keep up with the latest technological tendencies.

This chapter discusses the application of linear programming (LP) techniques to find the optimal solution of the economic dispatch (ED) problem without considering transmission losses. The ED problem is concerned with optimizing the power generated by several generating units. The objective is to find the optimal power produced by each unit to supply the required load at minimum total cost. The generation cost associated with each unit is usually in the form of a quadratic or cubic function of the power produced. To apply LP, these nonlinear cost functions have to be linearized. The optimal solution is then determined by LP based on the approximate linear model. Piece-wise linearization methodology is adopted in this chapter. To evaluate the performance of the linearization method, a comprehensive set of benchmark test problems is used. LP solutions of linearized ED problems are compared with several other techniques from the literature. The LP technique with piece-wise linearization shows an overall competitive advantage in terms of total cost, solution time, and load satisfaction.

Advanced manufacturing technologies (AMT) acquisition by maquiladoras (foreign-owned manufacturing companies) is a tendency that allows these companies to maximize their commercial benefits. However, it remains unclear how the AMT implementation impacts on their performance. In addition, this research studies 383 responses to a questionnaire about the AMT implementation in the Mexican maquiladora industry and reports an analysis with four latent variables associating obtained benefits after the AMT implementation—human resources, flexibility, production process, and commercial benefits—where their relationships are evaluated through six hypotheses using a structural equation model (SEM). Finally, the outcomes demonstrated that AMT benefits for human resources have a direct effect on flexibility, production process, and commercial benefits. However, the direct effect from human resources benefits, knowledge, and experience on commercial benefits are acquired through indirect effects, using flexibility and production process as mediator variables.

Innovation and technology are determinants of sustainably the competitive advantage of the company in Hong Kong. In order to deliver the message from industry users to R&D technologist to improve the rate of adoption of R&D technology into the logistics and supply chain industry in Hong Kong, a quantitative survey to collect the industry user concern and expectation of R&D technology is necessary. This chapter aims to develop a novel project management model to have a good grasp of factors influencing R&D project commercialization through a multi-perspective methodology. Through the quantitative survey methodology to collect the data from the industry side, the comprehensive data analysis will be conducted. The main concerns are whether the developed technology applied to the company can create the value for the company to solve the problem and add value.

In order to attain the expectations of the customers and to win their hearts and minds, manufacturing firms are looking towards customer-focused design approach for developing their products. This chapter is devoted to present a hybrid methodology for developing products with customer focus. The methodology is developed by using house of quality (HOQ) with the aid of other techniques, namely,

Nowadays a substantial share of the production processes of the world's apparel business is taking place in developing countries. In the apparel business, supply chain coordination needs resource and information sharing between business partners. Semantic web service computing (SWSC) provides numerous opportunities and value-added service capabilities that global apparel business requires to exchange information between distributed business partners. The ability to dynamically discover and invoke a web service is an important aspect of semantic web service-based architectures. This chapter describes the main features of an ontology-based web service framework, known as CSIA (collaborative service integration architecture) for integrating distributed business information systems in a global supply chain. The CSIA framework uses a hybrid knowledge-based system, which consists of structural case-based reasoning (S-CBR), rule-based reasoning (RBR), and an ontology concept similarity assessment algorithm.

Nowadays along with the rapid development of industrialization across the globe, the environmental and ecological impacts of products have become a serious issue. Taking into account purely the economic impacts of industrial decisions, and excluding their ecological impacts, make the human beings and animals more at risk to many threats such as global warming, ozone layer depletion, toxic environments, and natural resources depletion. To minimize the environmental effect, implementation of green supply chain management (GSCM) is much more essential for industries in the environmental and social point of view. The purpose of this chapter is to analyze barriers to an implementation of green supply chain management in a stone crushing plant of Southern India by using modified simple additive weighting (SAW) to rank approaches. Further, this study will help the small-scale industries to understand the factors affecting implementation of GSCM in their organizations.

Controlling inventories and achieving effective management of them can significantly improve the profits to any company. This chapter presents a case of study of an automotive supply company that has operational problems due to over inventories. An ABC study and the economic order quantity model (EOQ) for 942 raw materials are presented promptly in order to increase the rotation efficiently, the yield of their inventories, as well as to minimize the logistics costs and maximize medium and long-term operations.

Preface

Optimizing Current Strategies and Applications in Industrial Engineering provides a domain for professionals, academicians, researchers, policy makers, and practitioners working in the fields of industrial engineering, reliability engineering, TQM, management, and globalization to present their research findings, propose new methodologies, disseminate the latest findings, and to learn from each other's research directions. It focuses on applied research in the areas of industrial engineering with a bearing on inter and intra disciplinary research. This book contains full-length research manuscripts, insightful research, and practice notes, as well as case studies.

Optimizing Current Strategies and Applications in Industrial Engineering provides a forum for industrial engineering educators, researchers, and practitioners to advance the practice and understanding of applied and theoretical aspects of industrial engineering and related areas. The book contains empirical and theoretical research on the development, improvement, implementation, and evaluation of integrated systems in engineering. In addition, the book will especially be interested in those research studies that show a significant contribution to the area by way of intra and inter disciplinary approaches in industrial engineering. The target audience of this book is composed of professionals and researchers working in the field of industrial engineering including administrative sciences, management and education. Moreover, the book provides insights and support executives concerned with the management of expertise, knowledge, information and organizational development in different types of work communities and industries. The book contains fifteen chapters contributed by academic and industrial researchers from around the globe.

"New Perspectives on Industrial Engineering Education" discusses the adoption of the Internet of Things (IoT) in industry and the challenges for higher education. In this respect, the relationship between major concepts is discussed and at the same time, focus is placed on the presentation of some IoT enablers viewed as building blocks for higher education curricula. In order to ensure the required qualifications and to develop the necessary skills for current employees, management staff and students, academic engineering programs must undergo important changes. Some of these changes are also highlighted in this chapter.

"Linear Programming Based on Piece-Wise Linearization for Solving the Economic Load Dispatch Problem" discusses the application of linear programming (LP) techniques to find the optimal solution of the economic dispatch (ED) problem without considering transmission losses. The ED problem is concerned with optimizing the power generated by several generating units. The objective is to find the optimal power produced by each unit to supply the required load at minimum total cost. The LP technique with piece-wise linearization shows an overall competitive advantage in terms of total cost, solution time, and load satisfaction over several other techniques available in the literature.

"Role of Human Resources, Production Process, and Flexibility on Commercial Benefits From AMT Investments" considers how Advanced manufacturing technologies (AMT) acquisition and implementation impacts on the performance. The study evaluates huge responses to a questionnaire about the AMT implementation in the Mexican maquiladora industry, and reports an analysis with four latent variables associating obtained benefits after the AMT implementation: human resources, flexibility, production process, and commercial benefits, where it is their relationships are evaluated through six hypotheses using a structural equation model. The analysis demonstrates that AMT benefits for human resources have a direct effect on flexibility, production process, and commercial benefits.

"An Analysis of the Determinants of ITF R&D Projects Commercialization in Hong Kong's Logistics and Supply Chain Industry: Industry User Perspective" considers quantitative survey to collect the Industry User concern and expectation of R&D technology. This study illustrates a novel project management model to have a good grasp of factors influencing R&D project commercialization through a multi-perspective methodology.

"Kano-HOQ-GRA Hybrid Methodology for Customer-Driven Product Development" is devoted to present a hybrid methodology for developing products with customer focus. The methodology is developed by using House of Quality (HOQ) with the aid of other techniques namely Kano, Swing weighting method (SWM) and Grey relational analysis (GRA). HOQ serves a tool for mapping customers' perception and designers' conception during product development. The Kano helps to understand and categorize the customer need which is required to capture the voice of customers. The mapping of customer needs and design requirements in inter-relationship matrix of HOQ is carried by using SWM. An illustrative example is presented in this chapter to demonstrate the methodology.

"Role of R&D Practices for Effective Product Development Process in NPD" considers product quality and technological developments as performance attributes for development of comprehensive framework by structural equation modeling (SEM) approach. Primary data from 263 experts of Indian manufacturing industries has been collected for analysis purpose. This empirical research portrays the role of R&D practices along with its indirectly related success factors for effectively controlling PDP along with its sub-factors for developing high quality products with technological developments.

"Multi-Objective Territory Design for Sales Managers of a Direct Sales Company" presents a case study to organize the sales territories for a company in Mexico. The assignment problem is modeled as a mathematical program with two objective functions; minimize the maximum distance traveled by the manager, and minimizes the variation of the sales growth goals with respect to the national average. A heuristic and a metaheuristic based on simulated annealing were developed. The design of the heuristic generates good solutions for the distance objective. The metaheuristic produces better results than the heuristic, with a better balance between the objectives.

"Semantic Web Service for Global Apparel Business" describes the main features of an ontology-based web service framework, known as CSIA (Collaborative Service Integration Architecture) for integrating distributed business information systems in a global supply chain. The CSIA framework uses a hybrid knowledge-based system, which consists of Structural Case-Based Reasoning (S-CBR), Rule-Based Reasoning (RBR), and an ontology concept similarity assessment algorithm. The ability to dynamically discover and invoke a web service is an important aspect of semantic web service-based architectures.

"Study on Effect of Barriers in Green Supply Chain Management Using Modified SAW Technique: A Case Study" considers exposed analysis of barriers to an implementation of green supply chain management in Stone crushing plant of southern part of India by using modified simple additive weighting (SAW) to minimizing rank reversal approach. To minimize the environmental effect, implementation of green supply chain management (GSCM) is much more essential for industries in the environmental and social point of view.

"Improvement Proposal in Inventories at an Automotive Supply Company Settled in Mexico" presents a case of study of an automotive supply company which is TEAR two that has operational problems due to over inventories. An ABC study and the economic order quantity model (EOQ) for 942 raw materials are presented promptly in order to increase the rotation efficiently, the yield of their inventories, as well as to minimize the logistics costs, and maximize medium and long-term operations.

"A MILP and Genetic Algorithm Approach for a Furniture Manufacturing Flow Shop Scheduling Problem" proposes a MILP and Genetic Algorithm optimization models for the sequencing of jobs in a medium-sized factory, dedicated to the manufacturing of home furniture, where different categories and types of articles are produced and whose routes and manufacturing processing times vary widely. Different scenarios are considered for the objective function based on minimizing makespan and tardiness. The results of the optimization show important reductions in the productive system's usage times, oscillating between 10% and 20% with respect to a random initial sequence in the production plan. Improvements were similar in both techniques, the main difference being the solution time of each one.

"GUI Toolbox for Approximation of Fractional Order Parameters With Application to Control of pH Neutralization Process" proposes a MATLAB based GUI for the approximation of fractional-order operators. The toolbox is made up of four widely used approximation techniques namely Oustaloup, refined Oustaloup, Matsuda and curve fitting. The toolbox also allows numerical and stability analysis for evaluating the performance of approximated transfer function. To demonstrate the effectiveness of the developed GUI, a simulation study is conducted on fractional-order PID control of pH neutralization process. The results show that the toolbox can be effectively used to approximate and analyze the fractional-order systems.

"Lean Six Sigma Implementation Framework Using Resource-Based Theory Approach" presents a Resource Based theory (RBT) based Lean Six Sigma (LSS) framework. By reviewing the previous literature on RBT, LSS and Organizational strategy, this chapter builds an implementation framework that considers the firm resources, firm capability, competitive advantage, and LSS integration as the foundation of a firm's strategy. Application of RBT can add richness to LSS implementation within an organization, which will bring practical implications for forming the resource-based strategy in the organizations leading to competitive advantage.

"Bio-Inspired Meta-Heuristic Multi-Objective Optimization of EDM Process" deals with the optimization of EDM process parameters using two different bio-inspired optimization algorithms namely, Artificial Bee Colony algorithm and Whale optimization algorithm for single as well as multiple responses and compares the results.

"Hybrid Multi-Criteria Decision-Making Optimization Strategy for RP Material Selection: A Case study" presents a hybrid MCDM optimization strategy for optimal material selection of the RP process. The hybrid strategy consists of modified DEMATEL with TOPSIS where the extraction of criteria weights using modified DEMATEL strategy while ranking of RP alternatives considering the effect of beneficial and non-beneficial using TOPSIS strategy. A real-life case study on RP material selection is demonstrated to validate the proposed strategy.

The editor would like to thank all the authors for contributing their valuable research work as chapters for this book. The authors submitted the chapters and responded for the revisions on time which could make it possible to publish this book on time. The editor is thankful to all the reviewers who painstakingly reviewed all the submissions and the revisions, wherever needed. Finally, the editor is thankful to IGI Global, for allowing publication of the book, and also to the Book development editor and other team members at IGI Global for their motivation, encouragement, advice and support provided during organization of this edited volume that led to the publication of this book.

Prasanta Sahoo
Jadavpur University, India

Chapter 1
New Perspectives on Industrial Engineering Education

Cornel Turcu
University of Suceava, Romania

Cristina Elena Turcu
University of Suceava, Romania

ABSTRACT

The widespread deployment of inexpensive sensors, processors, embedded systems, etc., as well as the latest advances in data storage, analytics, cloud, etc. triggered significant changes in industrial engineering. This chapter aims to examine the adoption of the internet of things (IoT) in industry and the challenges for higher education. In this respect, the authors tried to explain the relationship between major concepts: IoT and industrial internet of things (IIoT). At the same time, they focused on the presentation of some IIoT enablers that could be viewed as building blocks for IIoT higher education curricula. In order to ensure the required qualifications and to develop the necessary skills for current employees, management staff, and students, academic engineering programs must undergo important changes. Some of these changes are also discussed in this chapter. Nevertheless, since there are so many uncertainties that lie between today and the future, higher education programs need to keep up with the latest technological tendencies.

INTRODUCTION

In recent years, the rapid evolution of information and communications technologies involved many changes in various fields, including industrial engineering. Thus, the widespread deployment of inexpensive processors, embedded systems, smart sensors, wireless sensor networks, but also the advances in data storage, analytics, cloud infrastructure, etc., enabled the rise of a new digital industrial wave, referred as the fourth industrial revolution (IR 4.0). And the biggest technological initiative for implementing this revolution is the Industrial Internet of Things (IIoT), considered at the heart of the 4th industrial revolution, according to various worldwide surveys. The adoption of IIoT could have a huge impact. Thus, Accenture estimates the Industrial Internet of Things (IIoT) could add $14.2 trillion to the

DOI: 10.4018/978-1-5225-8223-6.ch001

global economy by 2030 (Daugherty & Berthon, 2015). According to a McKinsey report (Manyika et al., 2015), this revolution is well under way, anticipating that by 2025, the percentage of factories adopting IIoT will reach 65%-90% in advanced economies and 50%-70% in developing economies. Several scientific papers and studies have focused their attention on this new paradigm, examining, among others, the multiple promised benefits, but also the demands and challenges encountered in developing it. In addition to the technological barriers, the uptake of the Internet of Things paradigm in industrial field is delayed by the so-called skills gap. Thus, one key to success would be the qualification and human resource development in the near future. Researchers estimate that, as former industrial revolutions, IR 4.0 will not only influence the industry itself, but also the labor market, by creating new jobs, and also causing the displacement of some existing ones. For example, the World Economic Forum, considering 15 important developed and emerging economies, anticipated the production of 2.1 million new jobs, offset by the elimination of 7.1 million jobs (WEF, 2016). According to an analysis from Deloitte (Giffi et al., 2015), over the next decade, there will be 3.5 million job openings in manufacturing, but only enough skilled labor to fill less than half of them. In order to reduce the skills gap, education is essential. Over the years, throughout the world, education in engineering has undergone important changes. But, in order to ensure the required qualifications and to develop the necessary skills for workforce, management and students as future workforce in the fourth industrial revolution, the education system must also evolve from Education 3.0 to Education 4.0. Also, in the future, companies will have to pay more attention to developing the skills of their employees, by re-training or further training on new technologies.

This chapter aims to examine the adoption of the Internet of Things in industry and the requirements for higher education. Currently, the pictures of the Internet of Things, Industrial Internet of Things and Industry 4.0 are still quite blurry. Although IoT, IIoT and Industry 4.0 are closely related concepts, they cannot be used interchangeably. In an attempt to understand these concepts, this chapter tries to clarify them. These concepts do not have precise and widely accepted definitions. The literature in the field proposes several definitions, some of them presented in this chapter. We also consider the related fields, such as big data, artificial intelligence, security technology, etc. Standardization plays an important role for the further development and spread of IIoT and we can assert that various standards regarding security communication, identification and others are used in IIoT. An IIoT platform is a future looking framework that offers various capabilities, essential to enabling the successful adoption of the Industrial Internet of Things. Some of the most important platforms are presented in this chapter. IIoT applications are being developed and/or deployed in various industries, including manufacturing, logistics, energy production, construction, mining, supply, transportation, etc., to name just a few. In fact, all enabling technologies and related fields could be viewed as building blocks for IIoT education, that have to be included in the higher education curricula. Traditional industrial engineering models are changing, as traditional industrial processes are being supplemented and optimized by the digital world. Higher education must adapt to the new ways in which people, machines, services and data can interact. Based on the trends so far, experts predict that in the context of IR 4.0, profound changes are needed in the major aspects of education, from content to delivery. To this extent, this chapter examines some problems and possible solutions. Thus, for example, in order to meet the changing requirements and to respond to the increased demand for a highly skilled workforce, in the future, new effective educational programmes will have to be developed and/or the existing academic programmes will have to be re-structured.

In this chapter, we use IIoT-HE as a shorthand to refer to a transformation journey that higher education institutions have to make in order to meet the needs of the IIoT.

BACKGROUND

The Industrial Internet of Things, considered as an industrial adaptation of the Internet of Things, is referred to under various names, such as Internet of Industrial Things (IoIT), "Industrial Internet" as GE terms it, "Internet of Everything" term proposed by Cisco, Rockwell Automation's "IoT Industrial Revolution", IBM's "Smarter Planet" or the European "Industry 4.0" or "Industrie 4.0" (original german term), respectively French "Industrie du Futur" (Industry of the Future).

Various studies estimate that IIoT can have a high impact in various domains and a high business potential, being viewed as "the biggest driver of productivity and growth over the next decade" that "will accelerate reinvention of sectors that account for nearly two-thirds of world production" (Daugherty & Berthon, 2015;). And innovation and digital competences are some of the key factors that make companies more competitive. Nevertheless, experts estimate that although the adoption of the IIoT might have massive economic benefits, these benefits are not guaranteed.

The widespread adoption of IIoT begs some key questions:

- What is Industrial Internet of Things? What is Industry 4.0? How does it compare with Industrial Internet of Things? What is Internet of Things? How does it compare with Industrial Internet of Things? What are the similarities and differences?
- Who are the key actors?
- Are companies prepared to implement the solutions offered by IIoT? What will be the adoption cycle of the Industrial Internet of Things from an end-user's standpoint? What are its implications for the industrial companies? What are the key challenges faced by industrial companies? What are their expectations and needs for the future?
- What are its implications for the industrial suppliers?
- Are workforces prepared for the Industrial Internet of Things adoption?
- What are the challenges and opportunities that will emerge in the next 5, 10, 20 years?
- Do governments put in place the right conditions to boost the application of IIoT across their economy, thus facilitating progress and gaining benefits?

The next sections try to answer these key questions along with raising others.

INDUSTRIAL INTERNET OF THINGS

IoT, IIoT and Industry 4.0

The Internet of Things, Industrial Internet of Things and Industry 4.0 are closely related concepts, but they cannot be used interchangeably, as sometimes happens. According to (Jasperneite, 2012, p. 27), "scientific research is always impeded if clear definitions are lacking, as any theoretical study requires a sound conceptual and terminological foundation. Companies also face difficulties when trying to develop ideas or take action, but are not sure what exactly for". Currently, there are no widely accepted definitions for the concepts of IoT, IIoT and Industry 4.0, despite the fact that the literature in the field proposes several, some of which presented below.

The Internet of Things is defined as:

- "Sensors and actuators connected by networks to computing systems. These systems can monitor or manage the health and actions of connected objects and machines. Connected sensors can also monitor the natural world, people, and animals" (Manyika et al., 2015).
- "A world-wide network of interconnected objects uniquely addressable, based on standard communication protocols" (Atzori et al, 2010)

We have to point out that the Internet of Things has not yet been consistently defined in academic literature (Atzori et al, 2010; Boos et al, 2013; Borgia, 2014).

The Industrial Internet of Things/Industry Internet of Things is defined as:

- "A network of physical objects, systems, platforms and applications that contain embedded technology to communicate and share intelligence with each other, the external environment and with people" (Daugherty & Berthon, 2015).
- "The use of internet of things technologies to enhance manufacturing and industrial processes" (Internetofthingsagenda, 2018)
- "Internet of things, machines, computers and people, enabling intelligent industrial operations using advanced data analytics for transformational business outcomes", by the (Industrial Internet Consortium, 2016). The Industrial Internet of Things or IIoT connect traditional OT - Operational Technology (such as, machinery, equipment, assets, monitoring and control systems) with people and traditional IT - Information Technology (for example, storage systems, computing technology, business applications, data analysis) based enterprise systems, forming larger end-to-end systems.

We have to introduce the Industrial Internet term, which is

- "A short-hand for the industrial applications of IoT, also known as the Industrial Internet of Things, or IIoT" (O'Halloran & Kvochko, 2015).
- "The unification of (industrial) machines and software", being viewed as "the foundation for system-wide optimization" (Bruner, 2013)

Industry 4.0/ Industrie 4.0

- "Represents a paradigm shift from "centralized" to "decentralized" production – made possible by technological advances which constitute a reversal of conventional production process logic. Simply put, this means that industrial production machinery no longer simply "processes" the product, but that the product communicates with the machinery to tell it exactly what to do. Industrie 4.0 connects embedded system production technologies and smart production processes to pave the way to a new technological age which will radically transform industry and production value chains and business models" (Sharma Asha-Maria, 2018).

But, the relations between these concepts are complex and still unclear.

The concept of Industry 4.0 refers to the current fourth generation of industry, focusing solely on the manufacturing industry scenario. In fact, Industry 4.0 is viewed as a subset of Industrial Internet of Things (Aazam, Zeadally, & Harras, 2018). And the IIoT is considered to be one of the drivers of Industry 4.0.

The literature related to IIoT or IoT mostly focuses on technology development, operation models, market trends, and case studies, and rarely on comparisons between Industrial Internet of Things and Internet of Things (non-Industrial). However, there are some points of view regarding the relationship between Industrial Internet of Things and Internet of Things. Some consider the IIoT as a subset of the more general IoT, while others believe that there are real distinctions between these two concepts. At the time this chapter is written, there is still no definitive answer.

We should emphasize the fact that some of the characteristics of the two aforementioned concepts are similar, as can be seen from the following:

1. Availability of connected devices

One of the primary characteristic of both IoT and IIoT is the availability of affordable and intelligent devices; their connectivity brings the true value of IoT and IIoT, ensuring monitoring, control, etc., although they are used for different purposes, as we can see below in the subsection dedicated to differences.

2. Architecture

The IIoT and IoT paradigms follow the [Collect | Store | Analyse | Share] architecture.

3. Technology involved

Another common characteristic of the IoT and IIoT paradigms is related to enabling technology. Both of them integrate, among others, embedded computing and communication technology. Sensor technology along with the collection, transmission, and processing of sensory data are used in both cases. And in order to make sense of and act on an unprecedented volume, velocity, variety, variability, veracity, and complexity of data, big data, advanced analytics and machine learning are used.

4. Challenges

IoT and IIoT face some common challenges, among which we can mention a few. Thus, there are various technical challenges, such as the multitude of available standards and technologies, managing and processing large amounts of data, interoperability assuring, scalability, etc. But the most important challenges are security and privacy protection, because issues related to cybersecurity and data privacy could alter the true potential of adopting these paradigms.

Also, the lack of understanding, the skill shift for users, lack of experience and expertise for developers, acceptance, capital, etc. are some other challenges that can be added to the list of challenges for both IoT and IIoT.

Although Industrial Internet of Things follows the same core definition of the Internet of Things, IIoT is different from the IoT, at least from the following perspectives:

1. Area of interest
    a. **IIoT:** Connects critical machines and sensors, usually with a high degree of complexity, in various industries and refers to industry-oriented applications, which tend to be "system centric" applications. The full potential of the Industrial IoT will lead to smart power grids,

smart healthcare, smart logistics, smart diagnostics, and numerous other "smart" paradigms (Thames & Schaefer, 2017).

b. **IoT:** Without the term "industrial", uses connected things (various devices) for consumer convenience and could be considered as Consumer IoT (CIoT), representing the class of consumer-oriented applications, which tend to be "human centric".

2. Infrastructure

A consumer IoT is essentially wireless, while the Industrial IoT has to deal with an installed base of countless industrial devices, either old or new, some of them linked by wires because of the potential interference induced by concrete walls and heavy iron pipes and machinery.

3. Things involved (connected)

IIoT uses a wide spectrum of devices, often sophisticated and complex, such as sensitive and precise sensors, actuators, industrial robots, motors, PLC controllers, assembly line components, 3D-printers, mining equipment, test equipment, healthcare devices and even automobiles, trains, planes, and many more (Thames & Schaefer, 2017).

IoT tends to involve a potentially higher volume of users and devices, but these devices are consumer-level, less expensive than in industry, such as consumer cell phone, various wearable devices (e.g., fitness tools, healthcare devices, etc.), home appliances, smart home sensors (e.g., thermometers) or smart home devices (smart thermostat, automatic pet feeders, etc.).

But, in particular, Industrial IoT devices can contain systems of IoT systems and IIoT's things could make IoT's things.

4. Communication capabilities

One of the most important characteristic of the (I)IoT paradigm is connectivity among things.

In the case of IIoT, that covers an extensive range of industries, there is a wide number of technologies used to connect machines and sensors and also, there are different standards that complicate the communication between various devices. For instance, in factories there is often a variety of automated processes, which require the system involved at the lower level or device flow to communicate with a higher-level, plant-wide control system. But, in many cases, industrial companies have inherited a variety of automated information and technology solutions, some of which being implemented over the years. These systems do not easily interoperate with each other.

In comparison, in IoT there is a small number of communication standards and technologies to choose from.

5. Access of the connected things

While the Internet of Things is providing Internet access to any 'thing', the Industrial Internet of Things restricts the 'things' to the scenario of industry (Aazam, Zeadally, & Harras, 2018).

6. Real-time requirement

Real-time requirement could be considered as the key differentiator. Time plays a key role in real-time systems, but the real-time feature has different meanings and measurement units in different contexts. Thus, real-time usually means a few seconds for today's Internet, respectively a sub-millisecond for the industrial field. IoT industrial solutions must ensure real-time acquisition, processing and analysis of (a large amount of) data to enable continuous monitoring, anomaly detection, warning and alarming, etc. The delay caused by the communication between the devices layer and an IoT platform and also the resulting traffic over the network could make the difference.

Usually, unlike in IIoT cases, the response times in IoT applications, do not have critical real time requirements, because they are dealing with less time-sensitive systems.

7. Dependency

This perspective considers the Industrial IoT as the infrastructure that must be built before IoT applications can be developed. In other words, to some extent, the IoT depends on the Industrial IoT.

Although, at this time, some viewed the Industrial Internet of Things as a part of the Internet of Things, the IIoT shouldn't be understood as a subset of the IoT.

Although the distinction between the IoT and the IIoT is still under debate among professionals, in the future, as the IIoT and the IoT will become real and useful to the world overall, maybe the lines between IIoT and IoT will be less visible and less significant.

The aim of the next section is to provide a brief presentation of the relevant IIoT enablers.

Enablers of the Industrial Internet of Things

In order for the Industrial Internet of Things to span all types of industries and to deliver its maximum economic impact, the development of IIoT enablers must be supported.

There are two categories of IIoT enablers: technical and non-technical. For starters, we are tacking the ones in the first category.

In order to implement efficient industrial systems, completely aligned to IIoT concepts, the development of all related enabling technologies (big data, cloud computing, machine learning, data analytics, security, etc.) and hardware systems (embedded systems, industrial robots, sensors, etc.) must be aimed at being integrated into IIoT systems.

Artificial Intelligence and Machine Learning

For things from Industrial Internet of Things it is essential to understand the context in which they operate in order to make autonomous decisions.

However, understanding the big data gathered in IIoT is currently a challenge, because the perception and interpretation of context may be fallible.

Big Data

Gathering data from an increased number of sources (such as things connected to the Industrial Internet of Things) is a serious challenge because IIoT success is dependent upon major key capabilities that need to be well understood and correlated. Indeed, the huge variety of tasks that can be accomplished come with specific costs. According to a survey undertaken by Accenture and GE (Accenture, 2014), 73% of world companies are already investing more than 20% of their overall technology budget on big data analytics, while more than 2 companies out of 10 are currently investing more than 30%.

Security Technologies

Legacy automation and control systems are not inherently secure. Security issues are viewed as one of the major obstacles to digital transformation. IIoT involves various hardware and software components working together, even under incomplete information. Also, the decision-making and control must be distributed amongst various parts. Consequently, in addition to hardware and software security, operational issues are also required to be considered for ensuring security. Industrial companies must invest in a holistic cybersecurity concept that addresses industrial cyber security best practices across people, process, and technology.

IIoT Platforms

As the IIoT has matured, "an IIoT platform market is emerging, along with a supporting ecosystem of technology and service providers. Today, IIoT platforms use next-gen technology to provide a platform-as-a-service (PaaS) that … enables flexible next-generation applications using both Cloud and Big Data analytics capabilities". According to (MFRF, 2018), the Global Industrial IoT Platform Market is expected to grow at USD ~650 million by the end of 2022 with ~23% of CAGR.

Although there is no one-size-fits-all platform, currently, there are many different options for selecting the IIoT platform. In order to choose an IIoT platform that best suits specific requirements, one ought to know the available platforms, as well as their performances and features.

Most IIoT platforms are characterized by the following:

- They are application enablement;
- They ensure connectivity between devices with different usage and technology characteristics;
- They enable monitoring, management, and control of connected devices;
- They offer support for IIoT data model, interoperability, collaboration, mobile, etc.;
- They have built-in industrial cyber security capabilities as part of the foundation platform;
- They allow advanced industrial analytics, such as, statistic, artificial intelligence and machine learning based models;
- They provide edge and cloud capabilities;
- Etc.

At this time, according to researchers, IIoT platforms are still in the early stages of maturity and none gained a dominant position in an industry sector. But, we should emphasize the fact that experts recommend the use of standards-based platforms in order to reduce costs and to enable innovation.

Table 1 from Appendix 1 presents a list of some IIoT platforms.

IIoT platforms have played and will play an essential role in supporting the development of IIoT applications, by decreasing the complexity of various phases of the application lifecycle, like, development, deployment, etc.

As IIoT becomes better defined and further developed, more powerful IIoT applications can and will be created and deployed, using or not an IIoT platform.

Standardization

As mentioned above, there is another category of IIoT enablers, namely, the non-technical ones. In fact, government action can be an important enabler. Thus, according to (Manyika, 2015), "governments can invest in research to address security risks, convene and fund multi-stakeholder centers to set standards and share information, model good security practices, and craft thoughtful rules to encourage security management and punish bad actors". Also, governments and other policy-making organizations, like industry groups, can promote and support the development of standards, best practices and processes of the Industrial Internet of Things, as an important step for IIoT to be widely supported and well-accepted.

Thus, for example, in the context of industrial standardization, we should emphasize the significant and important efforts carried out by the International Electrotechnical Commission (IEC), which "created many different Study Groups and Technical Committees on the subject and published a couple of white papers about IIoT and the smart factory with the aim of assessing potential global needs, benefits, concepts and pre-conditions for the factory of the future" (Sisinni, Saifullah, Han, Jennehag, & Gidlund, 2018). Also, the Industrial Internet Consortium (IIC) works with industry in order to propose standards for the Industrial Internet of Things. Although there is a vast number of emerging IIoT standards, IIoT standardization is still a work in progress.

Table 2 from Appendix 2 presents a short list with some examples of standards.

In fact, the IIoT enablers could be viewed as building blocks for IIoT education, that have to be included in a true 21st century IIoT-HE curricula.

But, the widespread adoption of IIoT is hampered by major challenges, some technical and others non-technical. Next section presents some of these challenges faced by the evolution of IIoT.

IIoT Challenges

The large-scale deployment of IIoT faces various challenges that need to be overcome. Some of these are technical challenges, related to the specific requirements of the industrial domain, such as requirements in interoperability, energy-efficient operations, real-time performance in dynamic environments, etc. Also, IIoT expansion is beset with several technical challenges related to the above presented IIoT enablers.

Besides technical ones, some other IIoT challenges are outlined in the following:

- The need to redefine industry practices;
- Redefining business models;
- defining new financial and governance models to share and using common data;
- The need for capital;
- A rapidly evolving employment landscape, etc.

IIoT has not been yet introduced across all the industrial processes, but IIoT applications in some areas illustrate the potential impact and opportunities of these new technologies.

IIoT Opportunities

Various reports anticipate a significant percentage of factories adopting IIoT, but the nature of these forecasts varies across different industries, countries and regions. The adoption of IIoT is expected to have a profound impact over a wide range of sectors. Experts estimate the economic value of the Industrial Internet of Things to reach 1.9 trillion dollars worldwide.

But, besides the estimated economic benefits, the switch to IIoT could offer various opportunities:

- **Business Opportunities**, for:
    - Enterprises across the industry spectrum –increase competitiveness, gain competitive advantage through new service enabled business models, more efficient operations and agile supply chain;
    - Individual business segments – warehousing and logistics, sales and services, etc. Experts estimate that IIoT will require a switch from the 'push into the market' of better products for customers to an individualized understanding of customers' needs through specialized, industry-specific solutions; shortly put, a 'pull from the customer'. Companies appreciate this switch as "opportunities for flexible customer integration and for boosting quality and efficiency" (Deloitte, 2015).
    - Technology suppliers. The market for IoT components and systems is projected to be above 30 percent a year by 2025 (Manyika, 2015). The supplier ecosystem of Industrie 4.0 solutions is expected to reach 420 Billion Euros in value by 2020 (FS, 2015).
    - Education providers - to engage in providing high qualifications to satisfy the increasing demand.
- **Research Opportunities**: the future development of the IIoT will require large scale multistakeholder investments for research and development activities:
    - Researchers to address current and anticipated IIoT challenges, for example, to enhance security, interoperability and reliability, to process and analyze IIoT data, to develop valuable knowledge and insights, etc.
    - Technology Providers: to adopt new technologies across different verticals.
- **Work Opportunities:** "IIoT also provides opportunities to enhance efficiency, safety, and working conditions for workers" (Sisinni, Saifullah, Han, Jennehag, & Gidlund, 2018).
- **Other Opportunities:** Are related to the new technologies used in IIoT systems, in order to merge digitization and automation.

In an attempt to emphasize some of the most important aspects of the IIoT, including the ones mentioned above, we consider Figure 1.

IIoT is a technology concept that is currently transforming and redefining virtually all markets and industries in fundamental ways. Also, many agreed that the development of IIoT will have significant and far-reaching effects on other areas, such as education.

Industrial Internet of Things and Higher Education

Industrial Internet of Things is no longer a future trend; this paradigm is wide and affects the entire economy. In fact, for many companies, it is now a key part of their business strategies. Currently, the

Figure 1. Some of the most important aspects of the IIoT

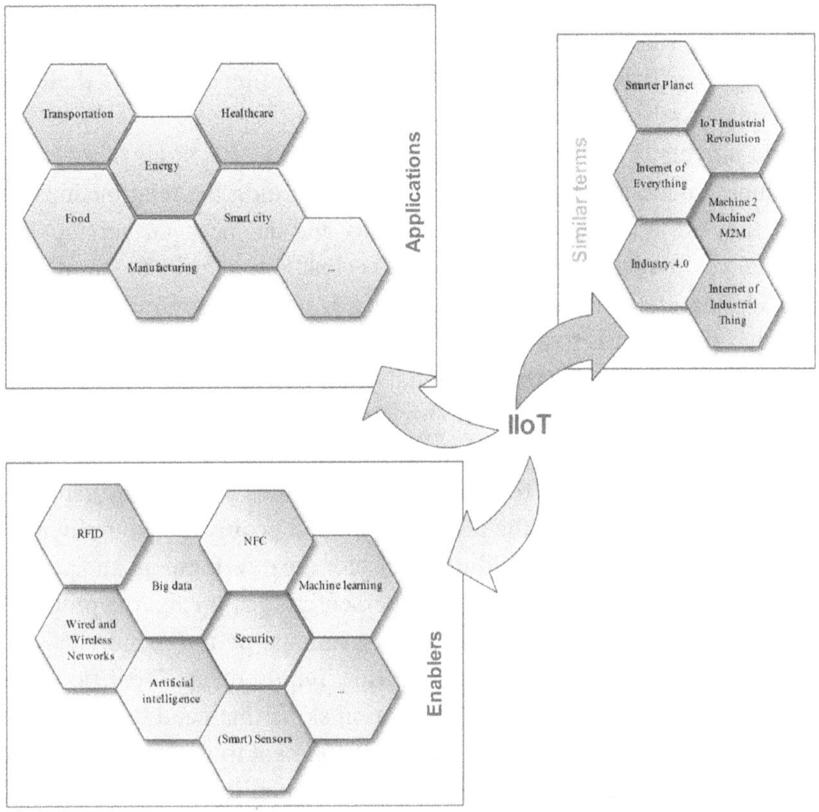

literature reports a significant number of different enabling solutions for the imminent IIoT era, already transferred into successful businesses. And, according to the various studies and reports, the IIoT applications will proliferate.

But, according to (Manyika, 2015) "across industries, the most fundamental cultural and organizational shift required for implementing the Internet of Things in corporations is to develop the skills and mindsets to embrace data-driven decision making".

Higher education offers solutions. However, the attempt of higher education to support the expansion of the IIoT raises various questions, some of which being outlined below.

Thus, there are certain questions to be tackled from the standpoint of the education provider regarding the changes that need to occur in the next 1, 5 or 10 years? What are the requirements in terms of type of equipment and lab space? What are some of the key end-user expectations in developed economies from an Industry 4.0 environment? "How will the increasing automation transform the future job market and skillsets required to succeed in the new economy" (O'Halloran & Kvochko, 2015)? What new digital workforce models will be created as a result of developing the Industrial Internet of Things? What changes will the Industrial Internet of Things (IIoT) bring to the field of higher education? What are the steps we need to take today, tomorrow, or four years from now to reach a set of goals available in five years from now? How will the equation between education providers and industry transform the Industrial Internet of Things age?

"In the context of technological developments, digitalisation, new requirements and expectations towards a changed employability and higher education as a mass phenomenon, higher education institutions will be likely to change their role and mode of operation entirely. More and more higher education institutions are opening up, in their business models, in their learning designs, in their access regulations and in the way they relate to the world of work" (Wipf, 2017).

Thus, the business-university collaboration is developing on a wide range of levels, such as, research and development, employment, technology transfer, etc. Regarding the relationships between employers and universities, they are often the focus of university and business partnerships (Pollard, E. et al., 2015). Employers' interest in this kind of relationships is clear, but in order to stimulate the involvement of universities, considering graduate employability as a key performance indicator for universities was a valid solution that has proven effective in practice.

Employers could even contribute to the curricular work of universities and/or co-design academic courses or laboratory activities. But the transition from higher education to employment proves a difficult task, whose success depends on several factors. Thus, making students 'work ready' is a challenge that could be alleviated not only through the efforts of universities and employers, but also through the efforts of students regarding training and the ability to respond positively to employment requirements.

The IIoT enablers could be viewed as building blocks for the development and provision of curricula and programmes for IIoT higher education. Also, it is essential to approach the problems not only from a technological viewpoint, but from a multidisciplinary vantage point (economics, social, integrating technology, etc.). Thus, for example, one of the industry requirements identified in various training needs assessment refers to strong social and collaboration skills that need to reinforce technical skills. "Universities need to use personalised employability development to equip students with attributes that meet this uncertain future – handling ambiguity, emotional intelligence, adaptability" (Edmonson & Ward, 2017). Of paramount importance will be providing lifelong learning opportunities to adults who will need to develop new skills and competencies throughout their careers (Broadband, 2017). According to (Broadband, 2017; Rainie, 2017) "the continuing education that is likely to be most relevant will provide adults with opportunities to upgrade skills inside their existing fields. It will also be increasingly important to support forms of education provision to (re)educate 'those whose skills are obsolete'. This need for updating existing skills, retraining and reskilling clearly implies the continued importance of responsive, flexible and equitable forms of …education".

Experts appreciate that it is critical that "broader and longer term changes to basic and lifelong education systems are complemented with specific, urgent and focused reskilling efforts in each industry" (WEF, 2016).

None of IIoT stakeholders, from businesses to governments, will provide solutions by themselves to close the gap, but together, through a concerted effort, academia, communities, and government can build in time a foundation to mitigate the skills gap.

FUTURE RESEARCH DIRECTIONS

An analysis undertaken on the Industrial Internet of Things education should also consider other aspects in addition to those presented in this chapter; nevertheless, due to space limitations, they have not been addressed here. Thus, in order to tackle the higher education in the context of Industrial Internet of Things, one can consider that the IIoT higher education could be essentially outlined by three dimensions:

1. **Scientific:** Universities are conducting research in the field, thus contributing to the expansion of knowledge horizons;
2. **Educational:** Through their teaching, universities actively support the dissemination of IIoT know-how amongst students and graduates, thus improving the stock of skilled human capital;
3. **Technological:** Universities support the transfer of their know-how to industry, centering on technology transfer.

A comprehensive study on how to develop an IIoT education system should take a holistic view of all IIoT-related educational perspectives.

If we consider only the educational dimension, we should emphasize the fact that, in order to build an enhanced environment for Industrial IoT education, the training needs in different domains have to be investigated more in deep. Also, academic and Higher Vocational Education and Training (HVET) programs, work-based education, virtual and remote laboratories for educational purposes, national and international educational policies and tools, etc. should be considered.

In the authors' opinion, the study should formulate answers to some key questions, such as:

- Is it possible to develop a strategy for the development of Industrial Internet of Things education?
- How to implement the outcomes from various IIoT-related studies and reports into the curricula offer?
- Will be possible to integrate international considerations into educational programmes in different countries?
- In the case of reskilling, it is possible to better use the accumulated experience of older employees and building a workforce with future proof skills?

Also, the study should compare different approaches in terms of both curricula and teaching methods. The obtained information, combined with the results of scientific studies, reports and policies about IIoT, will allow later to build-up the training process, to design training courses meeting the real needs of the labor market. Such a study of how to create an Industrial Internet of Things education should offer a crystal-clear plan centered on the good practices and recommendations to further build and consolidate effective education-job integration worldwide.

There are many uncertainties that lie between today and the future that higher education needs to explore. But, as (Johnson, 2017) emphasizes "these uncertainties are our potential for innovation and opportunity. Without these possibilities, the future would be fixed and closed, stamped with an expiration date and unable to be changed"… But, we "build the future — the way to prepare for the future is to invent it"… "People and organizations can all invent their futures" (Johnson, 2017). And one way is finding the right answers to the right questions and taking action.

CONCLUSION

This chapter aims to discuss some perspectives on industrial engineering education, exploring the Industrial Internet of Things and higher education in times of the fourth industrial revolution.

IIoT offers a chance to capitalize the opportunities presented by the fourth industrial revolution. However, to seize these opportunities, a better understanding of IIoT concepts is of paramount importance. In this respect, this chapter attempts to present the related concepts IoT, IIoT and Industry 4.0, which are often misused interchangeably; the focus would be on highlighting the similarities and the differences between the IoT and the IIoT. Next, the driving factors, challenges and opportunities of IIoT adoption are depicted.

In order to facilitate and support the effective extending IIoT, multistakeholder cooperation is essential. Thus, all stakeholders, from businesses to governments must intensify their efforts and escalate investments, but also must change their approach to education, skills and employment. Thus, for example, according to (WEF, 2016), "firms can no longer be passive consumers of ready-made human capital". Efforts aimed at filling the skills gaps require additional measures to be taken by higher education in order to anticipate and address in a timely manner the issues related to the changes occurring in the employment landscape and skills requirements. In order to design a true 21st century curricula and reform current education systems to better equip today's students to meet future skills requirements, education providers should collaborate and work closely with business and governments. This collaboration requires strong interdisciplinary partnerships between them for internships, training needs assessment, etc. All these are essential in addressing the complex challenges IIoT raises.

REFERENCES

Aazam, M., Zeadally, S., & Harras, K. A. (2018). Deploying Fog Computing in Industrial Internet of Things and Industry 4.0. *IEEE Transactions on Industrial Informatics*, 1–1. doi:10.1109/TII.2018.2855198

Accenture. (2014). *Industrial Internet Insights Report, For 2015, General Electric*. Accenture. Available at: https://www.accenture.com/ch-en/_acnmedia/Accenture/next-gen/reassembling-industry/pdf/Accenture-Industrial-Internet-Changing-Competitive-Landscape-Industries.pdf

Atzori, L., Iera, A., & Morabito, G. (2010). The internet of things: A survey. *Computer Networks, 54*(15), 2787–2805. doi:10.1016/j.comnet.2010.05.010

Boos, D., Guenter, H., Grote, G., & Kinder, K. (2013). Controllable accountabilities: The internet of things and its challenges for organisations. *Behaviour & Information Technology, 32*(5), 449–467. doi:10.1080/0144929X.2012.674157

Borgia, E. (2014). The Internet of Things vision: Key features, applications and open issues. *Computer Communications, 54*, 1–31. doi:10.1016/j.comcom.2014.09.008

Broadband. (2017). *Working Group on Education: Digital skills for life and work, Broadband Commission Working Group on Education, September 2017*. Available at: http://unesdoc.unesco.org/images/0025/002590/259013e.pdf

Bruner, J. (2013). *Industrial Internet*. O'Reilly Media, Inc.

Daugherty, P., & Berthon, B. (2015). *Winning with the Industrial Internet of Things: How to Accelerate the Journey to Productivity and Growth*. Dublín: Accenture.

Deloitte. (2015). *Industry 4.0. Challenges and solutions for the digital transformation and use of exponential technologies.* Deloitte AG. Available at: https://www2.deloitte.com/content/dam/Deloitte/ch/Documents/manufacturing/ch-en-manufacturing-industry-4-0-24102014.pdf

Edmondson, M., & Ward, A. (2017). *Tackling the disconnect between universities, Small businesses and graduates in cities and regions.* Gradcore. Available at: https://www.eurashe.eu/library/mission-phe/EURASHE_AC_LeHavre_170330-31_pres_EDMONDSON-WARD.pdf

FS. (2015). *Frost &Sullivan. Industry 4.0 Business Ecosystem to Change Dynamics in the Global Industrial Landscape.* FS.

Giffi, C., McNelly, J., Dollar, B., Carrick, G., Drew, M., & Gangula, B. (2015). *The skills gap in US manufacturing: 2015 and beyond.* Washington, DC: Deloitte and Manufacturing Institute. Available at https://www2.deloitte.com/content/dam/Deloitte/us/Documents/manufacturing/us-pip-the-manufacturing-institute-and-deloitte-skills-gap-in-manufacturing-study.pdf

IEC. (2017). *IEC International Standards. International Electrotechnical Commission.* Available at: https://webstore.iec.ch/home

IIC. (2017). *The Industrial Internet Reference Architecture Technical Paper.* Industrial Internet Consortium (IIC). Retrieved July 25, 2018, from http://www.iiconsortium.org/IIRA.htm

Internetofthingsagenda. (2018). *Industrial Internet of Things (IIoT) Definition.* IoT Agenda, Tech Target. Available at: http://internetofthingsagenda.techtarget.com/definition/Industrial-Internet-of-Things-IIoT

Jasperneite. (2012). Jasperneite, J., 2012: Alter Wein in neuen Schläuchen. *Computer & Automation, 12,* 24–28. Available at: http://www.ciit-owl.de/uploads/media/410- 10%20gh%20Jasperneite%20CA%20 2012-12_lowres1.pdf

Johnson, B. D. (2017). *How to invent the future, Trend, Analysis of the Facts, Numbers, and Trends Shaping the World.* Retrieved from http://trend.pewtrusts.org/-/media/post-launch-images/trend-magazine/summer-2017/trend_summer_2017.pdf

Kaa. (2018). Available at: https://www.kaaproject.org

Manyika, J., Chui, M., & Bisson, P. (2015). *The Internet of Things: Mapping the value beyond the hype.* San Francisco: McKinsey & Company. Available at https://www.mckinsey.de/sites/mck_files/files/unlocking_the_potential_of_the_internet_of_things_full_report.pdf

Manyika, J., Chui, M., Bisson, P., Woetzel, J., Dobbs, R., Bughin, J., & Aharon, D. (2015). Unlocking the Potential of the Internet of Things. *McKinsey Global Institute.* Available at: https://www.mckinsey.com/business-functions/digital-mckinsey/our-insights/the-internet-of-things-the-value-of-digitizing-the-physical-world

MRFR. (2018). Industrial Internet of Things Market Analysis 2018 To 2022. *Market Research Future.* Available to: https://www.marketresearchfuture.com/reports/industrial-iot-platform-market-2186

O'Halloran, D., & Kvochko, E. (2015). Industrial internet of things: unleashing the potential of connected products and services. In *World Economic Forum* (p. 40). Available at: http://www3.weforum.org/docs/WEFUSA_IndustrialInternet_Report2015.pdf

Pollard, E. (2015). *Understanding employers' graduate recruitment and selection practices. BIS Research Paper 231*. London, UK: Department for Business, Innovation and Skills. Available at https://derby.openrepository.com/derby/handle/10545/609154

Rainie, L. (2017). *Technological innovation? That's the easy part*. Trend Summer. Available at http://trend.pewtrusts.org/-/media/post-launch-images/trend-magazine/summer-2017/trend_summer_2017.pdf

Sharma Asha-Maria. (2018). *INDUSTRIE 4.0. Smart manufacturing for the future*. Germany Trade & Invest (GTAI). Retrieved May 3, 2018, from https://www.gtai.de/GTAI/Navigation/EN/Invest/Industries/Industrie-4-0/Industrie-4-0/industrie-4-0-what-is-it.html

Sisinni, E., Saifullah, A., Han, S., Jennehag, U., & Gidlund, M. (2018). Industrial Internet of Things: Challenges, Opportunities, and Directions. *IEEE Transactions on Industrial Informatics*, 1–1. doi:10.1109/TII.2018.2852491

Thames, L., & Schaefer, D. (2017). Industry 4.0: an overview of key benefits, technologies, and challenges. In Cybersecurity for Industry 4.0 (pp. 1–33). Springer.

WEF. (2016). The future of jobs: Employment, skills and workforce strategy for the fourth industrial revolution. *World Economic Forum*.

Wipf, A. (2017). *Professional Higher Education 4.0: A Change for Universities of Applied Sciences*. EURASHE. European Association of Institutions in Higher Education. Available at: https://www.eurashe.eu/library/mission-phe/EURASHE_AC_LeHavre_170330-31_portfolio.pdf

KEY TERMS AND DEFINITIONS

Big Data: The term is used for massive amounts of increasing data, structured or unstructured data, which can be processed and analyzed through various techniques/algorithms. As a concept, big data is defined around the five V's: volume (scale of data), velocity (analysis of streaming data), variety (different forms of data), veracity (uncertainty of data), and visibility (accessed from the disparate geographic location). But in the last years, another V is added: value.

Fast Data: This concept, which appeared soon after "big data," refers to the fast and efficient use of data to provide instant results and responses. This type of data is used when speed is important.

Industrial Internet of Things (IIoT): IIoT applies IoT concepts and technologies to the industry. The result will allow industrial things (devices, sensors, etc.) to communicate with other things, whether human beings or other devices.

Internet of Things (IoT): The concept depicts a world where different things, living and non-living entities, are connected to a single common network.

Real-Time Data: The term refers to the data acquired from the physical world and analyzed in a timely way.

Smart Data: This term refers to data, often obtained from big data and IoT that has value. Currently, there is an increasing focus on smart data instead of big data, big data being turned into smart data.

APPENDIX 1

Industrial Internet of Things platforms play an important role in supporting the development and implementation of IIoT applications, by reducing the complexity of various phases of the application lifecycle, like, development, deployment, etc. Some commercial and open source IIoT platforms are presented in Table 1.

Table 1. Some industrial internet of things platforms

Platform	Producer	Ref
AOS - Automation Operating System	ANT Industrial Software Systems	https://antsolutions.eu/aos-system-platform/
ARKTIC™	ARKADOS	https://www.solbrightgroup.com/platform/
AXON	Greenwave Systems	https://greenwavesystems.com/product/
AWS IoT	Amazon	https://aws.amazon.com/iot-1-click/
Ayla IoT Platform	Ayla	https://www.aylanetworks.com/
Azure IoT Suite	Microsoft	https://www.azureiotsolutions.com/Accelerators
Bosch IoT Suite 2.0	Bosch	https://www.bosch-si.com/iot-platform/bosch-iot-suite/homepage-bosch-iot-suite.html
Carriots	Carriots	https://www.carriots.com/
Cisco Jasper Control Center	Jasper	https://www.jasper.com/
Cloud der Dinge (Cloud of Things)	Deutsche Telekom	https://iot.telekom.com/plattformen/cloud-der-dinge/
Cloud of Things	Cloud of Things Ltd.	https://www.cloudofthings.com/
CloudPlugs	CloudPlugs	https://cloudplugs.com/
Connecthings IoT Platform	Connecthings	https://www.connecthings.com/product/
Cybus Connectware	Cybus	https://www.cybus.io/en/solutions-factory/
Cumulocity	Cumulocity	https://www.cumulocity.com/
Datonis	Altizon Systems	https://altizon.com/datonis-iiot-platform/
EcoStruxure	Schneider Electric	http://www.schneider-electric.com/
EpiSensor	EpiSensor	https://episensor.com/
GoFactory	GOFACTORY Inc.	http://go-factory.com/
IBM Watson Internet of Things Platform	IBM	http://www.ibm.com/internet-of-things/iot-solutions/watson-iot-platform/
ioTrust	Entrust Datacard	https://www.entrustdatacard.com/products/categories/iotrust
IzoT	Echelon Corporation	http://www.echelon.com/izot-platform
Kaa	CyberVision	https://www.kaaproject.org/
Kaa IoT Platform	KaaIoT Technologies	https://www.kaaproject.org
KEPServerEX	Kepware	https://www.kepware.com/products/kepserverex
Lumada	Hitachi Insight Group	https://www.hitachiinsightgroup.com/en-us/lumada-by-hitachi.html
M2M Intelligence Platform	M2Mi	http://www.m2mi.com/m2m-iot-platform
M2X	AT&T	http://www.moxa.com/IIoT/index.htm

continued on following page

Table 1. Continued

Platform	Producer	Ref
Meshify	Meshify	http://meshify.com/
Moxa	Moxa	http://www.moxa.com/IIoT/index.htm
OpenGate	Amplía Soluciones S.L	http://www.amplia-iiot.com/
Oracle IoT Cloud Service	Oracle	https://cloud.oracle.com/iot
Poolsure	Poolsure	http://poolsure.com/
Predix	General Electric (GE)	https://www.predix.io/
Salesforce IoT Cloud	Salesforce	http://www.salesforce.com/iot-cloud
SAP HANA Cloud Platform for IoT	SAP	http://go.sap.com/product/technology-platform/iot-platform-cloud.html
Telogis	Telogis	http://www.telogis.com/
Intel IoT Platform	Intel	http://www.intel.com/content/www/us/en/internet-of-things/iot-platform.html
Losant IoT Platform	Losant	https://www.losant.com/
thethings.iO	thethings.iO	https://thethings.io/
ThingWorx IoT Platform	PTC	http://www.thingworx.com/platforms
Verizon ThingSpace IoT Platform	Verizon	http://www.verizonenterprise.com/products/internet-of-things/
Worldsensing	Worldsensing	http://www.worldsensing.com/
Xively IoT Platform (by LogMeIn)	Xively	https://xively.com/platform/

APPENDIX 2

There is a vast number of emerging Industrial Internet of Things related standards, but IIoT standardization is still a work in progress. Some examples of standards applied in IIoT are presented in Table 2.

Table 2. Examples of standards applied in IIoT

Standard	is defined as	by	Reference
ISA100 Wireless	an open, universal IPv6 wireless network protocol	ISA100 Committee	https://isa100wci.org/
MTConnect	a standard that offers a semantic vocabulary for manufacturing equipment to provide structured, contextualized data with no proprietary format	MTConnect Institute	https://www.mtconnect.org
IEC62541	a solution to addresses connectivity issues	International Electrotechnical	https://webstore.iec.ch/publication/21987
OPC UA	"a protocol for industrial communication and has been standardized in IEC 62541. OPC UA interface provided a robust communication interface and a standardised information modelling capability that can be implemented on any hardware platform."		https://opcfoundation.org/about/opc-technologies/opc-ua/
IEC62657	an international standard that addresses the wireless communication networks in the context of industrial automation	International Electrotechnical Commission	https://webstore.iec.ch/publication/33125

Chapter 2
Linear Programming Based on Piece-Wise Linearization for Solving the Economic Load Dispatch Problem

Ahmad Al-Subhi
Yanbu Industrial College, Saudi Arabia

Hesham K. Alfares
https://orcid.org/0000-0003-4040-2787
King Fahd University of Petroleum and Minerals, Saudi Arabia

ABSTRACT

This chapter discusses the application of linear programming (LP) techniques to find the optimal solution of the economic dispatch (ED) problem without considering transmission losses. The ED problem is concerned with optimizing the power generated by several generating units. The objective is to find the optimal power produced by each unit to supply the required load at minimum total cost. The generation cost associated with each unit is usually in the form of a quadratic or cubic function of the power produced. To apply LP, these nonlinear cost functions have to be linearized. The optimal solution is then determined by LP based on the approximate linear model. Piece-wise linearization methodology is adopted in this chapter. To evaluate the performance of the linearization method, a comprehensive set of benchmark test problems is used. LP solutions of linearized ED problems are compared with several other techniques from the literature. The LP technique with piece-wise linearization shows an overall competitive advantage in terms of total cost, solution time, and load satisfaction.

DOI: 10.4018/978-1-5225-8223-6.ch002

INTRODUCTION

The economic dispatch (ED) problem is an essential short-term optimization problem in the power generation industry. Its aim is the determination of the optimal output power produced by several generators for meeting the required load at the lowest possible cost. Given several generation units, the objective of the ED is to determine the power output of each unit, in order to minimize the total cost while satisfying the power demand and other operational and technical constraints. ED is a crucial task for the planning and operation of electrical power generation plants. Hence, accurate and efficient estimation of the optimal generated power by each unit can significantly enhance the performance and reliability of the electrical power system. To achieve this purpose, various types of optimization methodologies have been presented in the literature. Ciornei and Kyriakides (2013) present a recent survey of ED literature, focusing on models and solution methods since 1991. They also provide a database of commonly used ED test problems, summarizing their features, the best results (costs) obtained, and the methods used to obtain them.

Recently, several new algorithms have been proposed to solve the ED problem. Moreover, some existing methods have been modified in order to meet the requirements of power system networks. The techniques differ in several aspects including time, complexity, ease of implementation, accuracy, and the software and hardware components involved. The techniques proposed in the literature to solve the ED problem can be categorized mainly into classical and novel techniques.

The classical techniques category includes Lambda-Iteration Algorithm (LIA), Gradient method, Newton's method, and Lagrange multiplier method. These techniques have been employed frequently to solve the ED problem. The novel techniques category includes new heuristic and artificial intelligence approaches that are becoming very popular in solving the ED problem. These techniques include Tabu Search (TS), Genetic Algorithms (GA), Simulated Annealing (SA), Particle Swarm Optimization (PSO), and Differential Evolution (DE). Other solution approaches include cost composite function, Quadratic Programming (QP), sequential approach with matrix framework, and Dynamic Programming (DP). In addition, hybrid methods, which are combinations of more than one method to solve ED problem, are also used.

LITERATURE REVIEW

Due to its practical and theoretical significance, the ED problem is an active area of continuing research. This is illustrated by a large number of ED papers that have been published in recent years. A recent literature search by the authors resulted in finding many works published on ED optimization using both classical and new methods. Some of the techniques used recently to solve ED problems include Mixed Integer Quadratic Programming (MIQP) (Absil et al., 2018), Gradient Method (Imen et al., 2013), Fast Lambda Iteration (FLA) (Zhan et al., 2014), Enhanced Lambda Iteration (ELI) (Singhal et al., 2014), and Artificial Neural Networks (ANN) (Momoh & Reddy, 2014).

Recent heuristic techniques used to solve the ED problem include PSO (Lin et al., 2015). PSO is one of the most commonly used heuristic algorithms used for ED applications. Abbas et al. (2017) perform a thorough literature survey of research papers that employed PSO to solve the ED problem. Other recently used heuristic techniques include Harmony Search (HS) algorithm (Al-Betar et al., 2018), Immune Log-Normal Evolutionary Programming (ILNEP) (Mansor & Musirin, 2017), Chaotic Bat Algorithm

(CBA) (Adarsh et al., 2016), Ant Lion Optimization (ALO) (Nischal & Mehta, 2015), Gravitational Search Algorithm (GSA) (Hota & Sahu, 2015), Flower Pollination Algorithm (FPA) (Vijayaraj & Santhi, 2016), GA (Oluwadare et al., 2016), stochastic Whale Optimization Algorithm (Mohamed et al., 2018) and Grey Wolf Optimization (GWO) (Tung & Chakravorty, 2015).

Hybrid techniques are also employed recently to solve the ED problem efficiently. For example, Rasoul et al. (2017) use a gradient search method with improved Jaya algorithm for an ED application with attention to environmental issues. Another example is given by Suppakarn (2017), who integrate Bee Algorithm (BA) with TS, obtaining better solutions than the standard BA heuristic. In addition, Wu et al. (2017) use hybrid Quadratic Programming (QP) and Compact Formulation Method (CFM) for ED with line losses and prohibited operating zones. Moreover, Ramadan & Logenthiran (2017) combine GA with Priority List (PL) algorithm for solving ED problems. A hybrid approach combining mixed-integer linear programming (MILP) and the interior point method (IPM) is proposed by Pan et al. (2018).

The ED problem has numerous applications in the area of Distributed Generation (DG) and microgrids. The objective of ED in the DG context is to optimize the power generated from various energy sources rather than only one conventional source. Examples of such sources include solar, wind, waterfall, fuel cell, and battery storage. Integration of these different energy sources, including their respective constraints, and trying to optimize the power generated from each source is a challenging task. For example, Cherukuri and Cortes (2017), Zhang et al. (2017), Tang et al. (2018), Meng and Wang (2017), Awan et al. (2017), Hasan and El-Hawari (2016), Shi et al. (2017) and Guo et al. (2017) show various approaches in microgrids environments to optimize the power generated from each DG. More recent DG applications are provided by Liang et al. (2018) who propose a hybrid Bat Algorithm for ED with random wind power. In addition, Ahmad et al. (2018), Nivedha et al. (2018), Bhattacharjee and Khan (2018), Ross et al. (2018), Chen and Yang (2018), Zhang et al. (2018) and Xia et al. (2018) also analyze the ED problem in the presence of multiple generation sources.

LP-based approaches are extensively used in formulating and solving ED problems. Hoke et al. (2013) utilize a quick and effective LP technique to solve an ED problem where there are several interconnected micro-grids. Those inter-dependent micro-grids contain different varieties of DGs such as conventional generators, wind turbines and energy storage. Computational experiments demonstrate the speed and accuracy of the LP technique. Jabr et al. (2000) develop a simplified homogeneous and self-dual (SHSD) LP interior-point algorithm and they apply it to the security-constrained economic dispatch (SCED) problem. The algorithm can detect infeasibility, which is common in SCED applications due to overloading usually occurring after an incident. Chamba and Ano (2013) propose an innovative LP-based hybrid methodology for solving the economic dispatch of both energy and reserve. Assuming the reserve requirement is determined in the optimization process, the method is used to calculate the optimal power flow and the reserve assigned to each unit. The hybrid methodology integrates LP with two alternative meta-heuristic algorithms: Evolutionary Particle Swarm Optimization (EPSO), and Mean-Variance Mapping Optimization (MVMO). Elsaiah et al. (2014) describe an LP-based method for solving the economic power dispatch problem in the presence of renewable energy sources. In their method, LP is utilized due to its flexibility, reliability and speed. Lazzerini and Pistolesi (2015) present an LP-driven multi-criteria decision-making approach for multi-objective ED in smart grids.

Generally, the complexity level of the ED optimization problem depends on the objective function as well as the system constraints. Applicable constraints include, for example, whether transmission line losses are considered or not. Other constraints that are typically added to the ED problem include valve point effect, prohibited zones, ramp rate limits, and restrictions on carbon emissions and environmental impact. With more constraints, the ED problem becomes more complicated and it needs more advanced solution techniques. For example, recent research works that analyze the ED problem with attention to carbon emissions are presented by Kuo et al. (2017), Bhongade and Agarwal (2016), Khan et al. (2016), and Goudarzi et al. (2017). Moreover, the problem of ED with constraints on valve point effect is studied by Wu et al. (2017), Bhui and Senroy (2016), Naveen et al. (2016) and Yang et al. (2016).

In this chapter, quadratic and cubic cost functions are approximated by linear functions, using either a single straight line or multiple-piece-wise linear segments. First, the LP model of the linearized cost function is formulated, and then large set of benchmark ED test problems is solved using LP. The chapter presents an original, compact formulation of the LP model for the linearized economic load dispatch problem, ignoring transmission losses and assuming equal widths of the piece-wise linear segments. Compared to previous works, this chapter presents a more comprehensive computational comparison of LP with other techniques for solving the economic load dispatch problem. This comparison is based on a larger number of solution algorithms and a bigger set of benchmark test problems. The chapter also discusses the advantage of incorporating several linear piece-wise segments instead of only a single straight-line cost function. Moreover, the chapter presents a graphical view of the LP solution convergence process as a function of the width of the piece-wise linear segments. All the computational analysis in this chapter is performed using MATLAB 7.10 software on a 2.4 GHz PC, using a 64-bit operating system.

Subsequent sections of this chapter are organized as follows. In the following section, the ED problem is briefly described. Next, the linearization approach is presented, and LP use for solving the ED problem is explained. Subsequently, comparative results are presented based on benchmark test problems. Finally, a summary is given, and conclusions and suggestions for future research are presented.

ECONOMIC DISPATCH PROBLEM

A typical power generation system has several power plants, and each power plant has several generating units. In any given time period, the total load (required power) from the system has to be met by the available generating units in different power plants. The ED problem determines the optimum power output of each generating unit, in order to minimize the total cost of power generated to satisfy the system load (Wood & Wollenberg, 2012). Consequently, the objective of the ED solution is to determine the optimal combination of power outputs for all generating units that will minimize the total cost while satisfying the load, capacity, and other operational constraints. The symbols used in describing and modeling the ED problem are listed below. The ED problem considered in this paper is based on the following assumptions:

Assumptions

- The objective is to minimize the total power generation cost.
- The required load must be satisfied exactly, without shortage or excess power.
- The capacity of each generator is specified by an upper limit and a lower limit on its power output.
- For each generator, power generation cost is a quadratic or a cubic function of the power output.
- The cost functions are approximated by piece-wise linear functions.
- The segment widths of the linearized cost functions are equal.
- The transmission losses are negligible.

Notation

P_i = power output of generator i.
$C_i(P_i)$ = total cost of generator i.
a_i, b_i, c_i = cost coefficients of generator i.
N = number of generators.
P_{load} = total required power (electrical load).
Z = total power generation cost.
$p_{i,min}$ = minimum power output of generator i.
$p_{i,max}$ = maximum power output of generator i.
K = number of segments in the piece-wise linear (PWL) solution.
W = width of the linearized segment.
s_{ik} = slope of segment k.
$p_{i,k-1}$ = start point of segment k.
$p_{i,k}$ = end point of segment k.
P_{ik} = PWL power values, measured from the start of segment k.

Figure 1 shows the configuration of a system consisting of N power-generating units connected to a single bus-bar serving a required electrical load P_{load}. The input to each unit, $C_i(P_i)$, represents the power generation cost of the unit. The output of each unit, P_i, is the electrical power generated by that particular unit. The total cost of this system is the sum of the power generation costs of all the individual units. The primary constraint on the operation of this system is that the sum of the output powers must be equal to the required load. The secondary constraint is the capacity limits specifying the minimum and the maximum power that can be generated from each unit.

Figure 1. ED problem with N generating units

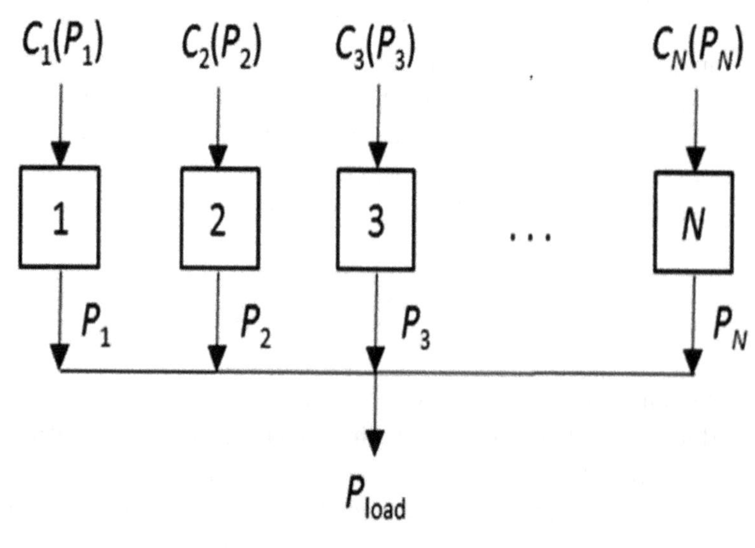

The ED problem can be mathematically represented by a concise LP model. First, the objective function Z is equal to the total cost of power generated for supplying the given load. The objective function Z is subject to several equality and inequality constraints. The problem is to minimize Z subject to the constraint that the sum of the powers generated must equal the required load. In addition, the power generated by any unit has to be within the maximum and minimum capacity limits of the given power generation units. The objective function of the LP model of the ED problem is presented below.

$$Minimize\ Z = \sum_{i=1}^{N} C_i(P_i) \quad (1)$$

Usually, the power generation cost is a quadratic function of the power output for each unit. In some cases, however, the cost function of a given generator is a cubic function of the power output. Assuming a quadratic relationship, the total cost (\$/h) for a given generator (i) is given by the following equation:

$$C_i(P_i) = a_i + b_i P_i + c_i P_i^2,\ i = 1, \ldots, N \quad (2)$$

The following equality constraint is known as the power balance equation, and it imposes equilibrium between total power generated and the total power required (system load). Ignoring transmission losses, this equilibrium is expressed in the following constraint:

$$\sum_{i=1}^{N} P_i = P_{load} \quad (3)$$

The following inequality constraints are known as the power generator's capacity bounds. Each generating unit i has a lower limit ($p_{i,\min}$) and an upper limit ($p_{i,\max}$) on power production, which are directly related to the generator design. These bounds are represented by the following constraints:

$$p_{i,\min} \le P_i \le p_{i,\max}, i = 1, \ldots., N \quad (4)$$

LINEARIZATION METHODOLOGY AND LINEAR PROGRAMMING

Linear Programming (LP) is an optimization technique to find the best solution in which the objective function and system constraints are represented by linear functions. As shown in equation (2), the cost function is a nonlinear function of the power generated. Usually, the cost is a quadratic function of the power, although it can be a cubic function in some cases. Given the nonlinear cost function for a generator i, this nonlinear (curve) function can be approximated by:

1. A single straight line, or
2. A series of straight-line segments.

Single-Segment Linearization Method

A quadratic or cubic cost function of the form shown in (2) is approximated by a single straight line. Therefore, for each generator i, the nonlinear cost function (2) is replaced by a straight-line function of the form:

$$C_i(P_i) = a_i + b_i P_i, \tag{5}$$

An example one-segment linearized cost function for generator i is illustrated in Figure 2.

Piece-Wise Linearization (PWL) Method

Of course, it is more accurate to approximate a nonlinear (quadratic or cubic) cost function by a series of straight-line segments. Piece-wise Linear (PWL) functions are several linear functions defined over pre-defined segments (intervals). The combination of these linear functions represents an approximation to one nonlinear function. Obviously, the accuracy of the linear approximation increases with a greater number of segments K. Usually, these segments are equal in horizontal width. Increasing the number of segments K decreases their width, W, and increases approximation accuracy. However, this also increases the computational time and effort. An example three-segment linearized cost function for generator i is illustrated in Figure 3.

In the PWL approach, all segment widths are usually considered to be equal. This equal-width convention is adopted in this chapter as it allows for easier calculation and simpler notation. Assuming K equal-width segments, the segment width is equal to:

Figure 2. One-segment linearized cost function

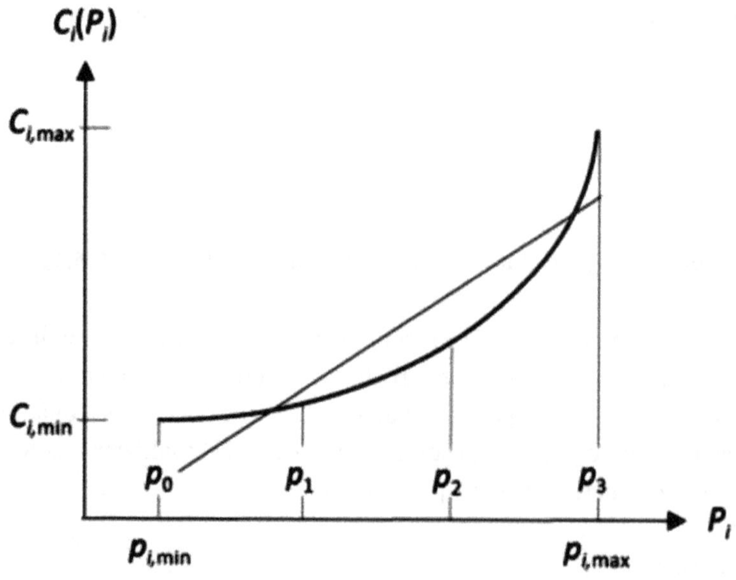

Figure 3. Three-segment linearized cost function

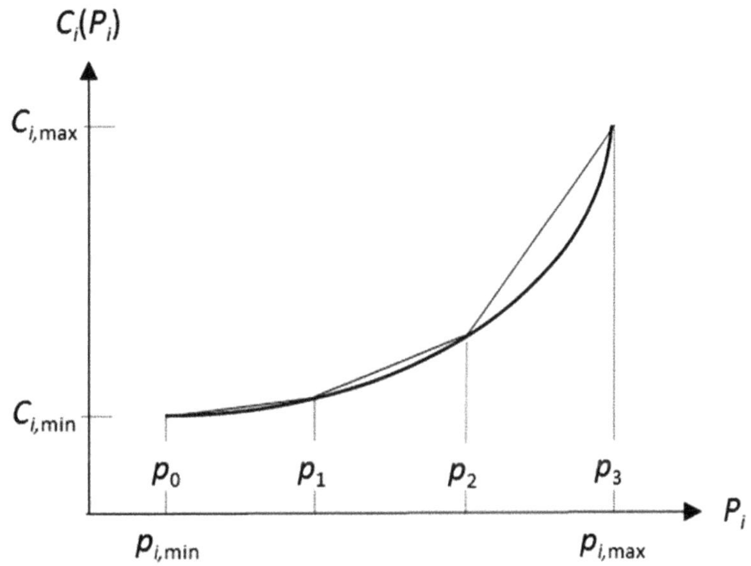

$$W = (p_{i,\max} - p_{i,\min}) / K \tag{6}$$

Obviously, increasing the number of segments K decreases the segment width W and improves the accuracy of the linear approximation, but it increases the computational time. The end point of segment k for generator i is calculated as follows:

$$p_{ik} = p_{i,\min} + k \times W, i = 1, \ldots., N, k = 0, \ldots., K \tag{7}$$

$$p_{i0} = p_{i,\min}$$

$$p_{ik} = p_{i,\max}$$

For each segment k, the slope of the linear segment s_{ik} is calculated by taking the difference between the cost values of the segment start and end points ($p_{i,k-1}$ and $p_{i,k}$, respectively) and then dividing the result by the segment width as follows:

$$S_{ik} = [C_i(p_{ik}) - C_i(p_{i,k-1})] / W \tag{8}$$

Now, assuming K equal-width segments, the cost function of generator i becomes:

$$C_i(P_i) = C_{i,\min} + s_{i1}P_{i1} + s_{i2}P_{i2} + ... + s_{iK}P_{iK} \tag{9}$$

Where

$$0 \le P_{ik} \le W, k = 1,2,...., K \tag{10}$$

$$P_i = P_{i,\min} + P_{i1} + P_{i2} + ... + P_{iK} \tag{11}$$

The cost function (9) is now made up of a linear combination of the P_{ik} values that can be optimally determined by LP. It is important to note that the fixed-cost constants $C_{i,\min}$ are ignored in the LP solution. However, these constants for all units must be added later to the optimum total cost value Z. Since all the units are on-line, then even if $P_i = 0$, the generator is running but not producing any power, and hence, the fixed cost of operation will be incurred for each generator.

The decision variables in the linearized problem are P_{ik}, where P_{ik} is measured from the start of segment k. For each segment k, the actual value of the corresponding variable P_{ik} is equal to:

$$P_{ik} = \begin{cases} \min(P_i, p_{i,k}) - p_{i,k-1}, \ if \ P_i > p_{i,k-1} \\ 0, \ otherwise \end{cases} \tag{12}$$

BENCHMARK TEST SYSTEMS WITH COMPARATIVE RESULTS ANALYSIS

In order to assess and compare the performance of the proposed LP-linearization technique, it was used to solve eleven benchmark ED test problems presented in recent literature (two 3-generator, one 6-generator, four 18-generator, one 20-generator, one 38-generator, and two 40-generator problems). In solving the ED test problems by LP, transmission losses were assumed negligible. All the test cases were simulated and run in MATLAB software environment. MATLAB is a numerical computing software for various mathematical and engineering applications including modeling, simulation, plotting, and matrix manipulations. Overall, two types of linearized LP solutions were compared with 14 other ED solution techniques from several recent references:

1. **The LP: Linear Solution**

This LP solution is based on single-segment linear approximation. Each nonlinear cost function is directly converted to a single linear function using the linear curve-fitting tool in MATLAB. This means that only one linear segment is considered for each unit. The linearized ED problem is then optimally solved by linear programming. The main purpose of this method is to show the advantage of using several piece-wise linear functions rather than a single linear function.

2. The LP: PWL Solution

This LP solution is based on multi-segment linear approximation using piece-wise linear cost functions. The nonlinear cost function of each generator is converted to a series of linear functions for different ranges of the power output. The linearized ED systems are then solved by the LP, assuming that all segment widths are equal for each generator.

Three-Generator Test System (Quadratic Function)

All the given parameters of this system are shown in Table 1. For each generator, the optimal power outputs and the corresponding total cost are shown in Table 2. Moreover, Table 2 compares the two LP solutions with the solutions obtained by Kumari and Sydulu (2009) using Simple Genetic Algorithm (SGA) and Fast Genetic Algorithm (FGA). In all tables, LP-Linear denotes the linear model of a single straight line fitted to the nonlinear cost curve. On the other hand, LP-PWL represents the LP solution for the linearization method using piece-wise linear cost functions. For each solution method, the balance error, i.e. difference between generated power and required power, is calculated as the left-hand side minus the right-hand side of equation (3). The last two rows of Table 2 show the LP segment width W in Megawatts (MW) and computation times, respectively.

As shown in Table 2, the solution obtained by LP-PWL provides good performance demonstrated by both low total cost and exactly meeting the required load. Although FGA method achieves the lowest cost, it has a balance error (deficiency) of 0.17 MW required load that is not met by the generators.

Table 1. Parameters of the 3-generator system (quadratic function)

Unit	P_{min} (MW)	P_{max} (MW)	a	b	c
1	100	600	561	7.92	0.00156
2	100	400	310	7.85	0.00194
3	50	200	78	7.97	0.00482

Table 2. The optimal solution for the 3-generator system (quadratic function)

Load (MW)	850			
Method	SGA (Kumari and Sydulu 2009)	FGA (Kumari and Sydulu 2009)	LP-Linear	LP-PWL
P_1 (MW)	393.26	393.09	400	393.18
P_2 (MW)	334.67	334.54	400	334.59
P_3 (MW)	122.25	122.2	50	122.23
P_{total} (MW)	850.18	849.83	850	850
Balance error	0.18	-0.17	0	0
Total cost ($/h)	8196	8192.7	8305.97	8194.356
Segment width	-	-	-	3
Time (seconds)	0.078	0.062	-	0.0465

Moreover, the computation time using LP-PWL is faster than the both SGA and FGA. The linear model with one segment (LP-Linear) gives higher cost when compared with other techniques. Figure 4 shows the convergence of the LP-PWL total cost, *Z*, as the segment width *W* decreases. As expected, the total cost decreases as the segment width gets smaller.

Three-Generator Test System (Cubic Function)

In this system, the power generation cost of each unit is a cubic function of its power output. The model is of the form:

$$C_i(P_i) = a_i + b_i P_i + c_i P_i^2 + d_i P_i^3,\ i = 1, \ldots, N \tag{13}$$

Table 3 shows the given parameters of the three generator units. All 3 generator units have cubic cost functions whose constants (a_i, b_i, c_i, d_i) are specified below.

Table 4 demonstrates the optimal power generation values for a load of 260 MW. A comparison of LP solutions is also shown with ANN and the classical Lagrange technique (CLT) by Tsekouras et al. (2013). Table 4 also shows that ANN has the least cost value, but it has a 0.22 MW deficiency. On the other

Figure 4. LP-PWL convergence for the 3-generator system (quadratic function)

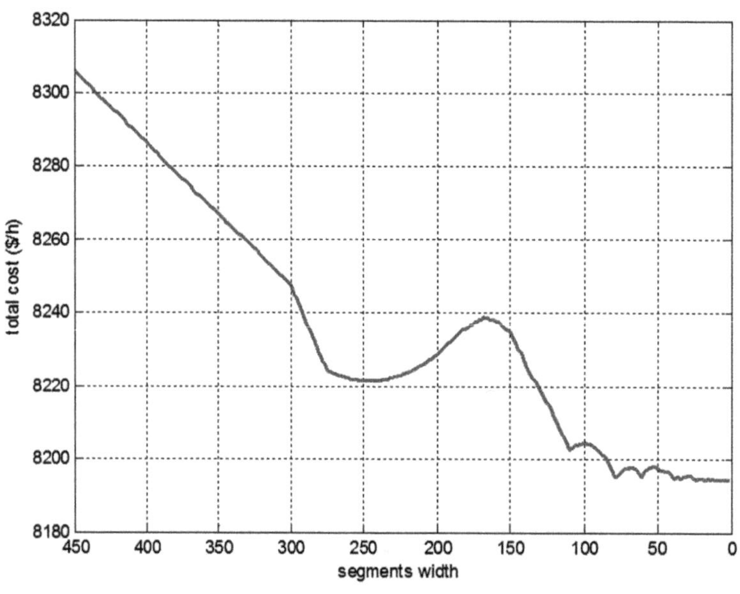

Table 3. Parameters of the 3-generator system (cubic function)

Unit	P_{min} (MW)	P_{max} (MW)	*a*	*b*	*c*	*d*
1	50	350	2700	100	0.03	0.0001
2	50	350	8760	75	0.1	0.00005
3	50	450	4800	90	0.03	0.00008

hand, the optimal LP-PWL satisfies the load requirement exactly with very low computation time, and it has a lower cost than CLT. The reduction of the LP-PWL solution's total cost with decreasing segment width is shown in Figure 5. The high cost obtained using the LP-Linear model is clearly demonstrated in the figure. Considering only one segment corresponds to the largest segment width, resulting in an inaccurate fit for the linearized model and leading to poor cost performance.

Six-Generator Test System

For this system, the given (input) parameters are displayed in Table 5. This system is known as the IEEE 30-bus test system. In Table 6, the optimal solutions found by LP are compared with the solutions by Newton's method (Wood & Wollenberg, 2012) and the Differential Evolution (DE) method (Sayah

Table 4. The optimal solution for the 3-generator system (cubic function)

Load (MW)	**260**			
Method	**ANN (Tsekouras et al. 2013)**	**ANN (Tsekouras et al. 2013)**	**LP-linear**	**LP-PWL**
P_1(MW)	49.99	50	50	50
P_2(MW)	109.27	108.78	160	108.8
P_3(MW)	100.52	101.22	50	101.2
P_{total}(MW)	259.78	260	260	260
Balance error	-0.22	0	0	0
Total cost ($/h)	40232	40254	43372.5	40253.8
Segment width	-	-	-	2
Time (seconds)	-	-	-	0.06

Figure 5. LP-PWL convergence for the 3-generator system (cubic function)

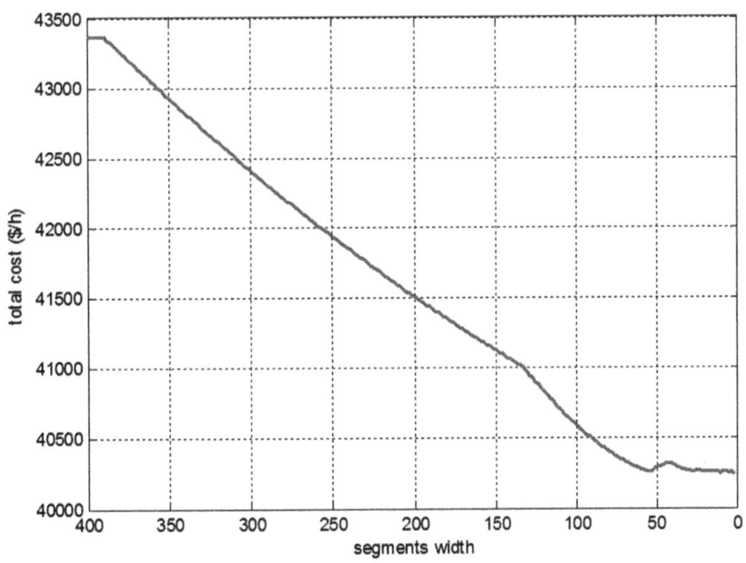

& Zehar, 2008). The LP-PWL solution's convergence process is shown in Figure 6. Among the four techniques listed in Table 6, LP-PWL has the minimum cost, shortest computation time, and highest accuracy in satisfying the required load.

Eighteen-Generator Test System

The given data for this system is shown in Table 7. All 18 generator units have quadratic cost functions whose constants (a_i, b_i, c_i) are specified below.

For four load values, the two LP solution techniques are compared in Table 8 with Lambda-Iteration Algorithm (LIA), binary GA (BGA), and real coded GA (RGA) from Damousis et al. (2003), and also with artificial bee colony (ABC) from Dixit et al. (2011). For a segment width of 0.25 MW, the LP computation times for all load values are in the range of 0.1-0.5 seconds. For a load value of 411.559 MW, the LP-PWL cost convergence process is shown in Figure 7. Table 8 shows that LP-PWL solu-

Table 5. Parameters of the 6-generator system

Unit	P_{min} (MW)	P_{max} (MW)	*a*	*b*	*c*
1	50	200	0	2	0.00375
2	20	80	0	1.75	0.0175
3	15	50	0	1	0.0625
4	10	35	0	3.25	0.00834
5	10	30	0	3	0.025
6	12	40	0	3	0.025

Figure 6. LP-PWL convergence for the 6-generator system

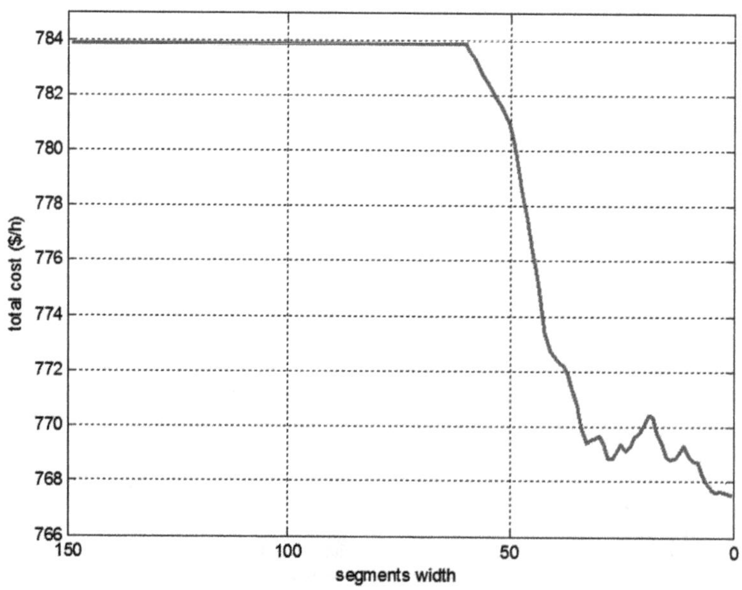

Table 6. The optimal solution for the 6-generator system

Load (MW)	283.4			
Method	Newton (Wood & Wollenberg, 2012)	(Sayah & Zehar, 2008)	LP-linear	LP-PWL
P_1(MW)	185.4	184.095	200	185.4
P_2(MW)	46.872	47.301	36.4	46.88
P_3(MW)	19.124	18.842	15	19.12
P_4(MW)	10	10.866	10	10
P_5(MW)	10	10.179	10	10
P_6(MW)	12	12.116	12	12
P_{total}(MW)	283.396	283.399	283.4	283.4
Balance error	-0.004	-0.001	0	0
Total cost ($/h)	767.6	767.78	783.90	767.6
Segment width	-	-	-	1
Time (seconds)	0.09	0.34	-	0.049

Table 7. Parameters of the 18-generator system

Unit	P_{min}(MW)	P_{max}(MW)	*a*	*b*	*c*
1	7	15	85.74158	22.45526	0.602842
2	7	45	85.74158	22.45526	0.602842
3	13	25	108.9837	22.52789	0.214263
4	16	25	49.06263	26.75263	0.077837
5	16	25	49.06263	26.75263	0.077837
6	3	14.75	677.73	80.39345	0.734763
7	3	14.75	677.73	80.393450	0.734763
8	3	12.28	44.39	13.19474	0.514474
9	3	12.28	44.39	13.19474	0.514474
10	3	12.28	44.39	13.19474	0.514474
11	3	12.28	44.390	13.19474	0.514474
12	3	24	574.9603	56.70947	0.657079
13	3	16.2	820.3776	84.67579	1.236474
14	3	36.2	603.0237	59.59026	0.394571
15	3	45	567.9363	56.70947	0.420789
16	3	37	567.9363	55.965	0.420789
17	3	45	567.9363	55.965	0.420789
18	3	16.2	820.3776	84.67579	1.236474

tion is comparable to the other techniques for the first two test problems and is superior for the last two problems. As expected, the highest cost among all approaches is obtained by the LP-linear model.

Twenty-Generator Test System

The input data for this system is shown in Table 9. For a 3600 MW required load, the optimal total cost and the optimal power outputs from all units are shown in Table 10. In Table 10, the LP solutions are compared with Simple Genetic Algorithm (SGA) and Fast Genetic Algorithm (FGA) from Kumari and Sydulu (2009). Although the FGA method has the least cost, it leads to a 1.4062 MW shortage in required load that is not supplied by the generators. On the other hand, LP-PWL provides the full power generation output to satisfy the given load at comparable time and cost. Figure 8 shows the convergence of the total cost for LP-PWL with a decreasing segments width.

Table 8. The optimal solution for the 18-generator system

	(Damousis et al., 2003)			(Dixit et al., 2011)		
Load	**LIA**	**BGA**	**RGA**	**ABC**	**LP-linear**	**LP-PWL**
411.559	29731.05	29733.42	29731.05	29730.8	29798.82	29731.07
389.898	27652.47	27681.05	27655.53	27653.3	27703.92	27653.75
346.576	23861.58	23980.24	23861.58	23859.4	24263.42	23855.29
303.254	20393.43	20444.68	20396.39	20391.6	20959.98	20386.22

Figure 7. LP-PWL convergence for the 18-generator system

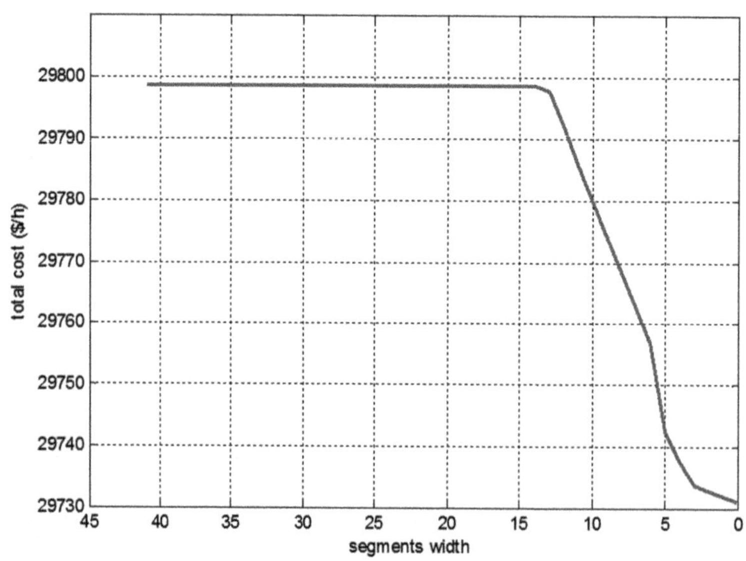

Table 9. Parameters of the 20-generator system

Unit	P_{min}(MW)	P_{max} (MW)	*a*	*b*	*c*
1	150	600	1000	18.19	0.00068
2	50	200	970	19.26	0.00071
3	50	200	600	19.8	0.0065
4	50	200	700	19.1	0.005
5	50	160	420	18.1	0.00738
6	20	100	360	19.26	0.00612
7	25	125	490	17.14	0.0079
8	50	150	660	18.92	0.00813
9	50	200	765	18.27	0.00522
10	30	150	770	18.92	0.00573
11	100	300	800	16.69	0.0048
12	150	500	970	16.76	0.0031
13	40	160	900	17.36	0.0085
14	20	130	700	18.7	0.00511
15	25	185	450	18.7	0.00398
16	20	80	370	14.26	0.0712
17	30	85	480	19.14	0.0089
18	30	120	680	18.92	0.00713
19	40	120	700	18.47	0.00622
20	30	100	850	19.79	0.00773

Figure 8. LP-PWL convergence for the 20-generator system

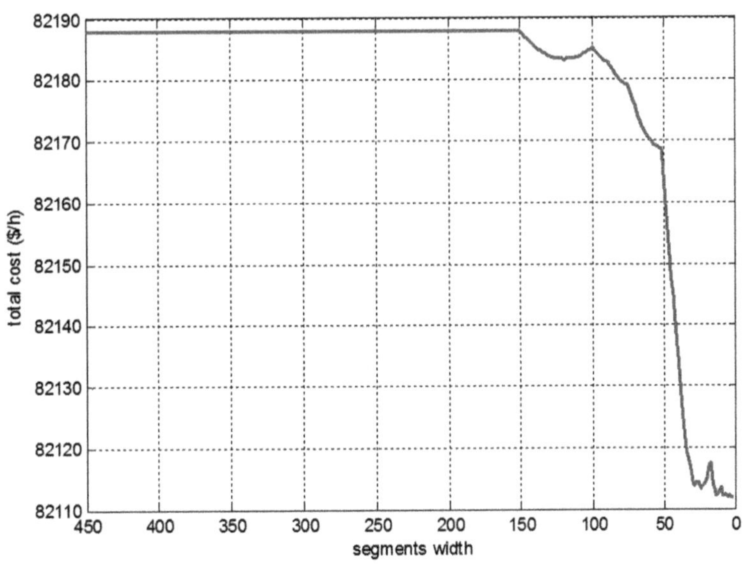

Table 10. The optimal solution for the 20-generator system

Load (MW)	3600			
Method	SGA (Kumari and Sydulu, 2009)	FGA (Kumari and Sydulu, 2009)	LP-linear	LP-PWL
P_1(MW)	600	600	600	600
P_2(MW)	200	200	200	200
P_3(MW)	75.146	74.3918	50	74.75
P_4(MW)	167.69	166.709	200	167.15
P_5(MW)	160	160	160	160
P_6(MW)	100	100	100	100
P_7(MW)	125	125	125	125
P_8(MW)	114.2	113.597	150	113.87
P_9(MW)	200	200	200	200
P_{10}(MW)	150	150	150	150
P_{11}(MW)	300	300	300	300
P_{12}(MW)	500	500	500	500
P_{13}(MW)	160	160	160	160
P_{14}(MW)	130	130	130	130
P_{15}(MW)	185	185	185	185
P_{16}(MW)	45.76	45.695	20	45.72
P_{17}(MW)	85	85	85	85
P_{18}(MW)	120	120	120	120
P_{19}(MW)	120	120	120	120
P_{20}(MW)	63.836	63.201	45	63.52
P_{total}(MW)	3601.632	3598.5938	3600	3600
Balance error	1.632	-1.4062	0	0
Total cost ($/h)	82145.85	82082.65	82187.943	82111.81
Segment width	-	-	-	3.8
Time (seconds)	1.076	0.094	-	0.2544

Thirty-Eight-Generator Test System

The input data for this system is shown in Table 11. In Table 12, the optimal LP solutions are compared with two other techniques from Kumari and Sydulu (2009), namely the Simple Genetic Algorithm (SGA) and the Fast Genetic Algorithm (SGA). Again, the FGA gives the least cost but it has a deficit of 6.63968 MW of power output from the 38 generators. The LP-PWL solution, however, gives accurate power generation to fully satisfy the required load, with comparable cost and short computation time. The LP-PWL total cost convergence process is shown in Figure 9.

Table 11. Parameters of the 38-generator system

Unit	P_{min}(MW)	P_{max} (MW)	*a*	*b*	*c*
1	220	550	64782	796.9	0.3133
2	220	550	64782	796.9	0.3133
3	200	500	64670	795.5	0.3127
4	200	500	64670	795.5	0.3127
5	200	500	64670	795.5	0.3127
6	200	500	64670	795.5	0.3127
7	200	500	64670	795.5	0.3127
8	200	500	64670	795.5	0.3127
9	114	500	172832	915.7	0.7075
10	114	500	172832	915.7	0.7075
11	114	500	176003	884.2	0.7515
12	114	500	173028	884.2	0.7083
13	110	500	91340	1250.1	0.4211
14	90	365	63440	1298.6	0.5145
15	82	365	65468	1290.8	0.5691
16	120	325	72282	190.8	0.5691
17	66	315	190928	238.1	2.5881
18	65	315	285372	1149.5	3.8734
19	65	315	271376	1269.1	3.6842
20	120	272	39197	696.1	0.4921
21	120	272	45576	660.2	0.5728
22	110	260	28770	803.2	0.3572
23	80	190	36902	818.2	0.9415
24	10	150	105510	33.5	52.123
25	60	125	22233	805.4	1.1421
26	55	110	30953	707.1	2.0275
27	35	75	17044	833.6	3.0744
28	20	70	81079	2188.7	16.765
29	20	70	124767	1024.4	26.355
30	20	70	121915	837.1	30.575
31	20	70	120780	1305.2	25.098
32	20	60	104441	716.6	33.722
33	25	60	83224	1633.9	23.915
34	18	60	111281	969.6	32.562
35	8	60	64142	2625.8	18.362
36	25	60	103519	1633.9	23.915
37	20	38	13547	694.7	8.482
38	20	38	13518	655.9	9.693

Table 12. The optimal solution for the 38-generator system

Load (MW)	7500			
Method	SGA (Kumari and Sydulu, 2009)	FGA (Kumari and Sydulu, 2009)	LP-linear model	LP-PWL
P_1(MW)	550	550	550	550
P_2(MW)	550	550	550	550
P_3(MW)	500	500	500	500
P_4(MW)	500	500	500	500
P_5(MW)	500	500	500	500
P_6(MW)	500	500	500	500
P_7(MW)	500	500	500	500
P_8(MW)	500	500	500	500
P_9(MW)	225.58	222.98	114	224.60
P_{10}(MW)	225.58	222.98	114	224.60
P_{11}(MW)	233.33	230.88	114	232.37
P_{12}(MW)	247.56	244.97	440	246.27
P_{13}(MW)	110	110	110	110
P_{14}(MW)	90	90	90	90
P_{15}(MW)	82	82	82	82
P_{16}(MW)	325	325	325	325
P_{17}(MW)	192.57	191.86	315	192.16
P_{18}(MW)	65	65	65	65
P_{19}(MW)	65	65	65	65
P_{20}(MW)	272	272	272	272
P_{21}(MW)	272	272	272	272
P_{22}(MW)	260	260	260	260
P_{23}(MW)	190	190	190	190
P_{24}(MW)	11.52	11.49	10	11.49
P_{25}(MW)	125	125	125	125
P_{26}(MW)	110	110	110	110
P_{27}(MW)	65.26	64.67	75	64.97
P_{28}(MW)	20	20	20	20
P_{29}(MW)	20	20	20	20
P_{30}(MW)	20	20	20	20
P_{31}(MW)	20	20	20	20
P_{32}(MW)	20	20	20	20
P_{33}(MW)	25	25	25	25
P_{34}(MW)	18	18	18	18
P_{35}(MW)	8	8	8	8
P_{36}(MW)	25	25	25	25

continued on following page

Table 12. Continued

Load (MW)	7500			
Method	**SGA (Kumari and Sydulu, 2009)**	**FGA (Kumari and Sydulu, 2009)**	**LP-linear model**	**LP-PWL**
P_{37}(MW)	31.84	31.75	38	31.74
P_{38}(MW)	29.87	29.78	38	29.79
P_{total}(MW)	7505.14	7493.36	7500	7500
Balance error	5.14	-6.64	0	-2E-06
Total cost ($/h)	10,824,602.1	10,809,797.2	10,926,825.32	10,818,260.9
Segment width	-	-	-	0.75
Time (seconds)	4.087	0.452	-	1.9135

Figure 9. LP-PWL convergence for the 38-generator system

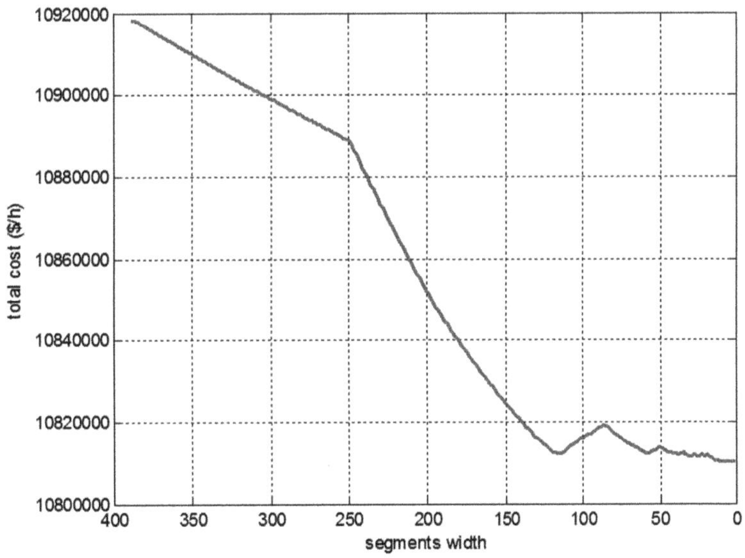

Forty-Generator Test System

The data for this system is shown in Table 13. Two alternative required loads are considered, either 9,000 MW or 10,500 MW. In Table 14, the LP solutions are compared with Variable Scaling Hybrid Differential Evolution (VSHDE), Hybrid Differential Evolution (HDE) SA, and GA solutions by Chiou (2007). For a load of 10500 MW, the LP-PWL cost convergence process with decreasing segment width is shown in Figure 10. The LP-PWL computation times (for two load values) are around 1.5 seconds using a segment width of 1 MW. As shown in Table 14, the lowest total cost for both loads is achieved by LP-PWL.

Figure 10. LP-PWL convergence for the 40-generator system

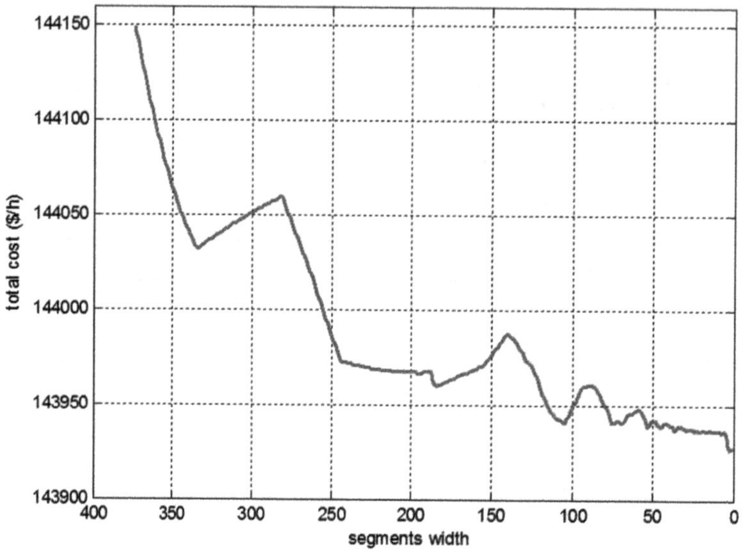

Table 13. Parameters of the 40-generator system

Unit	P_{min}(MW)	P_{max} (MW)	*a*	*b*	*c*
1	40	80	170.44	8.336	0.03073
2	60	120	309.54	7.0706	0.02028
3	80	190	369.03	8.1817	0.009420
4	24	42	135.48	6.9467	0.08482
5	26	42	135.19	6.5595	0.09693
6	68	140	222.33	8.0543	0.01142
7	110	300	287.71	8.0323	0.003570
8	135	300	391.98	6.999	0.004920
9	135	300	455.76	6.602	0.005730
10	130	300	722.82	12.908	0.006050
11	94	375	635.2	12.986	0.005150
12	94	375	654.69	12.796	0.005690
13	125	500	913.4	12.501	0.004210
14	125	500	1760.4	8.8412	0.007520
15	125	500	1728.3	9.1575	0.007080
16	125	500	1728.3	9.1575	0.007080
17	125	500	1728.3	9.1575	0.007080
18	220	500	647.85	7.9691	0.003130
19	220	500	649.69	7.955	0.003130
20	242	550	647.83	7.9691	0.003130
21	242	550	647.83	7.9691	0.003130

continued on following page

Table 13. Continued

Unit	P_{min}(MW)	P_{max} (MW)	*a*	*b*	*c*
22	254	550	785.96	6.6313	0.002980
23	254	550	785.96	6.6313	0.002980
24	254	550	794.53	6.6611	0.002840
25	254	550	794.53	6.6611	0.002840
26	254	550	801.32	7.1032	0.002770
27	254	550	801.32	7.1032	0.002770
28	10	150	1055.1	3.3353	0.52124
29	10	150	1055.1	3.3353	0.52124
30	10	150	1055.1	3.3353	0.52124
31	20	70	1207.8	13.052	0.25098
32	20	70	810.79	21.887	0.16766
33	20	70	1247.7	10.244	0.2635
34	20	70	1219.2	8.3707	0.30575
35	18	60	641.43	26.258	0.18362
36	18	60	1112.8	9.6956	0.32563
37	20	60	1044.4	7.1633	0.33722
38	25	60	832.24	16.339	0.23915
39	25	60	834.24	16.339	0.23915
40	25	60	1035.2	16.339	0.23915

Table 14. The optimal solution for the 40-generator system

	(Chiou, 2007)					
Load (MW)	**VSHDE**	**HDE**	**SA**	**GA**	**LP-linear model**	**LP-PWL**
10500	143,943.90	143,955.83	164,069.36	144,486.02	144,151.65	143,926.47
9000	121,253.01	121,266.40	135,229.69	121,839.72	122222.65	121,244.11

Comparison With Previous Computational Studies

Several other computational comparisons have been performed to evaluate ED solution methods. Table 15 summarizes the features of previous ED computational studies in comparison with this chapter. For each computational study, Table 15 shows the number of solution methods compared, the number of test problems, the maximum number of generators, and whether convergence analysis is performed. Clearly, this chapter presents the most comprehensive comparison of solution methods for the economic load dispatch problem.

Table 15. Comparison of ED computational studies

Paper	No. of methods	No. of problems	Max. No. of units	Convergence analysis
This chapter	16	11	40	Yes
(Kumari and Sydulu 2009)	2	7	38	Yes
(Tsekouras et al. 2013)	2	2	3	No
(Sayah & Zehar, 2008)	2	1	6	Yes
(Damousis et al., 2003)	3	4	18	Yes
(Dixit et al., 2011)	11	6	18	Yes
(Chiou, 2007)	4	2	40	No

Discussion and Implication of the Results

From the results presented so far, several observations can be made regarding the piece-wise linear LP approach (LP-PWL). First, the performance of LP-PWL is consistently excellent regardless of the ED problem size (i.e. number of generators). For example, LP-PWL achieves the least cost for all 6-generator problems and all 40-generator problems. Second, LP-PWL solutions provide the best load accuracy for all benchmark problems. This an impressive result, considering that LP-PWL is compared with 15 other recent ED solution techniques. Third, even in the instances where LP-PWL does not produce the lowest cost, its cost is very close to the least cost for the given problem. Fourth, the computational accuracy and performance of LP-PWL improves as the number of linear segments increases. Although this improvement is obtained at the expense of longer solution times, LP-PWL computation times are still very competitive.

The above observations have significant implications for real-life industrial applications. To production managers in electrical power plants, LP-PWL provides a useful alternative for solving their daily economic dispatch (ED) problems. Appropriate utilization of this method can lead to two important benefits: (1) lower power generation costs, and (2) accurate satisfaction of the required power load. Maximum benefit of using LP-PWL in power plants is obtained by applying on two rules. First, the LP-PWL method should not be used alone to solve the ED problem, but as one alternative among several solution techniques. The production manager should compare all the solutions produced by these methods and select the least-cost feasible solution. Second, LP-PWL should be used with different values of the number of linear segments K. For each power plant, a range of relevant values of K can be found by trial and error based on the typical data of the given plant.

CONCLUSION

This chapter presented linear programming (LP) approach for solving the economic load dispatch problem ignoring transmission losses. For each generating unit, the cost of power generation is usually a quadratic or a cubic function of the power produced. Therefore, in order to use LP, linearization of the cost functions is needed to transform the total cost (objective) function into a linear form. Two techniques were used in this chapter to replace the nonlinear cost function of each generator by a linear approximation. The first technique (LP-Linear) approximates the nonlinear cost function with a single straight line. The second technique (LP-PWL) uses a linear piece-wise approximation to replace the cost function with several straight-line segments. After linearizing all the cost functions, LP can be used to find the optimum solution of the given ED problem.

To evaluate the performance of the proposed linearization techniques, a large set of benchmark problems was solved using MATLAB. This set includes eleven test problems with different parameters, ranging in size from 3 to 40 generators. For each test problem, the two LP solutions were compared with the previous solutions by other techniques specified in the corresponding published references. In total, the two types of linearized LP solutions were compared with 14 other ED solution techniques from several recent references.

Based on the computational comparisons, the LP-Linear linearization technique has the worst cost performance. For all test systems, LP-Linear model has produced the most expensive solution when compared with other techniques. Clearly, converting the nonlinear cost function to a simple linear model (single straight-line segment) leads to higher cost values. This verifies the advantage of linearizing each nonlinear cost function using piece-wise linear segments instead of fitting it into one straight line.

The most important outcome of the computational experiments is the overall competitive advantage of the piece-wise linearization technique. This is demonstrated by the following facts. LP-PWL provided the highest generation accuracy (least power balance deficiency) in all 11 problems (100%) and the least cost in 5 problems (45%). Out of the 11 test problems, only 4 problems had previously published solution times; LP-PWL had the shortest computation times in 2 of these 4 problems (50%). The computational experiments included an analysis of the effect of the segment size for the piece-wise linearization method. As expected, reducing the segment size leads to better solutions (lower total cost) by the LP-PWL technique.

It is important to observe that even in the cases where the LP-PWL solutions did not provide the minimum total costs, these costs were still very close to the minimum values obtained by other methods in the literature. By extensive comparisons with 15 other techniques, it can be concluded that the piece-wise linearization LP technique (LP-PWL) has a significant competitive edge. This conclusion is based on the overall performance of the LP-PWL technique regarding total cost, demand (load) satisfaction, and computational time. Therefore, the PWL-LP method can provide significant economic advantages to power plant production managers if they use it to solve their ED problems.

Another important observation should be noted regarding the comparison of computational times of different solution techniques. The main difficulty faced in this study and similar comparative studies is the lack of a comprehensive and fair comparison of computation times for different solution methods. This is mainly due to two facts. First, computation times for most published algorithms are not reported in their corresponding references. Second, even if these times were reported, they would not be easy to compare due to the differences in hardware and software platforms. Computational times reported by different authors are obtained in different years based on different and fast-advancing hardware and software systems. Therefore, the differences in solution times must be taken with caution, as they could be attributed more to differences in computing technology than to differences in efficiency among the solution techniques.

There many possibilities for future research based on alternative extensions of this work. For example, the LP-PWL technique can be applied to other related problems such as ED problems with transmission losses, unit commitment problems, and other nonlinear optimization problems in power system. More constrains can be also added in the problem such as analyzing exponential cost functions rather than quadratic and cubic cost functions, considering prohibited zones for generator operations, and analyzing non-smooth cost functions with valve-point effects. Another possibility is to fine-tune the LP-PWL technique by optimizing the segment size, in order to maximize the balance between solution quality (total cost) and speed (computation time). A third possibility is to integrate the proposed LP-PWL method with other exact or heuristic solution techniques in order to improve the overall performance. Finally, additional relevant aspects can be considered in the ED problem, such spinning reserve requirements, tie line limits, non-smooth cost functions with multiple fuels, emission and pollution effect, and ramp up and down cases.

REFERENCES

Abbas, G., Gu, J., Farooq, U., Asad, M. U., & El-Hawary, M. (2017). Solution of an Economic Dispatch Problem through Particle Swarm Optimization: A Detailed Survey–Part I. *IEEE Access: Practical Innovations, Open Solutions*, *5*, 15105–15141. doi:10.1109/ACCESS.2017.2723862

Absil, P. A., Sluysmans, B., & Stevens, N. (2018, June). MIQP-Based Algorithm for the Global Solution of Economic Dispatch Problems with Valve-Point Effects. In 2018 Power Systems Computation Conference (PSCC) (pp. 1-7). IEEE. doi:10.23919/PSCC.2018.8450877

Adarsh, B. R., Raghunathan, T., Jayabarathi, T., & Yang, X. S. (2016). Economic dispatch using chaotic bat algorithm. *Energy*, *96*, 666–675. doi:10.1016/j.energy.2015.12.096

Ahmad, J., Tahir, M., & Mazumder, S. K. (2018). Dynamic Economic Dispatch and Transient Control of Distributed Generators in a Microgrid. *IEEE Systems Journal*, (99): 1–11.

Al-Betar, M. A., Awadallah, M. A., Khader, A. T., Bolaji, A. L. A., & Almomani, A. (2018). Economic load dispatch problems with valve-point loading using natural updated harmony search. *Neural Computing & Applications*, *29*(10), 767–781. doi:10.100700521-016-2611-2

Awan, F., Michalska, H., & Joos, G. (2017, June). Economic dispatch in microgrids using compromise solution method. In PowerTech, 2017 IEEE Manchester (pp. 1-6). IEEE.doi:10.1109/PTC.2017.7981250

Azizipanah-Abarghooee, R., Dehghanian, P., & Terzija, V. (2016). Practical multi-area bi-objective environmental economic dispatch equipped with a hybrid gradient search method and improved Jaya algorithm. *IET Generation, Transmission & Distribution*, *10*(14), 3580–3596. doi:10.1049/iet-gtd.2016.0333

Bhattacharjee, V., & Khan, I. (2018). A non-linear convex cost model for economic dispatch in microgrids. *Applied Energy*, *222*, 637–648. doi:10.1016/j.apenergy.2018.04.001

Bhongade, S., & Agarwal, S. (2016, January). An optimal solution for Combined Economic and Emission Dispatch problem using Artificial Bee Colony Algorithm. In *Power and Energy Systems: Towards Sustainable Energy (PESTSE), 2016 Biennial International Conference on* (pp. 1-7). IEEE.10.1109/PESTSE.2016.7516478

Bhui, P., & Senroy, N. (2016, March). A unified method for economic dispatch with valve point effects. In *Power Systems (ICPS), 2016 IEEE 6th International Conference on* (pp. 1-5). IEEE.10.1109/ICPES.2016.7584108

Chamba, M., & Ano, O. (2013). Economic dispatch of energy and reserve in competitive markets using meta-heuristic algorithms. *IEEE Latin America Transactions, 11*(1), 473–478. doi:10.1109/TLA.2013.6502848

Chansareewittaya, S. (2017, March). Hybrid BA/TS for economic dispatch considering the generator constraint. In *Digital Arts, Media and Technology (ICDAMT), International Conference on* (pp. 115-119). IEEE.

Chen, G., & Yang, Q. (2018). An ADMM-based distributed algorithm for economic dispatch in islanded microgrids. *IEEE Transactions on Industrial Informatics, 14*(9), 3892–3903. doi:10.1109/TII.2017.2785366

Cherukuri, A., & Cortés, J. (2017). Distributed coordination of DERs with storage for dynamic economic dispatch. *IEEE Transactions on Automatic Control.*

Chiou, J. P. (2007). Variable scaling hybrid differential evolution for large-scale economic dispatch problems. *Electric Power Systems Research, 77*(3), 212–218. doi:10.1016/j.epsr.2006.02.013

Ciornei, I., & Kyriakides, E. (2013). Recent methodologies and approaches for the economic dispatch of generation in power systems. *International Transactions on Electrical Energy Systems, 23*(7), 1002–1027. doi:10.1002/etep.1635

Damousis, I. G., Bakirtzis, A. G., & Dokopoulos, P. S. (2003). Network-constrained economic dispatch using real-coded genetic algorithm. *IEEE Transactions on Power Systems, 18*(1), 198–205. doi:10.1109/TPWRS.2002.807115

Dixit, G. P., Dubey, H. M., Pandit, M., & Panigrahi, B. K. (2011). Economic load dispatch using artificial bee colony optimization. *International Journal of Advances in Electronics Engineering, 1*(1), 119–124.

Elsaiah, S., Benidris, M., Mitra, J., & Cai, N. (2014, July). Optimal economic power dispatch in the presence of intermittent renewable energy sources. In *PES General Meeting| Conference & Exposition, 2014 IEEE* (pp. 1-5). IEEE.10.1109/PESGM.2014.6939903

Goudarzi, A., Swanson, A. G., Tooryan, F., & Ahmadi, A. (2017, February). Non-convex optimization of combined enviromental economic dispatch through the third version of the cultural algorithm (CA3). In *Power and Energy Conference (TPEC), IEEE Texas* (pp. 1-6). IEEE.10.1109/TPEC.2017.7868281

Guo, Y., Tong, L., Wu, W., Zhang, B., & Sun, H. (2017). Coordinated multi-area economic dispatch via critical region projection. *IEEE Transactions on Power Systems, 32*(5), 3736–3746. doi:10.1109/TPWRS.2017.2655442

Hasan, Z., & El-Hawary, M. E. (2016, October). Economic dispatch at peak load using load reduction for smart grid network. In *Electrical Power and Energy Conference (EPEC),* 2016 *IEEE* (pp. 1-5). IEEE.10.1109/EPEC.2016.7771744

Hoke, A., Brissette, A., Chandler, S., Pratt, A., & Maksimović, D. (2013, August). Look-ahead economic dispatch of microgrids with energy storage, using linear programming. In *Technologies for Sustainability (SusTech), 2013 1st IEEE Conference on* (pp. 154-161). IEEE.10.1109/SusTech.2013.6617313

Hota, P. K., & Sahu, N. C. (2015). Non-Convex Economic Dispatch with Prohibited Operating Zones through Gravitational Search Algorithm. *Iranian Journal of Electrical and Computer Engineering, 5*(6).

Imen, L., Mouhamed, B., & Djamel, L. (2013, November). Economic dispatch using classical methods and neural networks. In *Electrical and Electronics Engineering (ELECO), 2013 8th International Conference on* (pp. 172-176). IEEE.10.1109/ELECO.2013.6713826

Jabr, R. A., Coonick, A. H., & Cory, B. J. (2000). A homogeneous linear programming algorithm for the security constrained economic dispatch problem. *IEEE Transactions on Power Systems*, *15*(3), 930–936. doi:10.1109/59.871715

Khan, N. A., Sidhu, G. A. S., & Gao, F. (2016). Optimizing combined emission economic dispatch for solar integrated power systems. *IEEE Access: Practical Innovations, Open Solutions*, *4*, 3340–3348.

Kumari, M. S., & Sydulu, M. (2009). A fast computational genetic algorithm for economic load dispatch. *International Journal of Recent Trends in Engineering*, *1*(1).

Kuo, M. T., Lu, S. D., & Tsou, M. C. (2017). Considering Carbon Emissions in Economic Dispatch Planning for Isolated Power Systems. *Natural Gas*, *701*, 233–237.

Lazzerini, B., & Pistolesi, F. (2015, November). A linear programming-driven MCDM approach for multi-objective economic dispatch in smart grids. In *SAI Intelligent Systems Conference (IntelliSys)*, 2015 (pp. 475-484). IEEE.10.1109/IntelliSys.2015.7361183

Liang, H., Liu, Y., Shen, Y., Li, F., & Man, Y. (2018). A Hybrid Bat Algorithm for Economic Dispatch with Random Wind Power. *IEEE Transactions on Power Systems*, *33*(5), 5052–5061. doi:10.1109/TPWRS.2018.2812711

Lin, J., Chen, C. L., Tsai, S. F., & Yuan, C. Q. (2015). New intelligent particle swarm optimization algorithm for solving economic dispatch with valve-point effects. *Journal of Marine Science and Technology*, *23*(1), 44–53.

Manser, M. H., Musirin, I., & Othman, M. M. (2017). Immune Log-Normal Evolutionary Programming (ILNEP) for solving economic dispatch problem with prohibited operating zones. In *Industrial Engineering and Applications (ICIEA), 2017 4th International Conference on*. IEEE.10.1109/IEA.2017.7939199

Meng, W., & Wang, X. (2017). Distributed Energy Management in Smart Grid With Wind Power and Temporally Coupled Constraints. *IEEE Transactions on Industrial Electronics*, *64*(8), 6052–6062. doi:10.1109/TIE.2017.2682001

Mohamed, F., Abdel-Nasser, M., Mahmoud, K., & Kamel, S. (2018, February). Economic dispatch using stochastic whale optimization algorithm. In *Innovative Trends in Computer Engineering (ITCE), 2018 International Conference on* (pp. 19-24). IEEE.10.1109/ITCE.2018.8316594

Momoh, J. A., & Reddy, S. S. (2014, July). Combined economic and emission dispatch using radial basis function. In *PES General Meeting Conference & Exposition, 2014 IEEE* (pp. 1-5). IEEE.0.1109/PESGM.2014.6939506

Naveen, P., Chandel, A. K., Vedik, B., & Topwal, K. (2016, April). Economic dispatch with valve point effect using symbiotic oragnisms search algorithm. In *Energy Efficient Technologies for Sustainability (ICEETS), 2016 International Conference on* (pp. 430-435). IEEE.10.1109/ICEETS.2016.7583793

Nischal, M. M., & Mehta, S. (2015). Optimal load dispatch using ant lion optimization. *Int J Eng Res Appl*, *5*(8), 10–19.

Nivedha, R. R., Singh, J. G., & Ongsakul, W. (2018, January). PSO based economic dispatch of a hybrid microgrid system. In *2018 International Conference on Power, Signals, Control and Computation (EPSCICON)* (pp. 1-5). IEEE.10.1109/EPSCICON.2018.8379595

Pan, S., Jian, J., & Yang, L. (2018). A hybrid MILP and IPM approach for dynamic economic dispatch with valve-point effects. *International Journal of Electrical Power & Energy Systems*, *97*, 290–298. doi:10.1016/j.ijepes.2017.11.004

Ramadan, B. M., Logenthiran, T., Naayagi, R. T., & Su, C. (2016, November). Hybridization of genetic algorithm and priority list to solve economic dispatch problems. In Region 10 Conference (TENCON), 2016 IEEE (pp. 1467-1470). IEEE.

Ross, M., Abbey, C., Bouffard, F., & Joos, G. (2018). Microgrid economic dispatch with energy storage systems. *IEEE Transactions on Smart Grid*, *9*(4), 3039–3047. doi:10.1109/TSG.2016.2624756

Sayah, S., & Zehar, K. (2008). Using evolutionary computation to solve the economic load dispatch problem. *Leonardo Journal of Sciences*, *12*(12), 67–78.

Shi, L. B., Wang, R., & Yao, L. Z. (2016). Modelling and solutions of coordinated economic dispatch with wind–hydro–thermal complex power source structure. *IET Renewable Power Generation*, *11*(3), 262–270. doi:10.1049/iet-rpg.2016.0429

Singhal, P. K., Naresh, R., Sharma, V., & Kumar, G. (2014, May). Enhanced lambda iteration algorithm for the solution of large scale economic dispatch problem. In Recent Advances and Innovations in Engineering (ICRAIE), 2014 (pp. 1-6). IEEE.doi:10.1109/ICRAIE.2014.6909294

Tang, Z., Hill, D. J., & Liu, T. (2018). A Novel Consensus-Based Economic Dispatch for Microgrids. *IEEE Transactions on Smart Grid*, *9*(4), 3920–3922. doi:10.1109/TSG.2018.2835657

Tsekouras, G. J., Kanellos, F. D., Mastorakis, N. E., & Mladenov, V. (2013, September). Optimal Operation of Electric Power Production System without Transmission Losses Using Artificial Neural Networks Based on Augmented Lagrange Multiplier Method. In *International Conference on Artificial Neural Networks* (pp. 586-594). Springer.10.1007/978-3-642-40728-4_73

Tung, N. S., & Chakravorty, S. (2015). Grey Wolf Optimization for Active Power Dispatch Planning Problem Considering Generator Constraints and Valve Point Effect. *International Journal of Hybrid Information Technology*, *8*(12), 117–134. doi:10.14257/ijhit.2015.8.12.07

Vijayaraj, S., & Santhi, R. K. (2016, March). Multi-area economic dispatch using flower pollination algorithm. In *Electrical, Electronics, and Optimization Techniques (ICEEOT), International Conference on* (pp. 4355-4360). IEEE.10.1109/ICEEOT.2016.7755541

Wood, A. J., & Wollenberg, B. F. (2012). *Power Generation, Operation, and Control*. John Wiley & Sons.

Wu, Z., Ding, J., Wu, Q. H., Jing, Z., & Zheng, J. (2017). Reserve constrained dynamic economic dispatch with valve-point effect: A two-stage mixed integer linear programming approach. *CSEE Journal of Power and Energy Systems*, *3*(2), 203–211. doi:10.17775/CSEEJPES.2017.0025

Wu, Z. L., Wu, Q. H., Zhou, X. X., & Li, M. S. (2015, November). Hybrid quadratic programming and compact formulation method for economic dispatch with prohibited operating zones and network losses. In Innovative Smart Grid Technologies-Asia (ISGT ASIA), 2015 IEEE (pp. 1-6). IEEE.doi:10.1109/ISGT-Asia.2015.7386963

Xia, H., Li, Q., Xu, R., Chen, T., Wang, J., Hassan, M. A. S., & Chen, M. (2018). Distributed Control Method for Economic Dispatch in Islanded Microgrids With Renewable Energy Sources. *IEEE Access: Practical Innovations, Open Solutions*, *6*, 21802–21811. doi:10.1109/ACCESS.2018.2827366

Yang, P., Zhu, C., Zhao, L., Wang, M., Ning, X., & Liu, Y. (2016, November). Solving dynamic economic dispatch with valve point effect by a two-step method. In Region 10 Conference (TENCON), 2016 IEEE (pp. 546-550). IEEE.doi:10.1109/TENCON.2016.7848060

Zhan, J. P., Wu, Q. H., Guo, C. X., & Zhou, X. X. (2014). Fast lambda-Iteration Method for Economic Dispatch With Prohibited Operating Zones. *IEEE Transactions on Power Systems*, *29*(2), 990–991. doi:10.1109/TPWRS.2013.2287995

Zhang, L., Yuan, Y., Chen, B., & Su, D. (2017, May). A robust interval economic dispatch model accommodating large-scale wind power generation with consideration of price-based demand response. In *Control And Decision Conference (CCDC), 2017 29th Chinese* (pp. 6679-6684). IEEE.10.1109/CCDC.2017.7978379

Zhang, Y., Hajiesmaili, M. H., Cai, S., Chen, M., & Zhu, Q. (2018). Peak-aware online economic dispatching for microgrids. *IEEE Transactions on Smart Grid*, *9*(1), 323–335. doi:10.1109/TSG.2016.2551282

ADDITIONAL READING

Abdelaziz, A. Y., Ali, E. S., & Elazim, S. A. (2016). Implementation of flower pollination algorithm for solving economic load dispatch and combined economic emission dispatch problems in power systems. *Energy*, *101*, 506–518. doi:10.1016/j.energy.2016.02.041

Abdi, H., Dehnavi, E., & Mohammadi, F. (2016). Dynamic Economic Dispatch Problem Integrated With Demand Response (DEDDR) Considering Non-Linear Responsive Load Models. *IEEE Transactions on Smart Grid*, *7*(6), 2586–2595. doi:10.1109/TSG.2015.2508779

Al-Betar, M. A., Awadallah, M. A., Khader, A. T., & Bolaji, A. L. A. (2016). Tournament-based harmony search algorithm for non-convex economic load dispatch problem. *Applied Soft Computing*, *47*, 449–459. doi:10.1016/j.asoc.2016.05.034

Ali, E. S., & Elazim, S. A. (2018). Mine blast algorithm for environmental economic load dispatch with valve loading effect. *Neural Computing & Applications*, *30*(1), 261–270. doi:10.100700521-016-2650-8

Basu, M. (2015). Modified particle swarm optimization for nonconvex economic dispatch problems. *International Journal of Electrical Power & Energy Systems*, *69*, 304–312. doi:10.1016/j.ijepes.2015.01.015

Elattar, E. E. (2015). A hybrid genetic algorithm and bacterial foraging approach for dynamic economic dispatch problem. *International Journal of Electrical Power & Energy Systems*, *69*, 18–26. doi:10.1016/j.ijepes.2014.12.091

Elsayed, W. T., & El-Saadany, E. F. (2015). A fully decentralized approach for solving the economic dispatch problem. *IEEE Transactions on Power Systems*, *30*(4), 2179–2189. doi:10.1109/TPWRS.2014.2360369

Ghasemi, M., Ghavidel, S., Aghaei, J., Akbari, E., & Li, L. (2018). CFA optimizer: A new and powerful algorithm inspired by Franklin's and Coulomb's laws theory for solving the economic load dispatch problems. *International Transactions on Electrical Energy Systems*, *28*(5), e2536. doi:10.1002/etep.2536

He, X., Rao, Y., & Huang, J. (2016). A novel algorithm for economic load dispatch of power systems. *Neurocomputing*, *171*, 1454–1461. doi:10.1016/j.neucom.2015.07.107

James, J. Q., & Li, V. O. (2016). A social spider algorithm for solving the non-convex economic load dispatch problem. *Neurocomputing*, *171*, 955–965. doi:10.1016/j.neucom.2015.07.037

Kamboj, V. K., Bath, S. K., & Dhillon, J. S. (2016). Solution of non-convex economic load dispatch problem using Grey Wolf Optimizer. *Neural Computing & Applications*, *27*(5), 1301–1316. doi:10.100700521-015-1934-8

Kamboj, V. K., Bhadoria, A., & Bath, S. K. (2017). Solution of non-convex economic load dispatch problem for small-scale power systems using ant lion optimizer. *Neural Computing & Applications*, *28*(8), 2181–2192. doi:10.100700521-015-2148-9

Liu, D., Guo, J., Wang, W., Liu, C., Huang, Y., & O'Malley, M. (2014, July). A dynamic economic dispatch method considering with the uncertainty and correlation of wind power. In *PES General Meeting| Conference & Exposition, 2014 IEEE* (pp. 1-5). IEEE.10.1109/PESGM.2014.6938953

Lorca, A., & Sun, X. A. (2015). Adaptive robust optimization with dynamic uncertainty sets for multi-period economic dispatch under significant wind. *IEEE Transactions on Power Systems*, *30*(4), 1702–1713. doi:10.1109/TPWRS.2014.2357714

Mudumbai, R., & Dasgupta, S. (2014, February). Distributed control for the smart grid: The case of economic dispatch. In *Information Theory and Applications Workshop (ITA)*, 2014 (pp. 1-6). IEEE.10.1109/ITA.2014.6804224

Nguyen, T. T., & Vo, D. N. (2015). The application of one rank cuckoo search algorithm for solving economic load dispatch problems. *Applied Soft Computing*, *37*, 763–773. doi:10.1016/j.asoc.2015.09.010

Pradhan, M., Roy, P. K., & Pal, T. (2016). Grey wolf optimization applied to economic load dispatch problems. *International Journal of Electrical Power & Energy Systems*, *83*, 325–334. doi:10.1016/j.ijepes.2016.04.034

Prakash, T., Singh, V. P., Singh, S. P., & Mohanty, S. R. (2018). Economic load dispatch problem: Quasi-oppositional self-learning TLBO algorithm. *Energy Systems*, *9*(2), 415–438. doi:10.100712667-017-0230-3

Qu, B. Y., Zhu, Y. S., Jiao, Y. C., Wu, M. Y., Suganthan, P. N., & Liang, J. J. (2018). A survey on multi-objective evolutionary algorithms for the solution of the environmental/economic dispatch problems. *Swarm and Evolutionary Computation*, *38*, 1–11. doi:10.1016/j.swevo.2017.06.002

Rizk-Allah, R. M., El-Sehiemy, R. A., & Wang, G. G. (2018). A novel parallel hurricane optimization algorithm for secure emission/economic load dispatch solution. *Applied Soft Computing*, *63*, 206–222. doi:10.1016/j.asoc.2017.12.002

Shen, X., Zheng, J., Zhu, S., & Wang, X. (2014, July). Multi-scene security constrained economic dispatch with rational wind power curtailment in micro-grid. In *PES General Meeting| Conference & Exposition, 2014 IEEE* (pp. 1-5). IEEE.10.1109/PESGM.2014.6939482

Su, C. L., & Chuang, H. M. (2014, May). Economic dispatch of island power systems with distributed energy resources. In *Energy Conference (ENERGYCON), 2014 IEEE International* (pp. 880-886). IEEE.

Subathra, M. S. P., Selvan, S. E., Victoire, T. A. A., Christinal, A. H., & Amato, U. (2015). A hybrid with cross-entropy method and sequential quadratic programming to solve economic load dispatch problem. *IEEE Systems Journal*, *9*(3), 1031–1044. doi:10.1109/JSYST.2013.2297471

Tang, Y., Luo, C., Yang, J., & He, H. (2017). A chance constrained optimal reserve scheduling approach for economic dispatch considering wind penetration. *IEEE/CAA Journal of Automatica Sinica, 4*(2), 186-194.

Trivedi, I. N., Jangir, P., Bhoye, M., & Jangir, N. (2018). An economic load dispatch and multiple environmental dispatch problem solution with microgrids using interior search algorithm. *Neural Computing & Applications*, *30*(7), 2173–2189. doi:10.100700521-016-2795-5

Zhang, Y., & Giannakis, G. B. (2014, April). Efficient decentralized economic dispatch for microgrids with wind power integration. In *Green Technologies Conference (GreenTech), 2014 Sixth Annual IEEE* (pp. 7-12). IEEE.10.1109/GREENTECH.2014.12

Zhu, J. (2014, July). An optimal approach for smart grid economic dispatch. In *PES General Meeting| Conference & Exposition, 2014 IEEE* (pp. 1-5). IEEE. 10.1109/PESGM.2014.6938941

Zou, D., Li, S., Wang, G. G., Li, Z., & Ouyang, H. (2016). An improved differential evolution algorithm for the economic load dispatch problems with or without valve-point effects. *Applied Energy*, *181*, 375–390. doi:10.1016/j.apenergy.2016.08.067

Chapter 3
Role of Human Resources, Production Process, and Flexibility on Commercial Benefits From AMT Investments

Jorge Luis García-Alcaráz
https://orcid.org/0000-0002-7092-6963
Autonomous University of Ciudad Juarez, Mexico

Arturo Realyvásquez-Vargas
https://orcid.org/0000-0003-2825-2595
Instituto Tecnológico de Tijuana, Mexico

Emilio Jiménez-Macías
https://orcid.org/0000-0001-6749-4592
University of La Rioja, Spain

Liliana Avelar Sosa
Autonomous University of Ciudad Juarez, Mexico

Aide Aracely Maldonado-Macías
Autonomous University of Ciudad Juarez, Mexico

ABSTRACT

Advanced manufacturing technologies (AMT) acquisition by maquiladoras (foreign-owned manufacturing companies) is a tendency that allows these companies to maximize their commercial benefits. However, it remains unclear how the AMT implementation impacts on their performance. In addition, this research studies 383 responses to a questionnaire about the AMT implementation in the Mexican maquiladora industry and reports an analysis with four latent variables associating obtained benefits after the AMT implementation—human resources, flexibility, production process, and commercial benefits—where their relationships are evaluated through six hypotheses using a structural equation model (SEM). Finally, the outcomes demonstrated that AMT benefits for human resources have a direct effect on flexibility, production process, and commercial benefits. However, the direct effect from human resources benefits, knowledge, and experience on commercial benefits are acquired through indirect effects, using flexibility and production process as mediator variables.

DOI: 10.4018/978-1-5225-8223-6.ch003

INTRODUCTION

In 1965, the Mexican government launched the Border Industrialization Program (BIP), where a specific kind of manufacturing companies were born: maquiladoras. In addition, Mexican maquiladoras are usually foreign-owned companies that are established in the northern border of the country (García-Alcaraz, Maldonado, Iniesta, Robles, & Hernández, 2014); they import high levels of raw material and equipment on a tariff-free basis, as well as export the assembled or processed goods overseas (Hadjimarcou, Brouthers, McNicol, & Michie, 2013).

Moreover, in its beginning, the BIP were focused on solving the problem of increasing unemployment along the Mexico-US border. However, with the North American Free Trade Agreement (NAFTA), this maquiladora program became attractive for both American and Canadian companies (Cherniwchan, 2017). Additionally, the NAFTA guaranteed the importation of raw materials on a tariff-free basis, as well as their exportation at preferential tariffs among these countries, a as a result, American and Canadian companies could benefit themselves from assembling their products in Mexico with a low labor cost and a qualified work force.

In a specific and regional context, Ciudad Juárez - El Paso is the largest binational metropolitan area in the Mexico-US border, in fact, it is the seventh manufacturing center of North America (AMAC, 2013). In addition, with 326 maquiladoras in Ciudad Juarez (Mexico), the city owns 67.23% of the state's total and 6.42% of the country's total manufacturing industry (AMAC, 2013), since there are 5,074, where the manufacturing sector in Ciudad Juarez offers 222,040 direct jobs, representing 62.24% of the state's total and 9.90% of the country's total (AMAC, 2013). Also, most maquiladoras in Ciudad Juárez are automotive manufacturers (29%) and electric/electronics (25%) factories (AMAC, 2013).

Furthermore, the importance of studying the manufacturing industry in Ciudad Juárez can be justified by three facts. First, many studies about the growth of the Mexican maquiladora industry predict that by 2020 Mexico would provide 7.8% of the global gross domestic product (GDP) (García-Alcaraz et al., 2014). Second, this city yearly imports $22.6 USD billions in raw materials and exports $43 USD billions in assembled products. Third, maquiladoras are subsidiary factories owned by parent companies overseas, where there is a complex process of raw material importation and finished products exportation.

Those maquiladoras are equipped with adequate technological capacity at first, but eventually they will be provided with more machinery, equipment, and updated production methodologies, such are commonly known as advanced manufacturing technologies (AMT), which can be divided into different categories according to their applications as well as five categories by Small and Chen (1997):

- Design and engineering
- Processing, fabrication, and assembling
- Automated material handling
- Automated inspection and test equipment
- Information

As a matter of fact, maquiladoras and partners invest in AMT because they may obtain certain benefits along with the production system. Similarly, AMT may have a far-reaching impact on *Human resources*, commercial and marketing aspects, materials handling, *Flexibility*, as well as on the *Production process*; however, Jack and Laura (1994) indicate that the main benefit from AMT is the competitive advantage.

Moreover, several authors as Ordoobadi and Mulvaney (2001) classified AMT benefits in ten categories: increased *Flexibility*, increased quality, increased productivity, expanded usage of technology, promotion of strategic objectives, competitive strength, increased customer satisfaction, increased market opportunities, increased ease of operation, and improved employee relationships. However, Garcia-Alcaraz, Iniesta, and Juárez (2012) identified seven AMT categories, where advantages or benefits used a factor analysis: operative benefits, *Commercial benefits*, waste reduction, improved knowledge quality and management, improved product design, improved usage of layout, and improved inventory management.

Also, Hynek and Janecek (2009) report a list of twenty benefits expected by managers after the AMT implementation. For instance, the principal benefits expected are presented on the reduced cost and quality for final products. However, benefits as *Flexibility*, competitive advantages, and reduction of work in process are occupying the next places. Also, it is important to observe that improved workforce attitudes and management attitudes have the nineteenth and twentieth place respectively (the last two places).

Furthermore, by analyzing the previous paragraphs, it can be inferred that some AMT benefits trigger others. For example, without the appropriate management knowledge of *Human resources*, others AMT benefits may not be obtained, as quality and *Flexibility*. Despite extensive research on AMT implementation, it remains unclear how AMT benefits impact on one another, which prevents manufacturing companies from focusing their attention on those AMT factors that must be achieved. In order to fill this gap, this paper proposes a structural equation model (SEM) to quantify the relationships between four categories from AMT benefits: *Human resources*, *Production process*, *Flexibility,* and *Commercial.* In addition, it is assumed that *Human resources* are the most relevant benefits, because they motivate a better operator productivity, labor cost reduction, and engineering knowledge, since those benefits may develop a better *Flexibility* along with the *Production process*, as a result acquire *Commercial benefits*, which may be reflected as a financial income. In other words, *Commercial benefits* are seen as the ultimate goal of companies, having *Flexibility* and *Production process* as mediating variables.

Finally, results from this research may help managers and decision makers to focus their attention on certain benefits to optimize resources related to the AMT implementation, where they will identify each benefit with a priority in order to guarantee Commercial benefits associated with a higher financial income, which is what investors are looking forward to obtain.

BACKGROUND AND HYPOTHESES

In the past, operating and *Commercial benefits* were separately studied from other AMT advantages; however, current studies have managed to incorporate the human factor as a success variable, since AMT help operators to acquire knowledge and improve the skills that are necessary to effectively do their jobs. Consequently, it is now argued that human factors must be considered from the planning phase in the AMT implementation (Snell & Dean, 1992; Stahre, 1995).

In addition, *Human resources* are directly responsible from implementing AMT, and they are accountable for the subsequent benefits that are obtained in the *Production process* (Snell & Dean, 1992). Therefore, if operators and administrators are not convinced about the AMT benefits, they reject AMT plans in the production lines, it may be difficult for companies to obtain the expected benefits (Co, Patuwo, & Hu, 1998), as the AMT implementation process may be compromised (Stahre, 1995). Considering the role of *Human resources* in the *Production process*, the following first working hypothesis is proposed:

H_1. AMT benefits for *Human resources* have a positive and direct effect on AMT benefits for the *Production process.*

However, knowledge and experience obtained from the AMT implementation may be also reflected in the *Flexibility* process. On one hand, expansion *Flexibility* increases as machines and operators perform more and different activities (Pyoun & Choi, 1994), where modern AMTs help to have a reduced changeover and setup times, due to a high level in maintainability (Genchev & Willis, 2014), as a result, increasing *Flexibility* for companies. Also, this low changeover and setup times are key to achieve a small lot production, which may eventually lead to product customization (Azaiez et al., 2016; Huettemann, Gaffry, & Schmitt, 2016; Lafou, Mathieu, Pois, & Alochet, 2015). However, this process directly depends on employee capabilities and abilities to operate these new AMTs, therefore the following hypothesis is proposed:

H_2. AMT benefits for *Human resources* have a positive and direct effect on the *Flexibility* obtained from an AMT investment.

Furthermore, *Human resources* and senior management are the most valuable asset of any company as well as their main source of profits, although such relationships are not always direct. In order to understand the impact of human factors, employee knowledge and experience may help firms to stay competitive (Swink & Nair, 2007), have a competitive advantage in the market (Cook & Cook, 1994), provide quick responses to customer demands (Thomas, Barton, & John, 2008), and offer fast product deliveries (Garcia-Alcaraz et al., 2012). Also, if an AMT is applied to a product and system design, knowledge and experience gained from the AMT implementation guarantee companies an early market diffusion (Ranjan, Jha, & Pal, 2016), providing *Commercial benefits* to companies. Considering the impact of *Human resources* on *Commercial benefits*, the following hypothesis is presented:

H_3. AMT benefits for *Human resources* have a positive and direct effect on AMT *Commercial benefits.*

Additionally, firms that manage to increase *Flexibility* from an AMT implementation, automatically improve their *Production process* (Swink & Nair, 2007). As an example, reduction in lot sizes and production volumes implies that the time process and production costs have been diminished or that changeovers and setup times programs have succeeded (Esa, Rahman, & Jamaludin, 2015). Furthermore, expansion and production *Flexibility* may improve the capacity and layout (Lafou et al., 2015), since new and modern AMT machines are smaller in size but with similar capacities (Esmaeilian, Behdad, & Wang, 2016). Therefore, the following hypothesis is proposed:

H_4. *Flexibility* benefits from AMT have a positive and direct effect on AMT benefits for *Production process* indexes.

Even though the main goal for the AMT implementation is to maximize profits, it cannot be achieved without first improving other areas, such as *Flexibility* like production *Flexibility*, also the lot size reduction may help to acquire precise product requirements and focus on a particular group of customers (Prabhaker, Goldhar, & Lei, 1995), whereas improving the operating *Flexibility* from environmentally friendly machines, which improves the corporation image as well as ensures social acceptance (Kong,

Feng, Zhou, & Xue, 2016), consequently, increasing sales and maximizing profits. Likewise, since soft AMT increase *Flexibility* and reduce time in product design and prototyping, companies may benefit themselves from an early market diffusion (Singhry, Abd Rahman, & Imm, 2016). Therefore, the fifth hypothesis is established:

H_5. *Flexibility* benefits from AMT have a positive and direct effect on AMT *Commercial benefits.*

Moreover, most *Commercial benefits* from AMT implementation are the result of an improved *Production process*. For instance, improved quality may increase sales (Bülbül, Ömürbek, Paksoy, & Bektaş, 2013), whereas better production organization may reduce the product delivery times and processing errors, as a result improving quality (Bülbül et al., 2013). Similarly, production costs reduction might potentially increase sales, as long as the product quality and reliability and the positive corporation image are preserved. In fact, these three elements are reliable indicators for the successful AMT implementation. According to the previous information, the sixth hypothesis is proposed:

H_6. AMT benefits for the *Production Process* have a positive direct effect on AMT *Commercial benefits.*

Correspondingly, Figure 1 illustrates the six hypotheses proposed in this research, where each ellipse represents a latent variable while the arrow represents a hypothesis. In addition, latent variables are variables that are not directly observed but are rather inferred (through a mathematical model) from other variables that are presented (directly measured) and usually are called items (Kock, 2018; Nitzl, 2016). For example, the *Flexibility* latent variable can be integrated by different kinds of flexibilities, as expansion, volume, small lot size, among others.

In particular, it is important to mention that the term of *Commercial benefits* is being used instead of economic or financial benefits, since it is difficult to have access to financial data from companies. However, *Commercial benefits* are directly related to the economic benefits that are obtained after an adequate AMT implementation on industries.

METHODOLOGY

In order to quantify the relationships between the four latent variables from the AMT benefits as well as statistically validate the hypotheses, the following tasks are performed.

Stage 1: Survey Design

As a data collection instrument, a three-section questionnaire was designed to assess the AMT implementation and benefits among maquiladoras in Ciudad Juárez. In addition, the first section gathered sociodemographic information from participants, such as gender, company size, and industrial sector, whereas the second section evaluated the success in the AMT implementation, finally, the third section assessed the AMT benefits and integrated them into four latent variables as follows:

- *Human resources* (four items)
- *Production process* (nine items)

Figure 1. Proposed hypotheses

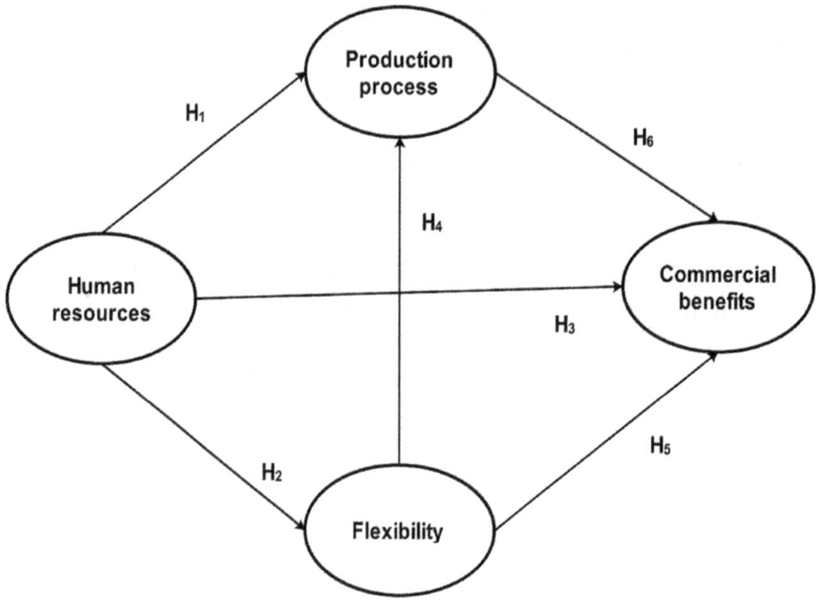

- *Flexibility* (seven items),
- *Commercial benefits* (seven items).

Finally, the fourth section has a list of barriers that managers have reported during the AMT justification and implementation. However, that information is not analyzed in this paper. Also, the survey is designed based on a basis literature review about AMT implementation. In addition, the list below reports the items in the survey that are included in each latent variable from the AMT benefits, which some authors have supported.

According to (Chung, 1996; Co et al., 1998; Corbett, 1988; Majchrzak, 1988; Slatter, Husband, Besant, & Ristic, 1989; Snell & Dean, 1992) *Human resources* benefits are:

- Increased operator productivity
- Labor cost reduction
- Increased engineering knowledge and expertise
- Increased administration expertise

According to (Abdul Ghani, Jayabalan, & Sugumar, 2002; Bourke & Roper, 2016; Khanchanapong et al., 2014; Koc & Bozdag, 2009; Percival & Cozzarin, 2010) *Production process* benefits are:

- Production costs reduction
- Process time reduction
- Increased product quality
- Increased plant capacity
- Improved capacity utilization

- Improved production organization
- Reduced machine setup times
- Increased machine reliability
- Improved space usage

According to (Azaiez et al., 2016; Genchev & Willis, 2014; He, Keung Lai, Sun, & Chen, 2014; Huettemann et al., 2016; Lafou et al., 2015; Pyoun & Choi, 1994) *Flexibility* benefits are:

- Machine size reduction
- Small lot production
- Machine *Flexibility*
- Process *Flexibility*
- Production volume *Flexibility*
- Expansion *Flexibility*
- Reprocessing and waste management

According to (Bourke & Roper, 2016; Efstathiades, Tassou, & Antoniou, 2002; Ordoobadi & Mulvaney, 2001; Schoute, 2011; Swink & Nair, 2007) *Commercial benefits* are:

- Competitive strength
- Increased market coverage
- Rapid response to customer needs
- Early market penetration
- Faster delivery times
- Lead time reduction
- Increased sales

Furthermore, the questionnaire is answered using a five-point Likert scale, where the lowest value (1) indicates that an AMT benefit has been never obtained, whereas the highest value (5) implies that an AMT benefit has been always obtained. In addition, the whole questionnaire is included in the appendix; however, in this research only some of the analyzed variables and items are presented. Also, a summary to illustrate the degree from the five points previously mentioned is presented in the following section:

1. Strongly disagree
2. Disagree
3. Neither agree nor disagree
4. Agree
5. Strongly agree

Stage 2: Survey Administration

The current survey is focused on maquiladoras stablished in Ciudad Juárez (Mexico), which have implemented AMT for at least 1 years to guarantee that they have reach and record benefits after their implementation. Therefore, the support of Maquiladora Association A.C (AMAC) was requested in order to reach such companies using a stratified sampling method and subsequently, respondents recommended other colleagues, as a consequence, the snowball sampling technique is applied when the manufacturing companies are not registered in the AMAC. In addition, the sample included AMT implementation leaders, and regarding the survey administration process, face-to-face personal interviews were performed from May to September 2017.

Stage 3: Data Registration and Questionnaire Validation

In the data registration process, a database was created with collected data using the statistical software SPSS 24®, where data is screened to detect missing values and outliers, which are replaced by the median value of the items, therefore, the latent variables are validated using the following indexes (Avelar-Sosa, García-Alcaraz, Cedillo-Campos, & Adarme-Jaimes, 2014):

- Cronbach's alpha and composite reliability for internal reliability, accepting values over 0.7 (Mani, Gunasekaran, Papadopoulos, Hazen, & Dubey, 2016), where both indexes are obtained through an iteration process, since removing items often increases reliability latent variables values (Avelar-Sosa et al., 2014).
- Average Variance Extracted (AVE) for convergent validity, accepting values over 0.5 (Yan & Azadegan, 2017).
- Predictive validity: Three predictive validity indexes are estimated; R^2, Adjusted R^2, and Q^2. In addition, as measures of parametric predictive validity, values over 0.02 for R^2 and Adjusted R^2 are accepted. Also, as measure of non-parametric predictive validity, values for Q^2 must be higher than 0 and similar to their corresponding R^2 values (Jacobs, Yu, & Chavez, 2016; Teixeira, Jabbour, de Sousa Jabbour, Latan, & de Oliveira, 2016).
- Variance Inflation Factors (VIF), used for measure internal collinearity of latent variables, setting 5 as the maximum value (Evermann & Tate, 2016).

Stage 4: Descriptive Analysis

A univariate analysis was conducted in order to find measures of central tendency as well as data dispersion for each item in the latent variables. In addition, since ordinal data was obtained through subjective assessments, the median or second quartile was used as a measure of central tendency, and the interquartile range (IQR) as a dispersion measure (Hair, Black, Babin, & Anderson, 2009). Also, the high median values indicated that an AMT benefit is always obtained in Mexican maquiladoras, while low median values implied that an AMT benefit was never acquired while for IQR (difference between the third and first quartiles), high values denoted low agreement among respondents regarding the median value of an item, whereas low values indicated high agreement regarding the median value of an item.

Stage 5: Structural Equation Model

A wide range of statistical tools, such as multiple regression analysis, discriminant analysis, and factor analysis, among others, allow to measure relationships between latent variables (Hair Jr, Anderson, Tatham, & Black, 2005). However, most of them cannot make simultaneous analyzes, as it is required in this research. Therefore, the structural equation model (SEM) technique is included to test simultaneously the hypotheses that were previously discussed.

Moreover, the SEM is considered a third-generation regression technique, combining factor analysis and lineal regression analysis (Liu, Cheng, & Wang, 2015), which is applied to study latent variables composed from observed variables that may change into a dependent or independent variables in subsequent analyses, as it is shown in Figure 1. In addition, SEM has been used to study purchaser-seller relationships by Ketkar, Kock, Parente, and Verville (2012), to analyze the impact of ERP systems on the supply chain performance by Su and Yang (2010), and to measure the effects of *Human resources* on Just in Time (JIT) implementation by García-Alcaraz et al. (2015).

In order to develop and test the SEM portrayed in Figure 1, the statistical software WarpPLS 5® was used, since it is widely recommended when analyzing small samples and non-normal or ordinal data, as in this research (Avelar-Sosa et al., 2014; García-Alcaraz et al., 2014), and based on Partial Least Squares Regression (PLS), showing a β value as a result of every relationship or hypotheses, where a P-value for a statistical significance test is presented. Also, it is the general underlying model of multivariate PLS (Kock, 2018; Schubring, Lorscheid, Meyer, & Ringle, 2016):

$$X = TP^{T} + E$$

$$Y = UQ^{T} + F$$

X is a nxm matrix of predictors, Y is a nxp matrix of responses; T and U are nxl matrices that are, respectively, projections of X (the X score, component or factor matrix) and projections of Y (the Y scores); P and Q are, respectively, mxl and pxl orthogonal loading matrices; and matrices E and F are the error terms, assumed to be independent and identically distributed in random normal variables. In addition, the decompositions of X and Y are done in order to maximize the covariance between T and U, where the β value is expressed in standard deviations because PLS allow to use different scales to measure or assess the non-observed variables or items.

Moreover, in the SEM model six efficiency indexes proposed by Kock and Lynn (2012) were estimated: Average Path Coefficient (APC), Average R-Squared (ARS), Average Adjusted R-Squared (AARS), Average block VIF (AVIF), Average Full collinearity VIF (AFVIF), and the Tenenhaus goodness of fit index (GoF).

On one hand, APC measured the general efficiency of the model, while ARS and AARS measure the predictive validity; in both cases, their corresponding P-values were estimated to determine the statistical significance of results at a 95% reliable level, thereby testing null hypotheses such as: APC=0, ARS=0 and AARS=0, versus alternative hypotheses: APC≠0, ARS≠0, AARS=0. On the other hand, AVIF and AFVIF allow to measure the collinearity between latent variables, setting 5 as the maximum value (Kock, 2011). Finally, the Tenenhaus GoF is estimated as the model fit measure, accepting values over 0.36.

In order to analyze the dependency measures between latent variables, three types of effects were estimated: direct, indirect, and total effects, all of them are statistically tested at a 95% reliable level (Hayes & Preacher, 2010).

- **Direct Effects:** They are represented in Figure 1 by arrows directly connecting two latent variables, direct effects express dependency between latent variables in standard deviation units. For each direct effect, a null hypothesis was tested $\beta_i=0$ against an alternative hypothesis $\beta_i\neq 0$; it was aimed to prove the alternative hypothesis, looking for P-values under 0.05.
- **Indirect Effects:** An independent latent variable may indirectly affect another dependent latent variable through a mediating variable, where one relationship may have many indirect effects, therefore, using at least two model segments. In addition, regarding the direct effects, the β value was estimated as a dependency measure, as well as its corresponding *P*-value to determine the statistical significance of each indirect effect.
- **Total Effects:** These are the sum of direct and indirect effects occurring in each relationship, where the total effects were validated in the same way as the direct and indirect effects.

Finally, for each dependent latent variable the R^2 values was estimated as a measure of the explained variance that can be decomposed to determine the effects' size that independent latent variables have over a dependent latent variable. Also, the R^2 value of a dependent latent variable is the sum of all the effects sizes from independent latent variables. In the present research, the value of R^2 was decomposed in order to identify which independent variable had the highest or lowest explanatory power in the variability of a dependent latent variable.

DISCUSSION AND RECOMMENDATIONS

This section is divided into different subsections according to the provided data.

AMT Implemented in Mexican Maquiladoras

After five months, 397 surveys were collected from Mexican maquiladoras located in Ciudad Juárez, where 14 were discarded since they contained more than 10% of missing values. In addition, only 383 valid questionnaires were analyzed to study the AMT implementation in the Mexican maquiladora industry, where the results are shown in Table 1 summarizing the types of AMT implemented in maquiladoras from Ciudad Juárez (Mexico), place them in a descending order according to their frequency of implementation in surveyed companies (column N). Also, median values (50th percentile) indicate whether the implementation of such AMT has been successful or not, while the IQR values indicate an agreement among respondents regarding the median values.

As an illustration, the first place ranked with 30.50% or 117 cases from the analyzed surveys; *Numerical control machines or processes such as NC, CNC, or DNC* seem to be the most frequent type of AMT implemented in Mexican maquiladoras from Ciudad Juárez (Mexico). In addition, considering the median value of the item, which is higher than 3, respondents considered the implementation of this AMTs as frequently successful, although the IQR value (2.05) indicates a moderate agreement on this matter.

Moreover, it is relevant to notice that the five following AMT are soft technologies rather than hard technologies, as it is the case of NC/CNC and DNC machines. In this sense, results showed that *Manufacturing resource planning (MRP II)* is the second most implemented AMT in maquiladoras, and its implementation is almost always successful (median value is 3.99). In addition, in this case, the IQR value is the lowest in the list, consequently, denoting great agreement among respondents regarding how effective this AMT is. Also, it is essential to mention that technologies associated to materials flow or planning have been implemented in maquiladora industries, because *Manufacturing resource planning (MRP II)* is occupying the second place, *Manufacturing resource planning (MRP)* the fourth place, and *Just in* time the fifth place. As a conclusion, this results may be normal due to the maquiladora nature, because they import raw materials and export finished products where supply chains and logistic operations have a key role for companies.

In the same way, particular attention must be focused on *Manufacturing resource planning (MRP)* and *Just in Time*, the fourth and fifth most popular AMT among Mexican maquiladoras, because according to participants and considering their median or 50th percentile values (3.98 and 3.92, respectively), the implementation of both MRP and Just in time frequently seems to be successful.

In addition, regarding the highest and lowest median values, results indicated that *computer-aided design (CAD)* is the most successful implemented AMT (median value is 4.09) among Mexican maquiladoras from Ciudad Juárez, whereas *lasers for materials processing* (LM) seem to be the least effective, as the median value is the lowest (3.37). Finally, regarding agreement, most participants agree on the success of *manufacturing resources planning (MRP II)* in the maquiladora industry, since this item showed the lowest IQR value in the analysis (1.55). However, there is a low agreement between participants regarding the usage of *other robots;* if they are actually effective and successful for the company; this item showed the highest IQR value in the analysis (2.85).

Table 1. Types of AMT implemented in Mexican maquiladoras

	N	Percentiles			IQR
		25th	50th	75th	
Numerical control machines or processes (NC/CNC or DNC)	117	2.63	3.84	4.68	2.05
Manufacturing resource planning (MRP II)	46	3.18	3.99	4.73	1.55†
Computer-aided design (CAD)	42	3.14	4.09‡	4.81	1.67
Manufacturing resource planning (MRP)	38	3.15	3.98	4.71	1.56
Just in time	34	2.91	3.82	4.65	1.74
Computer-aided process planning (CAPP)	18	2.84	3.82	4.66	1.82
Object-lifting robots	18	2.04	3.61	4.63	2.59
Automated materials handling systems (AMHS)	17	2.24	3.38	4.38	2.14
Automated auditing and test equipment (AITE)	15	2.83	3.80	4.64	1.81
Automated storage and retrieval systems (AS/RS)	13	2.21	3.41	4.44	2.23
Flexible manufacturing cells or systems FMC/FMS)	8	2.7	3.70	4.56	1.86
Computer-integrated manufacturing (CIM)	7	2.47	3.49	4.42	1.95
Other robots	5	1.74	3.48	4.61	2.87‡
Lasers for materials processing (LM)	5	1.94	3.37†	4.44	2.50

‡ Highest value, †Lowest value
Source: The authors

Descriptive Analysis of the Sample

Table 2 summarizes the results of the surveyed maquiladora sectors as well as the work experience from the respondents. Although 383 valid questionnaires were considered for analysis, only 369 participants answered these two sociodemographic questions. In addition, as it can be observed, the automotive industry was the most surveyed sector, with 186 reported cases; it was followed by the medical industry and machinery, with 48 cases. Regarding the work experience, it seems that few participants had little experience in the field of AMT implementation, as only 14 reported 1-2 years of work experience, representing only 3.79% of the sample. In contrast, most of the surveyed AMT implementation leaders had 2-3 years (40.65%) or 3-5 years (31.97%) of work experience, where both categories summed up a total of 268 participants resulting on 72.62% of the sample. Finally, these results demonstrated that the collected data about the AMT implementation in maquiladoras from Ciudad Juárez was retrieved from reliable respondents.

Additionally, Table 3 reports the genre of the surveyed respondents, as well as their job position, ranked in a descending order according to the number of interviewees in each category. On one hand, these results highlighted two facts: only 353 participants reported their job position and genre, where most AMT implementation leaders are male production managers with 156 responders. On the other hand, 188 production managers (male and female) were surveyed, representing 53.25% of the sample, followed by 121 production engineers (34.27%). In addition, both categories summed up a total of 87.53% from the sample. On the contrary, procurement manager was the least surveyed position, with 13 reported cases, therefore, confirming that the AMT implementation process in maquiladoras from Ciudad Juárez is usually performed in production departments.

Descriptive Analysis of the Items

Table 4 summarizes the descriptive analysis of the items related to benefits from AMT, illustrating the median value as the central tendency and the interquartile range as a deviation measure for and univariate analysis that helps to indicate the most significant obtained benefits. In addition, the items appear

Table 2. Surveyed sectors and work experience of participants

Sector	Work experience (years)					Total
	1-2	>2 and <3	>3 and <5	>5 and <10	> 10	
Automotive	2	67	73	32	12	186
Medical	2	21	13	7	5	48
Machinery	4	23	14	4	3	48
Electronics	4	16	9	9	3	41
Electric	2	18	6	6	0	32
Logistics	0	3	2	2	2	9
Aeronautic	0	2	1	1	1	5
Total	14 (3.79%)	150 (40.65%)	118 (31.97%)	61 (16.53%)	26 (7.04%)	369 (100%)

Source: The authors

Table 3. Work position and genre

Job position	Genre		Total
	Female	Male	
Production manager	32	156	188
Production engineer	23	98	121
Production supervisor	7	24	31
Procurement manager	2	11	13
Total	64	289	353

Source: The authors

in a descending order according to their median values for each category or latent variable analyzed. In general terms, the four most common AMT benefits for *Human resources* showed median values under 4 but over 3, indicating that they are frequently acquired after the AMT implementation. Also, the most important of these benefits according to participants it refers to *Increased engineering knowledge and expertise*, however, the *Increased administration expertise* for manager is occupying the last place, which indicates that AMT implemented in maquiladoras stablished in Ciudad Juarez are related to the *Production process*, such as, CNC machines.

Moreover, from a similar perspective, all *Flexibility* items presented median values like those from the previous category. In other words, all the analyzed *Flexibility* benefits are frequently obtained, but not always, because the median is under 4, but higher than 3. On one hand, participants reported *Expansion Flexibility* as the most common *Flexibility* benefit obtained, which indicates that AMT always allow an incremental growth, since new AMT may be effective in modern multitasking systems if they are appropriately installed. On the other hand, *Small lot production* was the lowest ranked item, implying that it is less frequently achieved, although if it is compared with the *Expansion flexibility* (first item) the difference in the median value is relatively small, only 0.23 units; this low value in *Small lot production* indicates that setup times are still and opportunity area to be improved, because currently it is necessary to have big lots of production, losing product customization.

As a matter of fact, special attention must be focused on the *Production process* latent variable and the benefits from AMT, in which some items presented median values over 4; the highest values reported in the whole analysis. In addition, these results indicated that *Improved space usage* is the most frequently benefit obtained, and it may be due to small, modern, and effective integration between soft and hard technologies through appropriate communication protocols, because nowadays modern machines are occupying not too much space for better layouts. On the other hand, *Reduced machine setup times* seems to be the least production benefit obtained, as its median value was the lowest in this category. Also, it is essential to remember that *Small lot production* in *Flexibility* benefits is occupying the last place in its category, and these two items are clearly related.

Finally, the analysis from the *Commercial benefits* latent variable has demonstrated that *Sales increased* are usually acquired after the AMT implementation, whereas the remaining benefits are less regular. In addition, it seems that AMT rarely allow the *Early market penetration*, although the median value of this item suggests that this benefit is still regular because the median values are higher than 3.

Table 4. Descriptive analysis of the items

Latent variables and items	Percentiles			IQR
Human resources	25th	50th	75th	
Increased engineering knowledge and expertise	3.08	3.88	4.64	1.56
Increased operator productivity	3.1	3.85	4.61	1.51
Labor cost reduction	2.91	3.78	4.61	1.7
Increased administration expertise	3	3.77	4.58	1.58
Flexibility				
Expansion *Flexibility*	3.08	3.87	4.64	1.56
Process *Flexibility*	3.07	3.84	4.61	1.54
Reprocessing and waste reduction	2.93	3.83	4.66	1.73
Production volume *Flexibility*	3	3.81	4.62	1.62
Machine *Flexibility*	2.93	3.8	4.63	1.7
Small lot production	2.77	3.64	4.47	1.7
Production process				
Improved space usage	3.25	4.11	4.79	1.54
Increased plant capacity	3.27	4.06	4.75	1.48
Increased product quality	3.29	4.04	4.71	1.42
Improved capacity utilization	3.23	4.01	4.71	1.48
Process time reduction	3.16	3.95	4.69	1.53
Increased reliability	3.15	3.94	4.68	1.53
Improved production organization	3.11	3.91	4.68	1.57
Production costs reduction	3.11	3.9	4.66	1.55
Reduced machine setup times	3.07	3.83	4.61	1.54
Commercial benefits				
Increased sales	3.22	4.02	4.74	1.52
Increased market coverage	3.39	3.99	4.71	1.53
Competitive strength	3.18	3.98	4.69	1.51
Lead time reduction	3.14	3.95	4.69	1.55
Faster delivery times	3.12	3.94	4.7	1.58
Rapid response to customer needs	3.06	3.84	4.62	1.56
Early market penetration	3.05	3.82	4.61	1.56

Source: The authors

DATA VALIDATION

Table 5 displays the results from the data validation process before any analysis or multivariate interpretation. In addition, findings indicate that all R^2 and Adjusted R^2 values are over 0.2, therefore, confirming parametric predictive validity of each latent variable or benefits. Similarly, all Q^2 values were under zero and similar to their corresponding R^2 and Adjusted R^2 values, which corroborated the non-parametric validity.

Moreover, regarding to the Cronbach's Alpha and the Composite Reliability index, results indicated that all latent variables had internal validity, since all values from both indexes were higher than 0.7. In addition, since, all AVE values were over 0.5, it is concluded that the four latent variables had enough convergent validity. Finally, all the VIF values were under the maximum value of 3.3, as a result, collinearity problems inside latent variables are discarded. In conclusion, results from the data validation process indicated that the latent variables data were appropriate and may be used to create the SEM.

Structural Equation Model

After validating the latent variables, the model proposed in Figure 1 was evaluated; in order to test the model's appropriateness from different statistical perspectives, the six model fit indexes described in the methodology section were estimated, therefore, the following results were obtained:

- APC=0.405, P<0.001
- ARS=0.572, P<0.001
- AARS=0.570, P<0.001
- AVIF=1.977, acceptable if <= 5, ideally <= 3.3
- AFVIF=2.882, acceptable if <= 5, ideally <= 3.3
- Tenenhaus GoF=0.628, small >= 0.1, medium >= 0.25, large >= 0.36

In the first place, the APC index demonstrated that all relationships between latent variables were statistically significant at a 99% reliable level, since its value in the analysis is equal to 0.405, and its corresponding *P*-value was lower than 0.001. Similarly, since the ARS and AARS values were over 0.2, it is concluded that the model as a whole has enough predictive validity. In addition, regarding to collinearity indicators, AVIF and AFVIF values confirmed that the model was exempted from collinearity problems; since they were lower than 3.3. Also, the Tenenhaus GoF value demonstrated enough model fit towards the data. Finally, considering these results, the model presented in Figure 2 were analyzed and interpreted, where each relationship between latent variables shows a β-value and a P-value, a basic measured dependency between latent variables, and the least determined statistical significance of effects. As it can be observed from Figure 2, the six effects corresponding to the hypotheses discussed in

Table 5. Validation of latent variables

Index	*Human resources*	*Flexibility*	*Production process*	*Commercial benefits*
R-Squared		0.396	0.59	0.729
Adjusted R-Squared		0.395	0.587	0.726
Composite Reliability	0.910	0.933	0.95	0.933
Cronbach's Alpha	0.869	0.915	0.941	0.914
Average Variance Extracted	0.718	0.664	0.679	0.700
Full collinearity VIFs	2.255	2.233	3.417	3.621
Q-Squared		0.397	0.59	0.728

Source: The authors

the introduction section were statistically significant at a 95% reliable level, since their corresponding *P*-values were under 0.05.

Direct and Size Effects

Direct effects help to test the hypotheses discussed in the introduction section, which are initially illustrated in Figure 1. In addition, each hypothesis has a β value and a P-value to test its statistical significance and according to it, in a general way, it can be stated that every relationship between the latent variables (benefits) is statistically significant because de P-values are under 0.05.

Furthermore, considering the results from the evaluated model in Figure 2, the following conclusions regarding the direct effects between latent variables are proposed:

H_1. There is enough statistical evidence to declare that the AMT benefits for *Human resources* have a positive and direct effect on AMT benefits for the *Production process*, since when the first latent variable increases in one unit its standard deviation, the second latent variable increases in 0.409 units.

H_2. There is enough statistical evidence to declare that AMT benefits for *Human resources* have a positive and direct effect on AMT benefits for *Flexibility,* since when the first latent variable increases its standard deviation in one unit, the standard deviation of the second latent variable increases in 0.630 units.

H_3. There is enough statistical evidence to declare that AMT benefits for *Human resources* have a positive and direct effect on AMT *Commercial benefits*, since when the first latent variable increases its standard deviation in one unit, the standard deviation of the second latent variable increases in 0.230 units.

Figure 2. Evaluated model

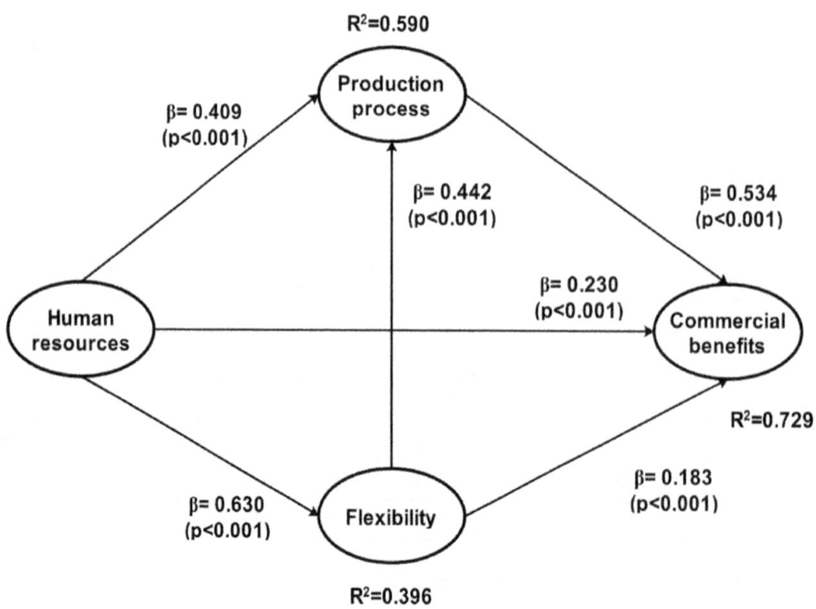

H_4. There is enough statistical evidence to declare that *Flexibility* benefits from AMT have a positive and direct effect on AMT benefits for *Production process*, since when the first latent variable increases its standard deviation by one unit, the standard deviation of the second latent variable increases in 0.442 units.

H_5. There is enough statistical evidence to declare that *Flexibility* benefits from AMT have a positive and direct effect on AMT *Commercial benefits*, since when the first latent variable increases its standard deviation by one unit, the standard deviation of the second latent variable increases in 0.183 units.

H_6. There is enough statistical evidence to declare that AMT benefits for the *Production process* have a positive and direct effect on AMT *Commercial benefits,* since when the first latent variable increases its standard deviation in one unit, the second latent variable increases in 0.534 units.

According to the β values, the following structural equations are proposed:

$$Production\ process = 0.409\ Human\ resources + 0.442\ Flexibility + Error$$

$$Flexibility = 0.630\ Human\ resources + Error$$

$$Commercial\ benefits = 0.183\ Flexibility + 0.230\ Human\ resources + 0.534\ Production\ process + Error$$

Regarding to the explained variance, Figure 2 portrays the R^2 value for each dependent latent variable. As it was previously discussed, R^2 reflects the size effects that independent latent variables have on a dependent latent variable. In addition, Table 6 presents the R^2 or size effects of decomposition for direct effects when a dependent latent variable is explained by two or more independent latent variables. For instance, *Production process* was explained in 59% (R^2=0.590) by *Flexibility* in 0.309 while *Human resources* in 0.281*;* however, the greatest size effect declares that *Flexibility* is more important than *Human resources* to explain the variability of *Production process* benefits, because the size effect is bigger.

From a similar perspective, the *Commercial benefits* latent variable was explained in 72.9% (R^2=0.729) by *Human resources* in 0.164, *Flexibility* in 0.12) and *Production process* in 0.437; however, since *Production process* showed the largest size effect, it is concluded that this latent variable is more significant to explain the variability of *Commercial benefits.* In other words, managers must focus their attention and resources to acquire benefits in the *Production process* after an AMT implementation, because that guarantees *Commercial benefits.* Also, those *Production process* benefits are represented by Improved

Table 6. Size effects

To	From			R^2
	Human resources	*Flexibility*	*Production process*	
Flexibility	0.396			0.396
Production process	0.281	0.309		0.590
Commercial benefits	0.164	0.128	0.437	0.729

Source: The authors

space usage, Increased plant capacity, Increased product quality, Improved capacity utilization, *Process time reduction, Increased reliability, Improved production organization, Production costs reduction,* and *Reduced machine setup times.*

Indirect Effects

Table 7 presents the total indirect effects between latent variables, and their corresponding *P*-values for the statistical significance test. In addition, this table also shows the size effects as a measure of the explained variance. In general, there are three indirect effects between latent variables, and all of them are statistically significant, since the *P*-values are under 0.05.

The following conclusions regarding the indirect effect are established:

1. The direct effect from *Human resources* on *Commercial benefits* are equal to 0.230. However, the total indirect effect given through *Production process* and *Flexibility* showed a higher value (0.482). Thus, its concluded that all expertise obtained by *Human resources* from AMT implementation should focus on increasing *Flexibility* and improving the *Production process*, since this strategy maximizes *Commercial benefits.*
2. As in the previous relationship, the direct effect from *Flexibility* on *Commercial benefits* was lower than the indirect effect (0.183 vs. 0.236), indicating that in an AMT implementation environment, if *Flexibility* is not reflected on the *Production process*, *Commercial benefits* may not be maximized.

Total Effects

Table 8 introduces the total effects found in the model, as well as their corresponding *P*-values and size effects. As a matter of fact, since *Human resources* had the highest effects on the other latent variables because it was placed on the top left corner of the model. In this sense, it is relevant to notice that the total effect from *Human resources* on *Commercial benefits* is equal to 0.712, and the former latent variable could explain up to 50.7% of the variability of the last variable.

Likewise, it was found that *Human resources* had the highest total effects on *Production process* (0.687), and it explained up to 47.2% of its variability (ES=0.472). In addition, these results statistically demonstrated that the *Human resources* knowledge and expertise obtained from the AMT implementation are key to guarantee a proper functioning of production lines, which would then translate into maximized profits.

Table 7. Total indirect effects

To	From	
	Human resources	*Flexibility*
Production process	0.278 (P<0.001) ES= 0.191	
Commercial benefits	0.482 (P<0.001) ES=0.343	0.236 (P<0.001) ES= 0.165

Source: The authors

Table 8. Total effects

To	From		
	Human resources	*Flexibility*	*Production process*
Flexibility	0.63 (P<0.001) ES=0.396		
Production process	0.687 (P<0.001) ES=0.472	0.442 (P<0.001) ES=0.309	
Commercial benefits	0.712 (P<0.001) ES=0.507	0.419 (P<0.001) ES=0.293	0.534 (P<0.001) ES=0.437

Source: The authors

Also, the total effects found between *Flexibility* and *Commercial benefits* are worth being highlighted, since an indirect effect was previously identified in this relationship through *Production process.* In this case, the total effects are equal to 0.419 units, where *Flexibility* explained up to 29.3% of the variance in *Commercial benefits*, since the SE=0.293.

FUTURE RESEARCH DIRECTIONS

Considering the study limitations related to geographical context and obtained results, future research will try to address the following issues:

1. Since most AMT are imported into Mexican maquiladoras, future studies will analyze the role of the Mexican government and tariff regulations in the AMT importation process.
2. In this research, it was took as basis the AMT classification proposed by Small and Chen (1997); however, future research will categorize these manufacturing technologies into soft and hard technologies, since the least have reported a positive impact along the production and distribution processes.
3. This research did not achieve full explanation for the variance of dependent latent variables (R^2); thus, further studies will pursue to integrate the product design and materials handling, since benefits from the AMT implementation are two possible impact elements, because many AMT have been positively associated with these benefits.

CONCLUSION

In this research, some effects were proposed, tested, analyzed, and measured between the four categories from the AMT benefits: *Human resources, Flexibility, Production process,* and *Commercial benefits* using a SEM. In addition, results from the model evaluation and the effects analysis between latent variables allow to provide the following conclusions regarding to the AMT implementation in the maquiladora industry from Ciudad Juárez:

- Even though AMT benefits are traditionally associated with financial performance, it is equally important for companies to focus on and ensure other AMT benefits, such as for *Human resources*, *Flexibility*, and *Production process*, since, only if these areas are improved, manufacturing companies in Ciudad Juárez may ensure their economic benefits from the AMT implementation.
- Senior managers of maquiladoras in Ciudad Juárez must attempt to guarantee *Human resources* benefits offered by the AMT implementation, since employees are directly responsible for generating *Flexibility*, *Production process,* and *Commercial benefits*; benefits that guarantee *Commercial benefits*.
- Senior and middle managers must make sure that everyone involved in the AMT implementation process – supervisors, operators, among others – uses AMT to their best, since when applying acquired knowledge and experience from the AMT implementation, employees contribute to increase *Flexibility* as well as improve the *Production process* benefits.
- The effects that AMT benefits for *Human resources* have on *Commercial benefits* are relatively low; however, if acquired knowledge and experience from the AMT implementation are applied into *Flexibility* and *Production process*, indirect and total effects are noticeably higher, which demonstrate their high influence as mediators variables.
- Senior and middle managers from maquiladoras established in Ciudad Juárez must strive to offer appropriate training every time new AMT are acquired, because as Waldeck and Leffakis (2007) argue, training is crucial for a successful AMT implementation process.

ACKNOWLEDGMENT

This research was supported by the Mexican National Council for Science and Technology (CONACYT) under the project Thematic Network in Industrial Process Optimization [grant number CONACYT-INS (REDES) 2018 - 293683 LAS].

Authors want to thank maquiladora industries established in Ciudad Juarez that answered the questionnaire in order to provide information regarding their investment practices in advanced manufacturing technologies, as well as to AMAC (Maquiladoras Association AC) for helping to connect with companies.

REFERENCES

Abdul Ghani, K., Jayabalan, V., & Sugumar, M. (2002). Impact of advanced manufacturing technology on organizational structure. *The Journal of High Technology Management Research, 13*(2), 157–175. doi:10.1016/S1047-8310(02)00051-2

AMAC. (2013). *Maquiladora Association AC - Maquiladora Overview 2016.* Retrieved from www.indexjuarez.org

Avelar-Sosa, L., García-Alcaraz, J., Cedillo-Campos, M., & Adarme-Jaimes, W. (2014). Effects of regional infrastructure and offered services in the supply chains performance: Case Ciudad Juarez. *DYNA - Colombia, 81*(186), 208-217.

Azaiez, S., Boc, M., Cudennec, L., Simoes, M. D. S., Haupert, J., Kchir, S., ... Tortech, T. (2016). Towards Flexibility in Future Industrial Manufacturing: A Global Framework for Self-organization of Production Cells. *Procedia Computer Science*, *83*, 1268–1273. doi:10.1016/j.procs.2016.04.264

Bourke, J., & Roper, S. (2016). AMT adoption and innovation: An investigation of dynamic and complementary effects. *Technovation*, *55-56*, 42–55. doi:10.1016/j.technovation.2016.05.003

Bülbül, H., Ömürbek, N., Paksoy, T., & Bektaş, T. (2013). An empirical investigation of advanced manufacturing technology investment patterns: Evidence from a developing country. *Journal of Engineering and Technology Management*, *30*(2), 136–156. doi:10.1016/j.jengtecman.2013.01.002

Cherniwchan, J. (2017). Trade liberalization and the environment: Evidence from NAFTA and U.S. manufacturing. *Journal of International Economics*, *105*, 130–149. doi:10.1016/j.jinteco.2017.01.005

Chung, C. A. (1996). Human issues influencing the successful implementation of advanced manufacturing technology. *Journal of Engineering and Technology Management*, *13*(3–4), 283–299. doi:10.1016/S0923-4748(96)01010-7

Co, H. C., Patuwo, B. E., & Hu, M. Y. (1998). The human factor in advanced manufacturing technology adoption: An empirical analysis. *International Journal of Operations & Production Management*, *18*(1), 87–106. doi:10.1108/01443579810192925

Cook, J. S., & Cook, L. L. (1994). Achieving Competitive Advantages of Advanced Manufacturing Technology. *Benchmarking for Quality Management & Technology*, *1*(2), 42–63. doi:10.1108/14635779410063329

Corbett, J. M. (1988). Ergonomics in the development of human-centred AMT. *Applied Ergonomics*, *19*(1), 35–39. doi:10.1016/0003-6870(88)90196-2 PMID:15676645

Efstathiades, A., Tassou, S., & Antoniou, A. (2002). Strategic planning, transfer and implementation of Advanced Manufacturing Technologies (AMT). Development of an integrated process plan. *Technovation*, *22*(4), 201–212. doi:10.1016/S0166-4972(01)00024-4

Esa, M. M., Rahman, N. A. A., & Jamaludin, M. (2015). Reducing High Setup Time in Assembly Line: A Case Study of Automotive Manufacturing Company in Malaysia. *Procedia: Social and Behavioral Sciences*, *211*, 215–220. doi:10.1016/j.sbspro.2015.11.086

Esmaeilian, B., Behdad, S., & Wang, B. (2016). The evolution and future of manufacturing: A review. *Journal of Manufacturing Systems*, *39*, 79–100. doi:10.1016/j.jmsy.2016.03.001

Evermann, J., & Tate, M. (2016). Assessing the predictive performance of structural equation model estimators. *Journal of Business Research*, *69*(10), 4565–4582. doi:10.1016/j.jbusres.2016.03.050

Garcia-Alcaraz, J. L., Iniesta, A., & Juárez, M. C. (2012). Benefits of advanced manufacturing technologies. *African Journal of Business Management*, *6*(16), 5524–5532. doi:10.5897/AJBM11.2777

García-Alcaraz, J. L., Maldonado, A. A., Iniesta, A. A., Robles, G. C., & Hernández, G. A. (2014). A systematic review/survey for JIT implementation: Mexican maquiladoras as case study. *Computers in Industry*, *65*(4), 761–773. doi:10.1016/j.compind.2014.02.013

García-Alcaraz, J. L., Prieto-Luevano, D. J., Maldonado-Macías, A. A., Blanco-Fernández, J., Jiménez-Macías, E., & Moreno-Jiménez, J. M. (2015). Structural equation modeling to identify the human resource value in the JIT implementation: Case maquiladora sector. *International Journal of Advanced Manufacturing Technology*, *77*(5), 1483–1497. doi:10.100700170-014-6561-5

Genchev, S., & Willis, G. (2014). A note on manufacturing flexibility as a firm-specific dynamic capability. *Manufacturing Letters*, *2*(4), 100–103. doi:10.1016/j.mfglet.2014.07.002

Hadjimarcou, J., Brouthers, L. E., McNicol, J. P., & Michie, D. E. (2013). Maquiladoras in the 21st century: Six strategies for success. *Business Horizons*, *56*(2), 207–217. doi:10.1016/j.bushor.2012.11.005

Hair, J. F., Jr., Anderson, R. E., Tatham, R. L., & Black, W. C. (2005). Analisis Multivariante (5th ed.). Pearson, Prentice-Hall.

Hair, J. F., Black, W. C., Babin, B. J., & Anderson, R. E. (2009). *Multivariate data analysis*. Upper Saddle River, NJ: Prentice Hall.

Hayes, A. F., & Preacher, K. J. (2010). Quantifying and Testing Indirect Effects in Simple Mediation Models When the Constituent Paths Are Nonlinear. *Multivariate Behavioral Research*, *45*(4), 627–660. doi:10.1080/00273171.2010.498290 PMID:26735713

He, Y., Keung Lai, K., Sun, H., & Chen, Y. (2014). The impact of supplier integration on customer integration and new product performance: The mediating role of manufacturing flexibility under trust theory. *International Journal of Production Economics, 147*(Part B), 260-270. doi:10.1016/j.ijpe.2013.04.044

Huettemann, G., Gaffry, C., & Schmitt, R. H. (2016). Adaptation of Reconfigurable Manufacturing Systems for Industrial Assembly – Review of Flexibility Paradigms, Concepts, and Outlook. *Procedia CIRP*, *52*, 112–117. doi:10.1016/j.procir.2016.07.021

Hynek, J., & Janecek, V. (2009). *Problems of advanced manufacturing technology benefits evaluation.* Paper presented at the 2009 International Conference on Intelligent Engineering Systems. 10.1109/INES.2009.4924746

Jack, S. C., & Laura, L. C. (1994). Achieving Competitive Advantages of Advanced Manufacturing Technology. *Benchmarking for Quality Management & Technology*, *1*(2), 42–63. doi:10.1108/14635779410063329

Jacobs, M. A., Yu, W., & Chavez, R. (2016). The effect of internal communication and employee satisfaction on supply chain integration. *International Journal of Production Economics*, *171*(Part 1), 60–70. doi:10.1016/j.ijpe.2015.10.015

Ketkar, S., Kock, N., Parente, R., & Verville, J. (2012). The impact of individualism on buyer–supplier relationship norms, trust and market performance: An analysis of data from Brazil and the U.S.A. *International Business Review*, *21*(5), 782–793. doi:10.1016/j.ibusrev.2011.09.003

Khanchanapong, T., Prajogo, D., Sohal, A. S., Cooper, B. K., Yeung, A. C. L., & Cheng, T. C. E. (2014). The unique and complementary effects of manufacturing technologies and lean practices on manufacturing operational performance. *International Journal of Production Economics*, *153*, 191–203. doi:10.1016/j.ijpe.2014.02.021

Koc, T., & Bozdag, E. (2009). The impact of AMT practices on firm performance in manufacturing SMEs. *Robotics and Computer-integrated Manufacturing*, *25*(2), 303–313. doi:10.1016/j.rcim.2007.12.004

Kock, N. (2011). Using WarpPLS in e-collaboration studies: Mediating effects, control and second order variables, and algorithm choices. *International Journal of e-Collaboration*, *7*(3), 1–13. doi:10.4018/jec.2011070101

Kock, N. (2018). *WarpPLS 6.0 User Manual*. Laredo, TX: ScriptWarp Systems.

Kock, N., & Lynn, G. S. (2012). Lateral collinearity and misleading results in variance-based SEM: An illustration and recommendations. *Journal of the Association for Information Systems*, *13*(7), 546–580. doi:10.17705/1jais.00302

Kong, D., Feng, Q., Zhou, Y., & Xue, L. (2016). Local implementation for green-manufacturing technology diffusion policy in China: From the user firms' perspectives. *Journal of Cleaner Production*, *129*, 113–124. doi:10.1016/j.jclepro.2016.04.112

Lafou, M., Mathieu, L., Pois, S., & Alochet, M. (2015). Manufacturing System Configuration: Flexibility Analysis For automotive Mixed-Model Assembly Lines. *IFAC-PapersOnLine*, *48*(3), 94–99. doi:10.1016/j.ifacol.2015.06.064

Liu, J., Cheng, Q., & Wang, J. (2015). Identification of geochemical factors in regression to mineralization endogenous variables using structural equation modeling. *Journal of Geochemical Exploration*, *150*, 125–136. doi:10.1016/j.gexplo.2014.12.021

Majchrzak, A. (1988). The human infrastructure impact statement (HIIS) A tool for managing the effective implementation of advanced manufacturing technology. *Jisuanji Jicheng Zhizao Xitong*, *1*(2), 95–102. doi:10.1016/0951-5240(88)90093-6

Mani, V., Gunasekaran, A., Papadopoulos, T., Hazen, B., & Dubey, R. (2016). Supply chain social sustainability for developing nations: Evidence from India. *Resources, Conservation and Recycling*, *111*, 42–52. doi:10.1016/j.resconrec.2016.04.003

Nitzl, C. (2016). The use of partial least squares structural equation modelling (PLS-SEM) in management accounting research: Directions for future theory development. *Journal of Accounting Literature*, *37*, 19–35. doi:10.1016/j.acclit.2016.09.003

Ordoobadi, S. M., & Mulvaney, N. J. (2001). Development of a justification tool for advanced manufacturing technologies: System-wide benefits value analysis. *Journal of Engineering and Technology Management*, *18*(2), 157–184. doi:10.1016/S0923-4748(01)00033-9

Percival, J. C., & Cozzarin, B. P. (2010). Complementarities in the implementation of advanced manufacturing technologies. *The Journal of High Technology Management Research*, *21*(2), 122–135. doi:10.1016/j.hitech.2010.05.002

Prabhaker, P. R., Goldhar, J. D., & Lei, D. (1995). Marketing implications of newer manufacturing technologies. *Journal of Business and Industrial Marketing*, *10*(2), 48–58. doi:10.1108/08858629510087373

Pyoun, Y. S., & Choi, B. K. (1994). Quantifying the flexibility value in automated manufacturing systems. *Journal of Manufacturing Systems*, *13*(2), 108–118. doi:10.1016/0278-6125(94)90026-4

Ranjan, S., Jha, V. K., & Pal, P. (2016). Application of emerging technologies in ERP implementation in Indian manufacturing enterprises: An exploratory analysis of strategic benefits. *International Journal of Advanced Manufacturing Technology*, 1–12. doi:10.100700170-016-8770-6

Schoute, M. (2011). The relationship between product diversity, usage of advanced manufacturing technologies and activity-based costing adoption. *The British Accounting Review*, *43*(2), 120–134. doi:10.1016/j.bar.2011.02.002

Schubring, S., Lorscheid, I., Meyer, M., & Ringle, C. M. (2016). The PLS agent: Predictive modeling with PLS-SEM and agent-based simulation. *Journal of Business Research*, *69*(10), 4604–4612. doi:10.1016/j.jbusres.2016.03.052

Singhry, H. B., Abd Rahman, A., & Imm, N. S. (2016). Effect of advanced manufacturing technology, concurrent engineering of product design, and supply chain performance of manufacturing companies. *International Journal of Advanced Manufacturing Technology*, *86*(1), 663–669. doi:10.100700170-015-8219-3

Slatter, R. R., Husband, T. M., Besant, C. B., & Ristic, M. R. (1989). A Human-Centred Approach to the Design of Advanced Manufacturing Systems. *CIRP Annals - Manufacturing Technology, 38*(1), 461-464. doi:10.1016/S0007-8506(07)62746-2

Small, M. H., & Chen, I. J. (1997). Economic and strategic justification of AMT inferences from industrial practices. *International Journal of Production Economics*, *49*(1), 65–75. doi:10.1016/S0925-5273(96)00120-X

Snell, S. A., & Dean, J. W. (1992). Integrated Manufacturing and Human Resource Management: A Human Capital Perspective. *Academy of Management Journal*, *35*(3), 467–504. doi:10.2307/256484

Stahre, J. (1995). Evaluating human/machine interaction problems in advanced manufacturing. *Jisuanji Jicheng Zhizao Xitong*, *8*(2), 143–150. doi:10.1016/0951-5240(95)00008-H

Su, Y., & Yang, C. (2010). A structural equation model for analyzing the impact of ERP on SCM. *Expert Systems with Applications*, *37*(1), 456–469. doi:10.1016/j.eswa.2009.05.061

Swink, M., & Nair, A. (2007). Capturing the competitive advantages of AMT: Design–manufacturing integration as a complementary asset. *Journal of Operations Management*, *25*(3), 736–754. doi:10.1016/j.jom.2006.07.001

Teixeira, A. A., Jabbour, C. J. C., de Sousa Jabbour, A. B. L., Latan, H., & de Oliveira, J. H. C. (2016). Green training and green supply chain management: Evidence from Brazilian firms. *Journal of Cleaner Production*, *116*, 170–176. doi:10.1016/j.jclepro.2015.12.061

Thomas, A. J., Barton, R., & John, E. G. (2008). Advanced manufacturing technology implementation: A review of benefits and a model for change. *International Journal of Productivity and Performance Management*, *57*(2), 156–176. doi:10.1108/17410400810847410

Waldeck, N. E., & Leffakis, Z. M. (2007). HR perceptions and the provision of workforce training in an AMT environment: An empirical study. *Omega, 35*(2), 161–172. doi:10.1016/j.omega.2005.05.001

Yan, T., & Azadegan, A. (2017). Comparing inter-organizational new product development strategies: Buy or ally; Supply-chain or non-supply-chain partners? *International Journal of Production Economics, 183*(Part A), 21-38. doi:10.1016/j.ijpe.2016.09.023

ADDITIONAL READING

Bai, C., & Sarkis, J. (2017). Improving green flexibility through advanced manufacturing technology investment: Modeling the decision process. *International Journal of Production Economics, 188*, 86–104. doi:10.1016/j.ijpe.2017.03.013

Cheng, Y., Matthiesen, R., Farooq, S., Johansen, J., Hu, H., & Ma, L. (2018). The evolution of investment patterns on advanced manufacturing technology (AMT) in manufacturing operations: A longitudinal analysis. *International Journal of Production Economics, 203*, 239–253. doi:10.1016/j.ijpe.2018.06.019

Eldressi, K. A. (2018). *A New Approach to Measure Intangibles in Economic Analysis of Advanced Manufacturing Technology Projects Reference Module in Materials Science and Materials Engineering*. Elsevier.

Ghaffar, S. H., Corker, J., & Fan, M. (2018). Additive manufacturing technology and its implementation in construction as an eco-innovative solution. *Automation in Construction, 93*, 1–11. doi:10.1016/j.autcon.2018.05.005

Janeček, V., & Hynek, J. (2015, 17-19 March 2015). Impact of advanced technologies utilization on manufacturing firms' efficiency in times of economic decline. Paper presented at the 2015 IEEE International Conference on Industrial Technology (ICIT). 10.1109/ICIT.2015.7125372

Lu, H.-P., & Weng, C.-I. (2018). Smart manufacturing technology, market maturity analysis and technology roadmap in the computer and electronic product manufacturing industry. *Technological Forecasting and Social Change, 133*, 85–94. doi:10.1016/j.techfore.2018.03.005

Nath, S., & Sarkar, B. (2017). Performance evaluation of advanced manufacturing technologies: A De novo approach. *Computers & Industrial Engineering, 110*, 364–378. doi:10.1016/j.cie.2017.06.018

Percival, J. C., & Cozzarin, B. P. (2010). Complementarities in the implementation of advanced manufacturing technologies. *The Journal of High Technology Management Research, 21*(2), 122–135. doi:10.1016/j.hitech.2010.05.002

Singh, H., & Kumar, R. (2013). Measuring the utilization index of advanced manufacturing technologies: A case study. IFAC Proceedings Volumes, 46(9), 899-904. doi:10.3182/20130619-3-RU-3018.00395

Wang, Ben. (2018). The Future of Manufacturing: A New Perspective. *Engineering*. doi:10.1016/j.eng.2018.07.020

Wieland, M., Steimle, F., Mitschang, B., Lucke, D., Einberger, P., Schel, D., . . . Bauernhansl, T. (2017, 4-8 Aug. 2017). Rule-based integration of smart services using the manufacturing service bus. Paper presented at the 2017 IEEE SmartWorld, Ubiquitous Intelligence & Computing, Advanced & Trusted Computed, Scalable Computing & Communications, Cloud & Big Data Computing, Internet of People and Smart City Innovation (SmartWorld/SCALCOM/UIC/ATC/CBDCom/IOP/SCI).

Xia, T., Dong, Y., Xiao, L., Du, S., Pan, E., & Xi, L. (2018). Recent advances in prognostics and health management for advanced manufacturing paradigms. *Reliability Engineering & System Safety*, *178*, 255–268. doi:10.1016/j.ress.2018.06.021

Yin, S., Cavaliere, P., Aldwell, B., Jenkins, R., Liao, H., Li, W., & Lupoi, R. (2018). Cold spray additive manufacturing and repair: Fundamentals and applications. *Additive Manufacturing*, *21*, 628–650. doi:10.1016/j.addma.2018.04.017

Zhou, H., & Leong, K. (2009). A comparative study of advanced manufacturing technology and manufacturing infrastructure investments in Singapore and Sweden. *International Journal of Production Economics*, *120*(1), 42–53. doi:10.1016/j.ijpe.2008.07.013

KEY TERMS AND DEFINITIONS

Advanced Manufacturing Technology: Represent a computer-controlled or micro-electronics-based equipment used in the design, manufacture or handling of a product. An example is a traditional CNC machine or an industrial robot.

Ciudad Juárez: A city in North Mexico, in Chihuahua state on the Río Grande, opposite El Paso, Texas. Currently has 1,469,000 (2018) population.

Commercial Benefits: Refers to any benefit that comes about as a result of the directors acting for the good of the company, which reflect positively on the company in the wider, commercial world.

Flexibility: Refers to the operational flexibility to alter production when demand varies from forecast. For example, if demand is strong, a company may profit from employees working overtime or from adding additional shifts.

Human Resources: People who make up the workforce of an organization, business sector, or economy. “Human capital” is sometimes used synonymously with “human resources”, although human capital typically refers to a narrower effect (i.e., the knowledge the individuals embody and economic growth). Likewise, other terms sometimes used include manpower, talent, labor, personnel, or simply people.

Investment: An investment is an asset or item acquired with the goal of generating income or appreciation after its use. In an economic sense, an investment is the purchase of goods that are not consumed today but are used in the future to create wealth. In finance, an investment is a monetary asset purchased with the idea that the asset will provide income in the future or will later be sold at a higher price for a profit.

Maquiladora: A foreign-owned factory in Mexico at which imported parts are assembled into final products for export to other countries.

Production Process: Mechanical or chemical steps used to create an object, usually repeated to create multiple units of the same item. Generally, involves the use of raw materials, machinery, and manpower to create a product.

Structural Equation Modelling (SEM): Statistical technique for building and testing statistical models, which are often causal models. It is a hybrid technique that encompasses aspects of confirmatory factor analysis, path analysis and regression analysis.

APPENDIX: THE SURVEY

Survey to Determine the Benefits of Implementing Advanced Manufacturing Technologies

Advanced Manufacturing Technologies (AMT) are a kind of physically hard and soft equipment. On one hand, hard technologies are classified as, robots, milling machines, lathes, inspection systems, computer vision, programmed equipment by numerical control, among others. On the other hand, soft technologies, are classified as, Just in Time, MRP, Kanban, software for designing, among others. In addition, these types of technology executed in production processes may provide several benefits; however, there is no general knowledge about which benefits are acquired when these AMTs are implemented. Therefore, in the following questionnaire, a list of benefits reported in the literature review is presented, in which a participant has to respond according to the scale that appears in the section below, where the respondent must judge in a subjective manner on a benefit that is obtained with the AMT implementation.

Types of Advanced Manufacturing Technologies

Directions: The following table displays a list of Advanced Manufacturing Technologies according to the United States Chamber of Commerce Chamber classification. Please indicate the level of implementation that is presented in your current company.

According to the previous classification, mention the technology that have been recently acquired in your company:

____

____

Table 9.

1	2	3	4	5
The Benefit is not expected	The Benefit is slightly expected	The Benefit is regularly expected	The Benefit is often expected	The Benefit has been completely expected
The Benefit has not been obtained	The Benefit has been slightly obtained	The Benefit has been regularly obtained	The Benefit has been often obtained	The Benefit has been completely obtained

Table 10. General information

Industrial Sector ☐ Machinary ☐ Electric ☐ Automotive ☐ Aeronautic ☐ Electronics ☐ Logistics ☐ Medical ☐ Other (Specify)
Hierarchical Job Position ☐ Production manager ☐ Production Engineer ☐ Procurement manager ☐ Production supervisor
Years of Experience: ☐ Less than 2 years ☐ 2-5 years ☐ 5-10 years ☐ More than 10 years
Gender ☐ Female ☐ Male

Table 11.

A. Stay-Alone Systems *1. Engineering and designing technologies*	1	2	3	4	5
Computer-aided Design (CAD)					
Computer-aided Process Planning (CAPP)					
2. Fabricating/Machining and Assembly Technologies					
NC/CNC or DNC Machines					
Materials Working with Laser (MWL)					
Object-lifting robots					
Other robots					
B: Intermediate Systems *3. Automated Material Handling Technologies*					
Automatic Storage and Retrieval Systems (AS/RS)					
Automatic Message Handling System (AMHS)					
4. Automated Inspection and Testing Systems					
Automatic Inspection Test Equipment (AITE)					
C: Integrated Systems 5 *5. Flexible Manufacturing Technologies*					
Flexible Manufacturing (Cells or Systems) (FMC/FMS)					
6. Computer-integrated manufacturing systems					
Computer Integrated Manufacturing (CIM)					
7. Logistic Related Systems					
Just in Time					
Material Requirements Planning (MRP)					
Manufacturing Resource Planning (MRP II)					

Directions: In the following section there is a list of benefits that the literature review reports, which are obtained by applying advanced manufacturing technologies in the production systems. Please, based on the most recently acquired technology, mention the degree where the benefits have been obtained. Mark with a ✓ as you consider it appropriate.

Principal Issues Faced in the AMT Implementation Process

Directions: In the following section, it is presented a list of the main issues reported in the literature review, which are obtained due to the advanced manufacturing technologies implementation. Please, based on the most recently acquired technology, mention the degree where these issues have been addressed. Mark with a ✓ or a ρ as you consider it appropriate.

Thank you for participating. If you wish to have a report about the results obtained, please write down an email.

Email: ____

Table 12. Obtained benefits

Design	Assessment				
Design time reduction	1	2	3	4	5
Reduction of time between product conceptualization and manufacturing	1	2	3	4	5
Quality of the design	1	2	3	4	5
Reduction of time between product conceptualization and manufacturing	1	2	3	4	5
Production					
Reduction in the cost production	1	2	3	4	5
Reduction in the time production	1	2	3	4	5
Increment in the product quality	1	2	3	4	5
Increment in the plant capacity	1	2	3	4	5
Improvement in the plant distribution	1	2	3	4	5
A better plant distribution has been achieved	1	2	3	4	5
Reduction of set-up time	1	2	3	4	5
Increment in the plant reliability	1	2	3	4	5
A better usage of the available space	1	2	3	4	5
Human Resources					
Increment in the operator productivity	1	2	3	4	5
Reduction in the workforce cost	1	2	3	4	5
A better engineering experiences is achieved	1	2	3	4	5
A better management experience is achieved	1	2	3	4	5
Comercials					
Helps to keep up with competitors	1	2	3	4	5
Market range is extended	1	2	3	4	5
Quick reaction towards consumer needs	1	2	3	4	5
Early market entry was achieved	1	2	3	4	5
Reduction in delivery times	1	2	3	4	5
Reduction in the time between order confirmation and delivery time	1	2	3	4	5
Increment in sales	1	2	3	4	5
Materials					
The level of stock in process was reduced	1	2	3	4	5
Reduction in products variety	1	2	3	4	5
Reduction in material handling	1	2	3	4	5
Reduction in the products and parts variety	1	2	3	4	5
Improvement on the stock rotation	1	2	3	4	5
Processes					
Reduction in the number of machines	1	2	3	4	5
Reduction in the size of the production lot	1	2	3	4	5
Machines flexibility	1	2	3	4	5
Process flexibility	1	2	3	4	5
Volume flexibility	1	2	3	4	5
Expansion flexibility	1	2	3	4	5
Reduction in rework and waste	1	2	3	4	5

Table 13.

Selection process	Assessment				
There were many alternative choices	1	2	3	4	5
Far away suppliers (in other countries)	1	2	3	4	5
Lack of economical resources	1	2	3	4	5
Financial issues (if it is financed)	1	2	3	4	5
Delivery time of the AMT	1	2	3	4	5
AMT guarantees from the supplier	1	2	3	4	5
Customs requirements for AMT importation	1	2	3	4	5
Disagreements in the internal selection process	1	2	3	4	5
Fear and risk of making financial investments	1	2	3	4	5
Difficulties in tracking technology (technological tracking)	1	2	3	4	5
Technical investment justification (what is it required for)	1	2	3	4	5
Implantation process					
Differences regarding design (the AMT is different from what was ordered)	1	2	3	4	5
Lack of knowledge to perform any installation	1	2	3	4	5
Installation manuals in a different language	1	2	3	4	5
It can be tuned or adjusted	1	2	3	4	5
Quality requirements are not clear	1	2	3	4	5
Special installation requirements are needed (energy, building)	1	2	3	4	5
Settings or communication interfaces issues	1	2	3	4	5
Operation					
Tranining for operators	1	2	3	4	5
Breach of guarantees	1	2	3	4	5
Frequent breakdowns	1	2	3	4	5
Expensive maintenance	1	2	3	4	5
Special maintenance personnel are required	1	2	3	4	5
High costs due to maintenance fees	1	2	3	4	5
Expensive and distant parts	1	2	3	4	5
Only a few operators know how to use maintenance machines	1	2	3	4	5

Chapter 4
An Analysis of the Determinants of ITF R&D Projects Commercialization in Hong Kong's Logistics and Supply Chain Industry:
Industry User Perspective

Siu Cheung Ho
City University of Hong Kong, Hong Kong

Kong Bieng Chuah
City University of Hong Kong, Hong Kong

ABSTRACT

Innovation and technology are determinants of sustainably the competitive advantage of the company in Hong Kong. In order to deliver the message from industry users to R&D technologist to improve the rate of adoption of R&D technology into the logistics and supply chain industry in Hong Kong, a quantitative survey to collect the industry user concern and expectation of R&D technology is necessary. This chapter aims to develop a novel project management model to have a good grasp of factors influencing R&D project commercialization through a multi-perspective methodology. Through the quantitative survey methodology to collect the data from the industry side, the comprehensive data analysis will be conducted. The main concerns are whether the developed technology applied to the company can create the value for the company to solve the problem and add value.

DOI: 10.4018/978-1-5225-8223-6.ch004

INTRODUCTION

According to NewGOV.HK (2017), it stated that "The SAR Government has been attaching great importance to the development of innovation and technology, and it was designated in 2009 as one of the six industries where Hong Kong enjoys clear advantages. It is a long-term investment to develop innovation and technology. Over the past decade, Hong Kong's research and development (R&D) expenditure has been increasing at an average annual growth rate of about 7%. The R&D expenditure by the public sector (including Government and higher education sectors) has increased at an average annual growth rate of 4.7%, from HK$5 billion in 2001 to HK$7.5 billion in 2010, accounting for 57% of the gross R&D expenditure." The research aims to break down technology barriers from initial stage of R&D to commercialization the R&D deliverables to the Industry. In addition, the research results will contribute towards the R&D design and adoption of unified the industry standard of ITF R&D technology adoption to the Logistics and Supply Chain Industry in Hong Kong.

Though quantitative survey study to understand the viewpoints of Industry Users for analysing the gap among 5 development stages of MSTAM model. The quantitative data analysis from demand side will be evaluated. The target people is a person who is working in Hong Kong Logistics and Supply Chain industry or has much working experience in this area.

Problem of ITF R&D Project Commercialization in Hong Kong's Logistics and Supply Chain Industry

ITF R&D project starting from science stage (science) develop the R&D technology to the industry (market) through basic research, applied research and experimental development to verify the developed technology is beneficial for the industry. (Ho and Chuah 2018). ITF R&D project involves many operations steps and conditions need to be fulfilled for monitoring whether ITF project deliverables are matching with industry requirement for further commercialization and adoption. The operation of ITF project has many control points such as progress report for the technical and administrative status of the project and audited account to check whether the funding used in a suitable area in complying the regulation. (Ho and Chuah 2018).

According to Ho and Chuah 2017 & Ho and Chuah 2018 research stated that many ITF projects were not successful completed and transferred the developed technology in the Hong Kong's Logistics and Supply Chain Industry. The common problems were:

- **Problem 1:** Control and Time Issue - Long period of time for approval in proposal stage
- **Problem 2:** Time and Control Issue – Project development process is too long (From proposal stage to project completion stage)
- **Problem 3:** Quality Issue - R&D project deliverable cannot match against industry demands
- **Problem 4:** Communication and Motivating Issue - Misunderstanding of the project expectation within each stakeholder in the project

Enabling Technology in Hong Kong' Logistics and Supply Chain Industry

Refer to Ho and Chuah 2018, the most popular R&D technology in Hong Kong's Logistics and Supply Chain Industry were 8 applications. Some examples of the ITF R&D project enabling research and commercial adoption in Hong Kong's Logistics and Supply Chain Industry are:-

1. Logistic and Transport
2. Airport Technology Initiative
3. Construction Industry
4. E-commerce Industry
5. Location-based Service (LBS)
6. Smart Community Service
7. Retail
8. Robotics Application

Table 1 summarizes some current enabling technologies for Logistics and Supply Chain Industry in Hong Kong. As can been seen, the ITF R&D enabling technologies can be applied in different environment and industries.

Table 1. Enabling technology in Hong Kong's logistics and supply chain industry

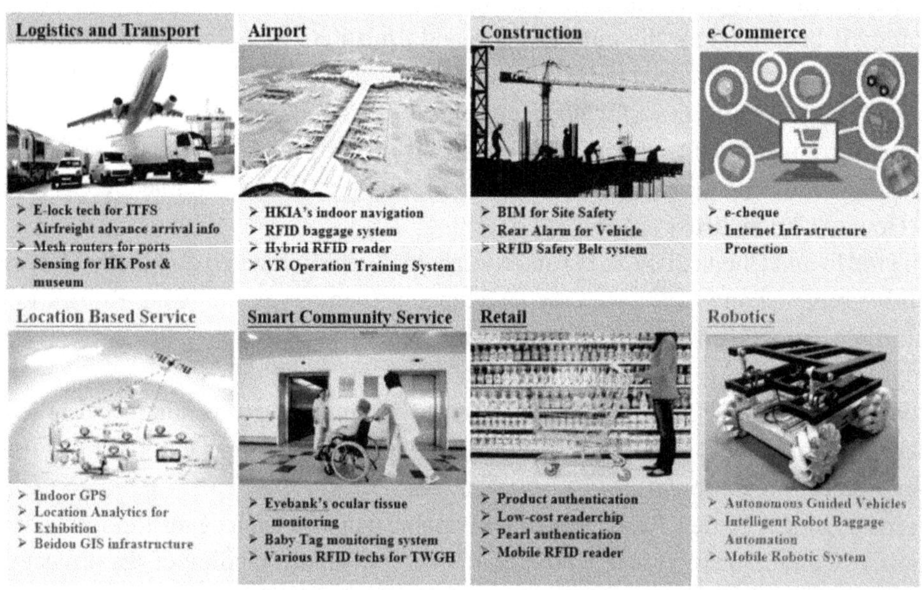

LITERATURE REVIEW, RESEARCH FOUNDATION AND RESEARCH METHODOLOGY

This chapter reviews the relevant quantitative, qualitative, and literature review articles and papers in technology development and commercialization category in previous study and research for helping to develop the survey methodology base on the industry-side perspective to list out which survey methods are suitable for this study. The basic R&D development process is reviewed for exploring the MSTAM model, hypothesis and gap analysis.

Industry Perspective (Demand-Side)

The research results of industry perspective are found in three papers using the qualitative and quantitative survey methodology, eleven papers using the quantitative survey methodology, two papers using the qualitative survey methodology, and two papers with data analysis the data through literature review. Fareena and Lillian (2000) examined the adoption of new technology at the level of middle managers, engineers, and technical personnel. Through qualitative and quantitative surveys the adoption of new technology within organizations is examined. To broaden the understanding of technology adoption at the individual level, the top management support was needed. Seo-Kyun et al. (2009) expressed learning and external networking is significant factors influence on innovation development. Investigate the R&D capabilities, technology commercialization and innovation performance of IT-related business SMEs; it was key process for transferring the resulting technologies to SME for enhancing their competitiveness in a systematic process. Hu and Mo (2006) collected the qualitative and quantitative data via deep interview and questionnaire methodology to analysis the core, important, minor, and auxiliary motivation factors for talent development. Base on the analysis results, the model of motivation factors of talent development in the industrial cluster were constructed. It seemed that qualitative and quantitative survey methodology was the basis data collection tool for evaluation.

Yuan and Dong (2015) proposed a risk assessment algorithm to identify the risk in 4 categories. These were 1) technical risk, 2) operational risk, 3) management risk and 4) emergency risk. Using Analytic Hierarchy Process (AHP) quantitative methodology to collect the group data for decision making, it helped to solve the accounting information system hierarchy risk assessment problem. Kimseng at al. (2016) examined the influence of knowledge management on product and process innovation in formal and non-formal R&D manufacturing firms. A set of questionnaires were related to manufacturing firms (small size to large size). The collected data indicated the formal and non-formal R&D firms tend in product and process innovation to determine innovation performance in the manufacturing industry. J. E. and R. T. (2011) proposed a commercialization model to define different R&D development process from basic R&D to fully commercial. Based on firm-level case study to collect the eighty enthusiasts comment, it explored the study of technology commercialization. The result suggested six issues which need to be considered as *"(i) Commercialization environment issues, (ii) Intellectual property issues, (iii) Technology management issues, (iv) Marketing strategy issues, (v) Innovation development issues, (vi) Finance issues*". Steven et al. (2002) examined the different roles of established and new firms in a disruptive technology commercialization. Using quantitative survey data collected from 72 Microelectromechanical Systems (MEMS) manufacturing firms to prove that market-pull strategies are preferred. Jiancheng and Quan (2003) collected the manufacturing firm feedback through the questionnaire. The data collected from 1997 to 1999. A questionnaire was created to collect the response from the industry.

The response rate is 65.9% and involved 17 industries. This monitoring framework can be identified the project risk in the ongoing R&D project. Ya-lan and Ling-ling (2007) proposed hygiene factors and motivation factors to align with user focus, system focus and satisfy users to keep use the same search engine respectively. Through the questionnaire to collect the user feedback, it could help the researcher to guide the designers to decide the system requirements of search engines. Ikram and Muhammad (2013) conducted the industrial survey via software engineer to identify de-motivational and motivational factors and evaluated which critical factors to cause the on-going R&D project termination. Stavros and Anastasios (2014) discussed the "*Attitudes towards Using*" (ATU) and "*Intention to Use*" (ITU) mobile-based assessment from the perspective of the self-determination theory of motivation. Using survey questionnaire to collect the user feedback for analysing and identifying three basic psychological, it needed for competency, autonomy and relatedness. Stavros and Anastasios (2014) defined the main external variables into 4 factors as perceived usefulness (PU), perceived case of use (PEOU), attitude towards usage (ATU), and behavioural intention to use (BIU) to identify what factors influenced the user behavioural intention to use a mobile-based assessment. This research investigated the factors of influence the acceptance of mobile-based assessment through the questionnaire survey to collect users comment through quantitative survey methodology. Henri et al. (2014) depicted how teachers' motivation and cultural barriers. Through questionnaire survey to collect the data from teachers, the data could help to predict how the extracted factors influenced the motivation of teachers to share and collaborate. Külli et al. (2016) collected the IT student feedback via quantitative survey. The outcome was helped to know the IT student interest and behavior for enrolling them in IT studies and increasing student interest in IT studies. Consequently, quantitative data could help to comprehend the demand side point of view for problem solving.

According to the survey result, Chun-Hsien et al. (2015) defined government sponsored innovation was crucial drive force for knowledge commercialization in emerging China market. Gathered primary data from multiple respondent groups and high-tech industrial policy experts, they were participated in the in-depth interviews to provide their insight and comment for the study. Nor et al. (2012) proposed the SME technology adoption need to be collected the information from internal and external to identify the significant factors for analyzing the uniqueness of SMEs characteristics. To understand the need in deep of supply side and demand side, in-depth interview is essential.

Gerald and John (1997) went through the literature review to deploy the individual cognitive abilities with tasks in each phase of the innovation process. Ibrahim Iskin (2011) through the literature to define the demand side management perspective to point out the critical adoption barrier in the energy efficient technology, also, a comprehensive list of adoption barriers, drivers, and policy tools was identified. Summarized the captioned findings, literature review is a good reference for helping to develop the thesis.

Research Methodology: Quantitative Study of Industry User

The main research survey was conducted by using an online self-administrated questionnaire being posted on Survey Monkey website. The purpose of this research is finding out the solution for increasing the opportunity of ITF R&D project deliverables commercialization and adoption to the industry. Quantitative survey is to collect the empirical data from industry user working in Hong Kong Logistics and Supply Chain Industry and their concern on the adoption of the new technology deliverables in their company. The initial target response size of this questionnaire is around 100.

Table 2. Related industry survey studies on R&D and commercialization

Author	Methodology	Key Findings and Insights
Fareena and Lillian (2000)	Qualitative Survey & Quantitative Survey	Through qualitative and quantitative survey to examine the adoption of new technology within organizations. To broaden the understanding of technology adoption at the individual level.
Seo-Kyun et al. (2009)	Qualitative Survey & Quantitative Survey	Investigate the R&D capabilities, technology commercialization and innovation performance of IT-related business SMEs. Transferring the resulting technologies to SME for enhancing their competitiveness.
Hu and Mo (2006))	Qualitative Survey & Quantitative Survey	Collecting the qualitative and quantitative data via deep interview and questionnaire methodology, The core, important, minor, and auxiliary motivation factors of talent development were separated out.
Yuan and Dong (2015)	Quantitative Survey	Using AHP quantitative methodology to collect the group data for decision making, it helped to solve the accounting information system hierarchy risk assessment problem.
Kimseng at al. (2016)	Quantitative Survey	A set of questionnaires were related to manufacturing firms (small size to large size). The collected data indicates the formal and non-formal R&D firms tend in product and process innovation to determine innovation performance in the manufacturing industry.
J. E. and R. T. (2011)	Quantitative Survey	Based on firm-level case studied to collect the eighty enthusiasts comment, the study of technology is explored for commercialization.
Steven et al. (2002)	Quantitative Survey	Using quantitative survey data collected from 72 MEMS manufacturing firms to prove that market-pull strategies are preferred.
Jiancheng and Quan (2003)	Quantitative Survey	The R&D projects data collected from manufacturing firms in Beijing from 1997 to 1999. A questionnaire was created to collect the response from the industry. The response rate was 65.9% and involved 17 industries. The key concerns of industry have been obtained.
Ya-lan and Ling-ling (2007)	Quantitative Survey	Through the questionnaire to collect the user's expectation, it helped to guide the designers to decide the system requirements of search engines.
Ikram and Muhammad (2013)	Quantitative Survey	Based on an industrial survey conducted from software engineer, the de-motivational and motivational factors are identified.
Stavros and Anastasios (2014)	Quantitative Survey	This study discussed the "*Attitudes towards Using*" (ATU) and "*Intention to Use*" (ITU) mobile-based assessment from the perspective of the self-determination theory of motivation. Using quantitative survey, the questionnaire was developed to collect the user feedback for analysis.
Stavros and Anastasios (2014)	Quantitative Survey	A questionnaire survey from the undergraduate students had been collected for investigating the impact factors of the acceptance of mobile-based assessment.
Henri et al. (2014)	Quantitative Survey	This research depicted how teachers' motivation and cultural barriers. Through questionnaire survey to collect the data from teachers to predict how the extracted factors influenced the motivation of teachers to share and collaborate.
Külli et al. (2016)	Quantitative Survey	The IT student's feedback via quantitative survey was collected. The outcome was helped to the IT students to know their interest and behavior to enroll them in IT studies.
Chun-Hsien et al. (2015)	Qualitative Survey	The primary data from multiple respondent groups and high-tech industrial policy experts were collected. These in-depth interviews could help to obtain their insight and comment for understanding the industry side expectation and concerns.
Nor et al. (2012)	Qualitative Survey	Through interview with SMEs' owner and managers, their feedback and concern for technology adoption was identified.
Gerald and John (1997)	Literature Review	The literature review of technical innovation was to deploy the individual cognitive abilities with tasks in each phase of the innovation process. The reason of unsatisfactory delays in the innovation process was determined.
Ibrahim Iskin (2011)	Literature Review	This research study used the literature review on adoption of energy efficient technologies from demand side perspective to identify the taxonomy of adoption barriers, drivers and policy tools.

THEORETICAL FRAMEWORK OF ORGANIZATIONAL ADOPTION

Refer to ITF IP (2017), ITF is to provide financial support for R&D projects to contribute and promote the innovative technology in Hong Kong and upgrade the value of enabling technology for the industry to enhance the competitive advantage of Hong Kong. ITF R&D project has some critical issue in the development process, it affects the R&D deliverables productizing or commercializing to the industry in the effective way. Through the literature review, there are 4 factors are important in the R&D development process. These factors can be used in ITF R&D projects for enhancing the commercialization success to the industry. These factors are described as below.

Control Factors

Mats and Jeffrey (2000) proposed specification based prototyping system to simulation and debugging the software specifications for control system. Detail software system specification in early development stage has many advantages for helping the developer to evaluate and address poorly understood aspects between the development side and user side. Controlling the in and out process, the system guaranteed consistency between the prototype and the requirements specification was addressed in advance. The problem of inconsistent will be eliminated. Chonyacha and Nathasit (2009) addressed the issue of technology adoption within an organization. Through the Technology Acceptance Model (TAM) to identify the key factors influenced the adoption process of technology in the company such as function performance, acquisition cost, ease of use, operating cost, reliability, serviceability and compatibility. Therefore, a systematic control system from development initial stage to technology adoption in the organization is needed. Abderrahim et al. (2013) proposed a control and time approach to manage the project execution through the statistics methodology to improve "*Interoperability Degree*" of a quality factor of information systems.

Time Factors

Steven et al. (2002) examined a variety of variables that were significant in adoption of technology and the determinant of new firms select the primarily disruptive technologies. The result found that the new firms prefer market-pull strategies from innovation process beginning with technology creation to the end of the user adoption and application. The results proved that "*Time to market for new firms is one-fourth that for established firms. These results suggest that new firms have two advantages in commercialization of disruptive technologies—flexibility in marketing strategy and much shorter times to market*". Therefore, speeding up the time of technology development, it could increase the commercialization opportunity to the market. Abderrahim et al. (2013) used the control and project planning theory to prioritize each project task duration and consequence tasks in a sequence to minimize the risk. Reducing the project implementation time for accelerate the advancement of the project operations. Time factor is important.

Quality Factors

Herry et al. (2016) examined different variables influence the project success and found that the key success factor is the quality of the project. It involved scope, time and cost factors in the project management mechanism and those factors are interactive to the project quality. Naoki Takada (2016) suggested high and low project autonomy promoted the technological outcome integrate the project to the firm. The conceptual framework was to contribute the understanding of the dynamic of innovation through collaboration and its success factors. The technological performance and integration between project and firms are the key monitoring process in the framework to control the project quality for commercialization. Svetlana and Anton (2015) analyzed the information technology using in the process of Good Manufacturing Practice (GMP) to test and improve the quality of products. In order to fulfill the GMP regulation, the quality control and quality assurance is necessary to ensure the quality of the product. As a consequence, quality and control factors are correlated in the development or production process.

Motivating Factors

Chuang and Wu (2007) evaluated the hygiene and motivation factors of search engine from the user perspective. This research survey collected 758 people to answer. The results showed that "*Motivation factors are those more additional services of a search engine and make users willing to keep using the same search engine*". The motivating factor could help to align with the user focus and improve the loyalty of user willing to use it. Ikram and Muhammad (2013) proposed motivational factors and de-motivational factors to find out what reason to cause software projects failure. Through the questionnaire to collect the information from software engineer and combined with literature review to list out the key motivational and de-motivational factors to realize the software engineer concern and expectation in the software projects. In–depth understanding the motivating factors of users, it could be greatly helped in the project management.

To sum up the literature review, the R&D deliverables need to be enhanced; the quality factors are core value of the R&D deliverables. The factor of control, time, and motivating are all related and interactive. Those factors will interact and exchange the information between them and create the value for the R&D deliverables. Therefore, synthesize the above factors, the basic R&D project management mechanism is proposed in Figure 1.

Figure 1. Basic R&D project management mechanism

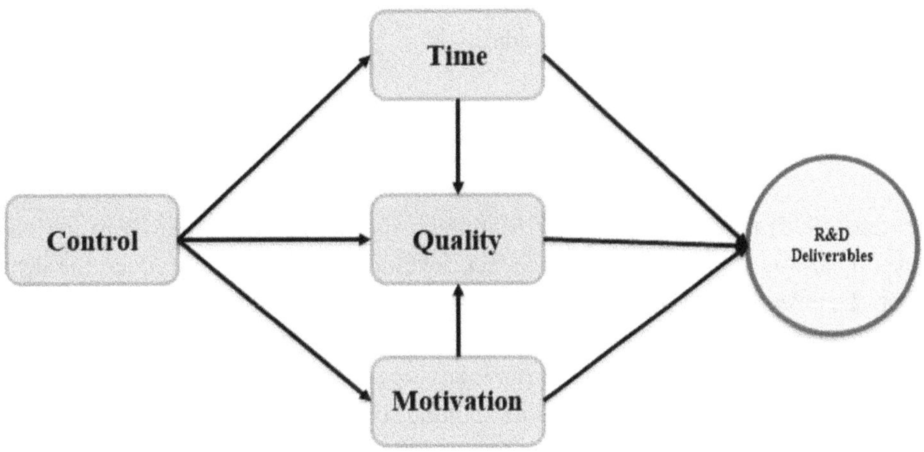

ITF R&D DELIVERABLES COMMERCIALIZATION HYPOTHESIS

To determine whether the elements is workable for enhancing commercialization success of ITF R&D project to the Hong Kong's Logistics and Supply Chain Industry, the following hypotheses are proposed and tested in a case study and the detail was stated in Ho and Chuah (2018).

Uda and MIMOS (2002) proposed the project development stage should be included technical requirement and product requirement such as manpower, product specification, project tasks, equipment, pilot schedule and customer's requirement. Technical requirements are an essential part in the initial stage of product development for covering the customer's requirements and need. Bruno at al. (2016) suggested that definition of system requirements should be started in early initial stage. Technical and managerial factors were correlated with system life cycle, it affected the requirements quality. Joko and Anggi (2017) conducted a multiple case studies method to collect the industry requirements of Enterprise Resource Planning (ERP) system in six micro, small and medium enterprises. The findings showed that the functional aspects collection in earlier development stage might help to comprehend the requirement of a company's integrated the new information system in the company. These lead the author first hypothesis is:

Hypothesis 1: Industry requirement collection can help to develop the R&D projects (Link with E1 and E2)

Hennann Loeh (2006) proposed that the supplier should participate in the project and product development stage to share the responsibility to take the monitoring role in the project for ensuring the project or product integration smoothly in the company. Frequent collaboration between the customer and supplier helped to sort out the project issues in advance for the developer to eliminate those issues in the project development process. Uda and MIMOS (2002) also suggested that "*Strategic partnership and contract manufacturing are initial sales and marketing effort.*"

Product Specification (PS) was the core testing requirements elements of the product for testing and examining indictor for identifying the performance of the product. The limitation and product performance need to be identified in the product specification and stated in the initial product development stage such as specifying the product acceptance testing level and worst case problem, and product limitation for aligning both agreed acceptance level between developer and customer. (Lauren et al. 2017)

Summarized the literature, the second hypothesis is:

Hypothesis 2a: Industry user provides real market situation to help the R&D technologist explore the new idea of R&D project (Link with E2 and E3)

John E. Ettlie (1995) proposed that Early Manufacturing Involvement (EMI) in the New Product Development (NPD) was significant in the product development process to reduce the design change for speeding up the development progress. The research finding indicated that if there are fewer design change requests for developing a new product, it was more likely to successfully commercialize the new product to the market. Thomas Åstebro (2004) examined some characteristics of technological entrepreneur to develop the R&D projects. The characteristics include several dimensions and the key factors, these were user need and market demand. GAO and Ping (2009) proposed a multi-criteria sorting method to evaluate the supplier performance for supplier selection in the early stage of new product development. The multi-criteria sorting method was based on the supplier's perspectives of quality, cost, delivery, development potentiality, technical capabilities, and cooperation capability. Early supplier involvement in new product development was suggested. Luo and Bin (2016) examined the impact of design collaboration among supplier, designer, and customer. The supplier and the designer provided the technical design and creative design respectively in fashion supply chain product development process. Both design efforts proved that market demand could be increased. Therefore, the design collaboration with supplier and designer should be considered to join in on the new product development for enhancing the market demand and product performance. Base on those literatures, the third hypothesis is:

Hypothesis 2b: Industry user involve in the R&D projects in the project initial stage, it can speed up the R&D development process (Link with E2 and E4)

L. Zhang and M. Swirski (2002) suggested that "*The key to successful specification management is to build project capability to enable the dynamical process of validating customers' requirements and supplier solutions*". The flexible, comprehensive procedure, and product specification management need to be established to balance the changes and project schedule. Venlakaisa et al. (2009) validated ESI could be provided a benefit to suppliers and customers. Improving and sharing the information about the process methods could reduce mistakes in the development process. Earlier communication could help the designers to do the work in right time, quantity, and process for reducing the rejected parts, production process, and time. The product development process in the foundry industry should be provided with clear information for matching with the customer requirements for building long-term collaboration. Qing et al. (2011) analyzed a Transaction Memory System (TMS) in term of early supplier involvement that helping the team in the new product development process to improve the communication and interaction between customer and supplier. Effective knowledge sharing was needed for new product development. Following the literature review, the basic R&D product specification was essential elements for enhancing the communication and understanding align with each party in the new product development process. The forth hypothesis is proposed to;

Hypothesis 3: Create a basic R&D product specification in the R&D project initial stage; it can help all stakeholders easy to know the R&D deliverables (Link with E3 and E5)

Pek-Hooi and Edward (2005) examined the performance impact for research capability between internal research capability and external sources of innovation in the corporate. A survey was conducted with management executives and engineer. The statistical findings showed that acquiring external sources of innovation and increasing the number of partners have positive influences over the innovative performance of the corporate. C.L. Zhao et al. (2009) proposed ESI (Early Supplier Involvement) adapted in manufacturing industries, an additional value in the development process and aligns the enterprise goals with supplier collaborative goals can be enhanced. The proposed elements were examined to solve the problems of the inefficient and ineffective coordination between teams and systems. Chun-Hsien Wang et al. (2015) proposed a co-innovation network connected with the commercializing of scientific technologies and academic knowledge together, it could help to transfer the technology knowledge from academic to the industry. The innovation process and understanding of academic side and industry side could be aligned. It helped to promote the high-tech technology to the market in effective way. Summarized the literature review, the fifth hypothesis is

Hypothesis 4: Industry users' participation in the R&D projects can enhance the project value of the R&D deliverables (Link with E4 and E5)

Floortje Blindenbach-Driessen (2015) examined the moderating role to improve the communication and relationship to difference functional teams in the organization. The contribution of the moderating role was a good mechanism and more functionally organized. The finding proved that "*Cross-functional knowledge is important for all innovations*". Souhaib et al. (2015) evaluated the Requirement Engineering (RE) techniques and challenge in software industry. The research result shows that the influence between both depended and independent variables was positive for the software project development. Blen et al. (2017) proposed an educational prototype sharing system by mobile devices to share the information for improving the communication with students. The systematic flow chart showing the priority of the system and the basic operation flow of the system was discussed to demonstrate the beneficial of frequency sharing. Chokri et al. (2017) described the prototype and experimental detail of electric vehicle. Through market analysis for developing the prototype function and feature, it should be developed the electrical and mechanical features of the prototype. Information sharing in the development process was needed to eliminate the risk of operation. To summarize the captioned literature, the sixth hypothesis is proposed to:

Hypothesis 5a: Regular project status update of the R&D project, it can reduce the communication issue among all stakeholders (Link with E5 and E6)

Mats and Jeflrey (2000) proposed that "*The capability to dynamically analyze, or execute, the description of a software system early in a project has many advantages*". The specification based prototype could be tested and evaluated in advance in the development process. It helped the analyst to address the aspects of the product issue that is not verified in detail. The development was involved different researchers for development. To reduce the complexity and communication, specification-based prototyping was an approach to eliminate many of the current drawbacks with rapid prototyping for stimulating and debugging of formal software specifications for control system. Yan et al. (2010) presented an evaluation index system for determining the commercialization potential of emerging technologies and technology commercialization success factors, economic factor, qualification factor, consistent factor and social factor. The index system listed the key performance indicator to identify the essential ratio of each factors, it helped to eliminate the uncertainties and complexities in the early development stage. Ferraz, et al. (2015) stated an innovative prototype for blood pre-transfusion test in a short time. The preliminary result was positive and could be characterized as commercial added value. It also presented a user-friendly interface with prototyping sample for users. The literature review pointed out that the preliminary test result and prototype sample were essential modules in the early development process. The seventh hypothesis is proposed to

Hypothesis 5b: The preliminary test result and prototype sample can help the industry user easy to understand the product for future commercialization and adoption (Link with E5 and E7)

P.M. Przybysz et al (2013) proposed a service productivity model to identify the critical tasks in the project development schedule and evaluate the co-relation among each tasks in three different dimensions as potential dimension, process dimension, and outcome dimension. The model defines the relations between task characteristics, skills and abilities to prior evaluate of the productivity of different team compositions possible to reduce the project risk in advance. Jumie and Stanley (2013) through the Quality Function Deployment (QFD) matrices model to analyze the relationships between project stakeholders and illustrate how QFD variances could impact decision-making of the project for facilitating stakeholder-value prioritization, requirement analyses, and conflict resolutions through QFD model to define the different variances in the early development stage. Mirza Ral et al. (2014) expressed that "*Dependency among non-functional requirements (NFRs) is one of the major issues to handle for delivering quality software*". To solve the conflict of NFR in the early development stage, it helped to design the software with better management. The interdependence relation of NFRs had 9 key elements such as performance, security, scalability, accountability, availability, integrity, usability, efficiency, and reliability to identify the relation where NFR help to ensure another NFR. It would reduce the conflict and communication barrier in the software development process to build the quality outcome. R. Shahzadi et al. (2014) through the questionnaires and interviews to collect the data from project stakeholders to analyze the conflict in the distributed projects. The results indicated three elements as team members, process, and technology. These were main factors causing conflict in the distributed projects. To align the different standards and common technical language among the member of project, the research results proved that conflict management was a good tool for improving the communication and eliminating the risk of project failure of the project. Therefore, the eighth hypothesis is assumed as follow.

Hypothesis 5c: In the R&D project initial stage, this thesis suggests that define project risk in advance such as administrative and technical issue, it can eliminates the risk of project failure (Link with E5 and E8)

Chen et al. (2007) used Dependency Structural Matrix (DSM) theory for research foundation to explore an algorithm for optimizing the tasks of the project and evaluate the risk, cost, and co-relationship of the project tasks to reduce the unnecessary activity rework in the project progress. The optimization algorithm and risk evaluation model for complex project should be taken into account of activity inter-dependencies, rework, and learning effect. Byung Chun SHIN (2010) presented some of preliminary test results of the prototype of urban maglev train to test the performance of the product. The prototype sample evaluation included two parts as 1) R&D for technical improvements and 2) Demonstration line construction. The prototype sample could be help the R&D technologist to test the performance in advance before mass production and simulate the critical tasks for evaluating the performance results in the early project development stage such as system engineering, performance test, guideway facilities, site selection, line design, system integration planning, test and commissioning. Ville et al. (2013) proposed a new prototype sample to test the usability for the new product development and combine with the field test for product evaluation. The research result was able to detect the key usability problems and reduce the project risk of failure in the early development stage. The prototype sample was a crux of the matter of the new product success. The user could be identified the function of the product and usability evaluation in the early stage of the development process that could help to collect the real and actual data from the end user. The product performance could be find tuned in the prototype sample stage. Qazi et al. (2015) utilized a prototype device to test the performance of the W-Fi enabled communication in homes. The preliminary test result was a good indicator for the technologist for analyzing the data in the development process for improving the functionality of the final product. The above literature review was proven that the prototype sample was a crux of the matter of new product development. To summarize the finding, the ninth hypothesis is proposed to:

Hypothesis 6: The preliminary results and prototype sample of the R&D product can help the industry user easy to know and how to use the R&D deliverables (Link with E6 and E7)

Lenina SVB (2016) suggested inventor and researcher should take care of the right of Intellectual Property (IP) and be briefed with the importance of IP protection for them. The researcher and inventor should be briefed with the process of filing patent procedure. Helandi et al. (2016) expressed that IP management was an essential element in the new product development and proposed to protect technology process from research and development to admit, format protection, request protection, commercialization technology and the final stage of technology transfer to the developer. Qing and Shawn (2017) proposed a cluster intellectual property protection management mechanism for mass innovation spaces. Intellectual property was inextricably linked to different development of innovation in mass innovation spaces. The IP protection mechanism was to protect the right for the developer. Jian and Qing (2017) also agreed the intellectual property management was important for protect creations and inventions for industrial, scientific, academic fields. A preliminary study of the key factors of IP management at art institutions, including IP strategies, IP creation, IP protection, and IP applications are proposed. Megumi and Koichiro (2017) conducted a survey to inventors for collecting their views and acceptance of the harmonization in terms of promoting invention. The inventors' point of view of IP was important for harmonization of intellectual property systems. The results indicated the acceptance of harmonization intellectual property systems by inventors was positive. The finding indicated that IP protection is important. The tenth hypothesis is proposed as follow.

Hypothesis 7: Intellectual Property (IP) registration can protect the rights of the developer and facilitate further commercialization (Link with E8 and E7)

Yayoi Hirose (2012) evaluated the cultural gap between technology researchers and potential user of technology transfer and commercialization. The finding showed that human engagement was a key factor to transfer the technology to the market effectively by using patent strategy and marketing strategy. Ruowei et al. (2013) suggested an information presentation system by using mobile phone and lap top device for interactive sharing of the information between the student and teacher. A survey had been conducted and the result found that this system had a significant impact on promoting communication with each other which was around 2 times in class and 3 times after class.

The adoption of information technology innovation was a complex process. Michael and Stephen (2013) stated that the process was involved three main entities such as *"(i) promoters, who facilitate the adoption process, (ii) inhibitors, who oppose the adoption of technology, and (iii) imitators who are influenced by both promoters and inhibitors."* Through what if analysis experiments to calculate the adoption rate of adopter population and potential adopters for effective management of the technology adoption process in this complex environment. To avoid inhibitor copy and phase the technology to the market and strip the technology from the original researcher, the effective promotion was needed for attracting the users to know the developed technology and adopt in their company. The eleventh hypothesis are raised to

Hypothesis 8a: The promotion activity can attract the industry users to know the R&D product value for commercialization (Link with E7 and E9)

S. Hatcher and M. Samuels (2000) reviewed the traditional procurement operations of new weapons system involved a complex testing requirement and condition. The onsite technical evaluation and operational test are suggested to eliminate the risk of final installation of the product. Konstantinos et al. (2008) demonstrated the value of virtual and augmented reality web-based platform to provide the textile products for potential customer to review the product and feature in a virtual environment before the final production stage. The web-based platform addresses the needs of collaborative product development among all stakeholders and aligned the understanding of the product concept in the same direction for them. The information and operation process for each stakeholder involved in the development project could be synchronized together with same pace. The end user and developer through this platform to review the prototype sample in advance. The project time, product quality, communication could be improved. Kawtar et al. (2015) also used a small scale testing in pilot for reviewing the product performance. It encouraged industrial integration in the workplace to ensure the installation is in good condition for improving the performance of the final product. Gilbert et al. (2016) through an industry trial to evaluate the system validation and traceability data in two organizations, the final results were positive and proven trial implementation in the real workplace could increase the performance of the system integration. Therefore, the twelfth element is

Hypothesis 8b: The R&D project trial implementation in the real industry premises can help the industry user to adopt the R&D deliverables (Link with E7 and E10)

Suleiman et al. (2002) through the questionnaire to collect the data from small and large firms, the research support was important for commercialization of disruptive and sustaining technology to the firm. Xianhua et al. (2009) proposed five roles in cluster knowledge system for facilitating the technology transfer and adoption from technologist side to the industry. The roles were 1) Technological Gatekeepers (TGs) has strong technology knowledge and connection between external knowledges sources and transfer knowledge to local enterprises; 2) Active Mutual Exchangers (AMEs) has external links and responsible a bridge between the local knowledge absorber and the external sources; 3) Weak Mutual Exchangers; (WMEs) plays a role to balances the absorption and transfer knowledge; 4) External Stars (ES) has a strong link to external sources; 5) Isolated Firms (IFs) is an active role of knowledge system. All of these roles are to support the technology transfer and adoption to the industry. Liliana et al. (2016) identified the key performance indicators to evaluate the project management mechanism. The stakeholders were elicited to provide the feedback in order to enhance the project management framework and coaching approach need to be conducted to ameliorate the rollout and adopt the product. The thirteen hypothesis is based on captioned findings to summarize as below:

Hypothesis 9: Technical support of R&D technologist can help the industry user easy to adopt the R&D deliverables in the company (Link with E9 and E10)

Chonyacha and Nathasit (2009) proposed Technology Acceptance Model (TAM) to evaluate the factor, which could help the company to adopt the new technology into the company. For those factors influencing technology adoption in an organization, it might be impacted through seven dimensions such as functional performance, acquisition, ease of use, reliability, compatibility, serviceability and compatibility. Through the case study to identify the main factors of technology adoption were customer requirements, quality of human resources, management support, and change management skills, clarification of project scope, and individual commitment and communication. J.E. Amadi-Echendu and R.T. Rasetlola (2011) suggested linear approach and functional approach to technology commercialization. The Linear models interacted between activities and linkages similar as project management approach, "*where the outcomes are reasonably determined a priori*". For the functional approach was to ensure the commercialization process on track and fulfil the market requirements.

The performance of electric vehicles through different experiments and evaluation is to test the product performance. A systematic examination process and detail project operations tasks were performed. A basis product specification was described in detail for standardizing the evaluation level of adoption. (T.W. Ching and T. Xu 2014). Ryohei et al. (2015) evaluated the correlative metrics of conflict management of project to understand how to solve the critical issues from large scale of project. The results suggested that "*most of conflicts are resolved by simple way*". Detail and enough information should be provided for conflict resolution. Therefore, comprehensible and understandable product specifications were key elements for industry to adopt the technology into the company. The fourteen hypotheses are proposed to:

Hypothesis 10: Clear and detailed R&D product specification can speed up the industry user to adopt the R&D deliverable (Link with E10 and E11)

J.E. Amadi-Echendu and R.T. Rasetlola (2011) suggested the firm should consider the influencing factors of technology commercialization. One of the factors of marketing strategy issues that the high score obtained was a technology company need to have marketing activities in place. W.K Wan and S. Monsef (2012) introduced four distinct elements in the new product development process via open innovation. These were planning stage, development stage, marketing stage and commercialization stage. From new product proposal evaluation to product idea generation and development tasks planning, then conduct the approval from the customer, and finally the product cost or service could be adjusted to match with the market requirements. Yinghua et al. (2013) proposed computer resources market model to define different stakeholders and their requirements in different layers such as physical resources layer, resources management layer, and market management layer for supply-demand matching. The results could be proved that the price of resources will be regulated. According to the captioned literature review, new product development involved different activities and function to promote the developed product to the market effectively. That promotion activities involved multiple stakeholders and tasks, so clear marketing plan was needed. The final hypothesis is proposed as below.

Hypothesis 11: Clear marketing plan can promote the R&D deliverables to the market effectively. (Link with E11 and E12)

RESEARCH MODEL DEVELOPMENT

A project has many uncertain factors to cause the project terminate or failure. Starting from the project commencement to project completion, there are a sequence tasks linking together and most of the tasks are correlated. Figure 2 is the basis project network template. The basis project tasks break down structure and operations flow has been shown the "before-after" relation. (Project Network 2015)

The relationship between R&D project management and gap analysis are discussed in the author's previous research (Ho and Chuah, 2017). To combine the section 3 findings, the critical factors of control, time, quality and motivation are discussed for enhancing the R&D deliverables commercialization to the market in the effective way. The relationship among each factor in the R&D project are to be aware of the intrinsic value of tasks of operations are linked in countless ways.

In the previous research from Ho and Chuah (2017), the foundation of MSTAM model, the model framework and proposition are developed. Summarized the section 3 findings, the detail of MSTAM model is listed in the Figure 3. Ho and Chuah (2017) modified STAM model (Phaal et al. 2011) to develop a new model, MSTAM (Market, Science, Technology, Application, and Market) to study the market need first before start the development process The different between STAM and MSTAM are showed in Table 3.

MSTAM refers to a data-driven improvement cycle used for improving, optimizing and stabilizing R&D processes and enhancing the commercialization. The MSTAM R&D cycle is the core tool used to drive ITF R&D projects. MSTAM can be used as the framework for other improvement applications. MSTAM model is an abbreviation of the five development steps it comprises: Market, Science, Technology, Application and Market. All of the MSTAM process steps are required and always proceed in the given order.

Industry User Survey: Quantitative Study

The main research survey was conducted by using an online self-administrated questionnaire being posted on Survey Monkey website. The purpose of this research is finding out the solution for increasing the opportunity of ITF R&D project deliverables commercialization and adoption to the Industry. Quantitative survey is to collect the empirical data from Industry User working in Hong Kong Logistics and Supply Chain Industry and their concern on the adoption of the new technology deliverables in their company. The initial target response size of this questionnaire is around 100.

Figure 2. Project network template

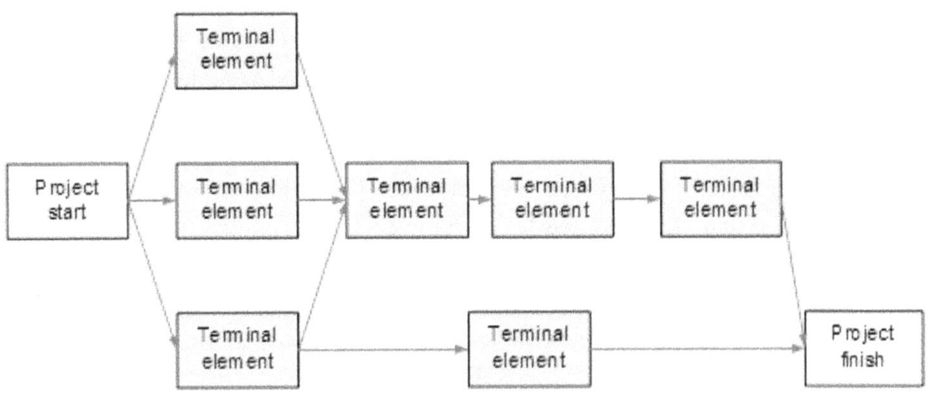

Figure 3. MSTAM Model and Hypothesis Development

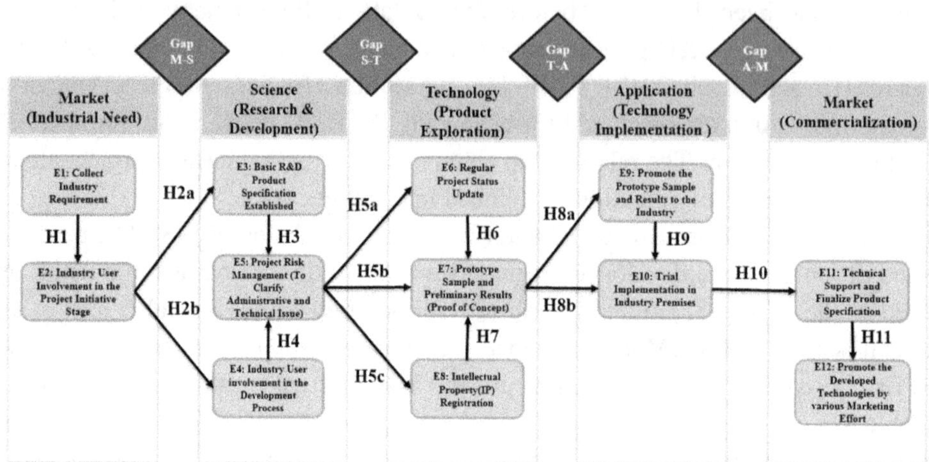

Quantitative Analysis

The questionnaire (See Appendix A in detail) was emailed to around 1100 industry users. 184 people in total participated in the survey, with 133 fully complete responses and 51 partially completed or incomplete responses. The results are based on 133 complete responses. The target people, who is working in the Hong Kong's Logistics and Supply Chain Industry or has an extensive working experience in this area. The questionnaire is designed in three parts; 1) Participant Detail; 2) Industry User's Point of View of R&D Projects; and 3) Challenge of R&D Technology Adoption and Commercialization in the Corporate.

The second and third parts of the question through the quantitative responses from the industry user base on their experience or what their believe to answer the questionnaire for evaluating the ratings of importance for these determinants of influencing ITF R&D project deliverables commercializing and productizing to the industry efficiently. The mean score, standard deviation and determinants are listed in Table 4. The mean score range from a minimum of 3.76 to a maximum of 4.26 and the standard deviation minimum range is around 0.28 to 0.33.

1. **Quality Factors:** The highest mean score factor is 4.26 for the item of R&D technology can provide the new value of the company. The lowest mean score factor is 3.76 for the item of R&D technology is the critical successful factor for the company.
2. **Control Factors:** The highest mean score factor is 4.22 for the item of Trial implementation or pilot test in the company can help the user adopt the R&D deliverables. The lowest mean score factor is 3.81 for the item of concerning about risk of failure of the R&D technology adoption.
3. **Motivating Factors:** The highest mean score is 4.05 for the item of R&D technology can extend the business to the market. The lowest score factor is 3.77 for the item of R&D technology promotion activities are the main factor of R&D deliverables commercialization.
4. **Time Factors:** The highest mean score is 4.11 for the item of concerning about the development time in the R&D process. The lowest score factor is 3.92 for the both items of it is the right time for us to adopt the R&D technology in the company and this is the right time to introduce R&D technology to our top management for consideration.

Table 3. STAM Model and MSTAM Model comparison chart

	Work / Tasks	STAM Model	MSTAM Model
Market	Collect Industry requirement	N	Y
	Industry user involvement	N	Y
	Demonstrator / Gatekeeper	N	Y
Market to Science	Industry user provide real market information	N	Y
	Industry user involve in the R&D project initial stage	N	Y
Science	Basic R&D product specification establishment	N	Y
	Demonstrator / Gatekeeper	Y	Y
	Demonstrating scientific principles	Y	Y
	Convince potential funders	Y	Y
	Project Risk Assessment	N	Y
	Project administrative and technical support	N	Y
	Communicate with scientific community	N	Y
	Communicating with Industry user	N	Y
Science to Technology	Regular project update	N	Y
	Preliminary test result and prototype sample	N	Y
	Define project risk	N	Y
	Applied science and technology (feasibility) demonstrators	Y	Y
	Visualizing potential future applications	Y	Y
	Demonstrating potential to scale-up the physical size of the science	Y	Y
Technology	Demonstrator / Gatekeeper	Y	Y
	Regular Project Status Update and Unification Schedule Management	N	Y
	Demonstrating technical feasibility of potential future applications	Y	Y
	Prototype Sample and Preliminary Results (Proof of Concept)	N	Y
	Pre-Intellectual Property (IP) Registration	N	Y
Technology to Application	Promotion	N	Y
	Trail Implementation	N	Y
Application	Demonstrator / Gatekeeper	Y	Y
	Trail Implementation	N	Y
	Promotion	N	Y
	Demonstrating technical feasibility of a specific application	Y	Y
	Demonstrating commercial feasibility of a specific application	Y	Y
	Demonstrating potential to scale-up and reproduce in volume	Y	Y
	Demonstrators to convince potentials investors	Y	Y
Application to Market	Product specification	N	Y
	Price-performance	Y	Y
Market	Demonstrator / Gatekeeper	Y	Y
	Promotion	N	Y
	Prototype Sample	Y	Y
	Technical Support and Finalize Product Specification	N	Y
	Promote the Developed Technologies by various Marketing Effort	N	Y

The mean score range in the survey is 3.76 to 4.26 and the Standard Deviation (SD) range is 0.28 to 0.33. There are 14 items over 4 mean score and 10 items mean score rating is below 4, but it has 7 items scored around 3.9. These are 58.3% and 41.7% respectively.

According to the research results for the quality factors influencing R&D project, the Industry User totally agree those factors (Table 4) are the key elements in the ITF R&D project. There are 8 items in this area and 7 items scored over 4 and 1 item scored 3.76. (See Figure 4)

For the control factors influencing R&D project, the industry user think those factors are reasonable and acceptance (Table 4). A All 8 items in this area are supported and 4 items scored over 4 and 4 items scored below 4.. (See Figure 5)

The motivating factors influencing R&D project, all response are positive. (Table 3) The average scored is 3.9 and the maximum scored item is 4.05 and the minimum item scored is 3.77. The detail is shown in Figure 6.

The time factors influencing R&D project, all response are positive (Table 3). The average scored is 4.01 and the maximum scored item is 4.11 and the minimum item scored is 3.92. The detail is shown in Figure 7.

For the third parts of the question, the mean score, standard deviation and determinant factors are listed in Table 5. The mean score range from a minimum of 4 to a maximum of 4.26 and the standard deviation minimum range is around 0.28 to 0.32. The results are positive and supported to the hypothesis. All items scored over 4.

The research results for the third parts (Challenge of R&D Technology Adoption and Commercialization in the Corporate), the industry user totally agree those factors and 100% scored over 4 (Table 5). As such, all factors what industry user concern and understanding the following elements are the critical components in the R&D project. The detail is stated in Figure 8.

Determinants Factor for R&D Deliverables Commercialization

The qualitative in-depth interview with the representative of industry experts and quantitative survey data were collected from industry user to test whether the proposed elements, hypotheses, and MSTAM model are correct. The qualitative and quantitative survey helps to identify the industry user point of view and collect their expectation. Base on their experience and what their belief to answer the questionnaire. In-depth interview has been conducted to seek six industry experts to share their comment to deeply understand the industry need. The quantitative online questionnaire was emailed or sent the message to around 1100 industry user for seeking their feedback. This is an initial study for ITF R&D project and the data only collected 184 responses from the industry user. The response rating is 16.73%. The findings are summarized as below:

Industry User Comment Summary

The quantitative survey were collected 184 responses, but only 133 people were fully completed the questionnaire. Therefore, the quantitative results will be evaluated base on 133 people comment. The findings were summarized as follow:

1. Both top level and senior management level people were scored 63.91% together.

Table 4. Industry User's Point of View of R&D Projects

Industry User's Point of View of R&D Projects	Mean Score	SD
Quality Factors		
R&D technology can provide the new value of the company	4.26	0.30
R&D technology can improve company productivity.	4.20	0.31
R&D technology can support the plurality of applications in logistics and supply chain industries.	4.07	0.31
R&D technology can increase the company competitive advantage in the market.	4.23	0.31
R&D technology is the critical successful factor for the company.	3.76	0.30
R&D technology can support the company long term growth.	4.26	0.31
R&D technology result quality is the key concern for user adoption.	4.08	0.31
Concerning about the quality of R&D technology result whether suitable for the company operations	4.02	0.30
Control Factors		
R&D technology can guide the company to a new transformation.	3.98	0.30
Concerning about technical support of R&D implementation weather enough.	3.93	0.32
Concerning about risk of failure of the R&D technology adoption.	3.81	0.32
Detail R&D product specification and clear guidance can help the user to use the R&D product.	4.16	0.32
R&D technology process control mechanism is main factor of R&D deliverables commercialization.	3.95	0.31
Trial implementation or pilot test in the company can help the user adopt the R&D deliverables.	4.22	0.32
Regular review meeting with the R&D technologist can help the user easy to know the R&D technology.	4.11	0.30
It would be desirable to adopt R&D technology for Hong Kong Logistics and Supply Chain Industries.	4.13	0.31
Motivating Factors		
R&D technology can extend the business to the market.	4.05	0.31
Hong Kong Government funding support for R&D technology can help the SME to build up core capability of the company.	3.98	0.32
R&D technology results can enhance the interactivity between internal and external users.	3.90	0.32
R&D technology promotion activities are the main factor of R&D deliverables commercialization.	3.77	0.29
Time Factors		
R&D technology can support the company long term growth.	4.09	0.31
Concerning about the development time in the R&D process.	4.11	0.31
It is the right time for us to adopt the R&D technology in the company.	3.92	0.33
This is the right time to introduce R&D technology to our top management for consideration.	3.92	0.32

*Five-point scale: 1 = Strongly Disagree, 2 = Somewhat Disagree, 3 = Neutral, 4 = Somewhat Agree, 5 = Strongly Agree

2. High educational level for the respondents scored 90.23%. This percentage included Doctoral Degree, Master Degree, and Bachelor Degree.
3. 69.17% respondents are working in over 100 staffs companies.
4. Working experience over 15 years in the industry scored 55.64%
5. Total years of working experience over 15 years scored 74.44%

Figure 4. Quality Factors influencing R&D Project (Industry View)

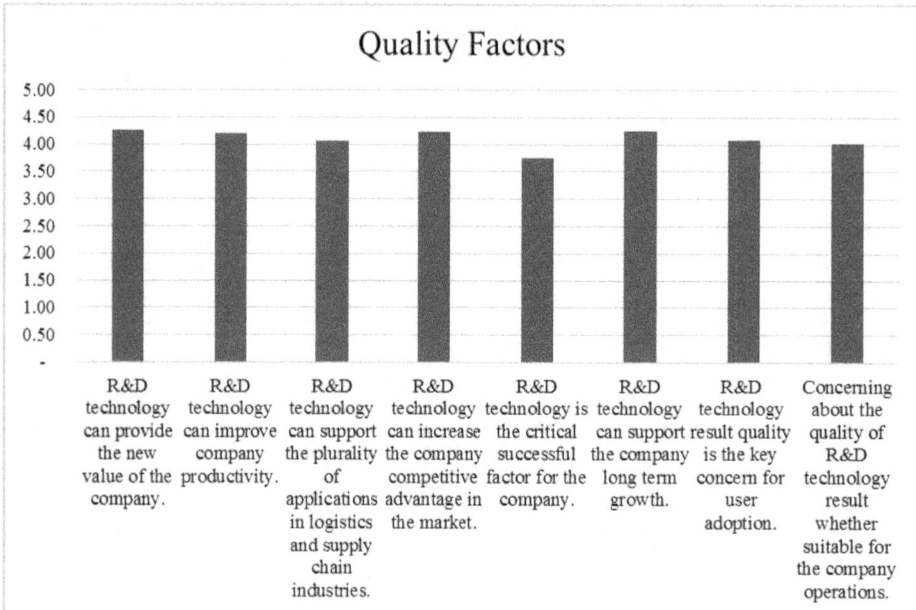

Figure 5. Control Factors influencing R&D Project (Industry View)

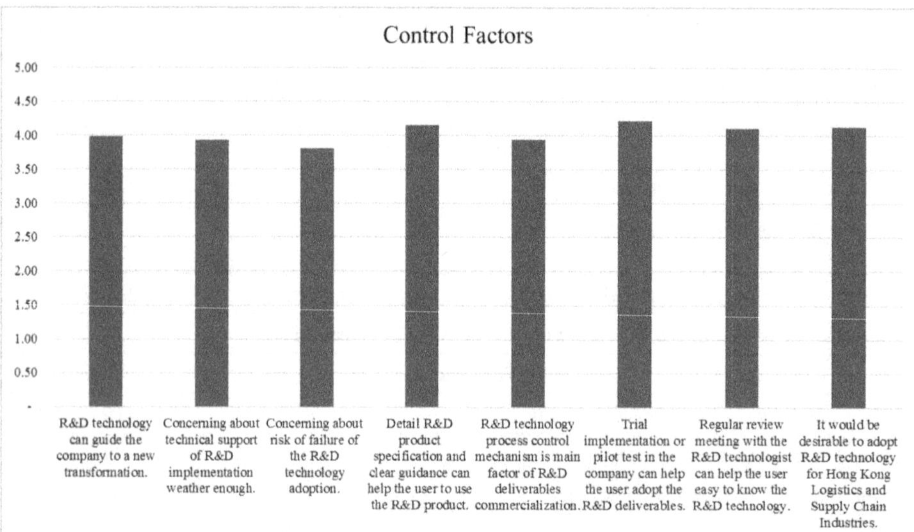

6. Haven't participated in the R&D projects scored 57.14%
7. Believe R&D project is beneficial for company growth scored 95.49%
8. Not believe R&D project is beneficial for company growth scored 4.51% and most respondents are operational level

Figure 6. Motivating Factors influencing R&D Project (Industry View)

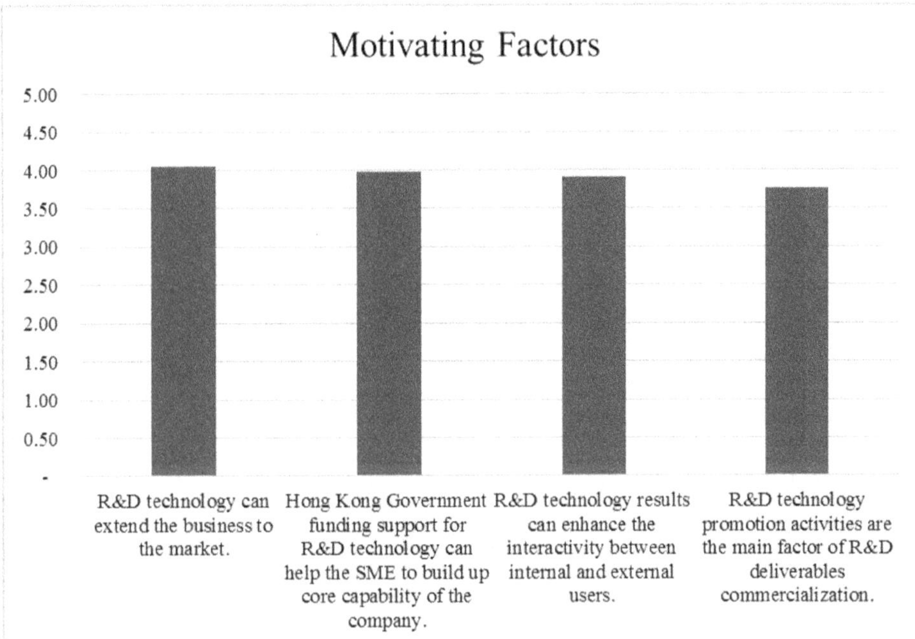

Figure 7. Time Factors influencing R&D Project (Industry View)

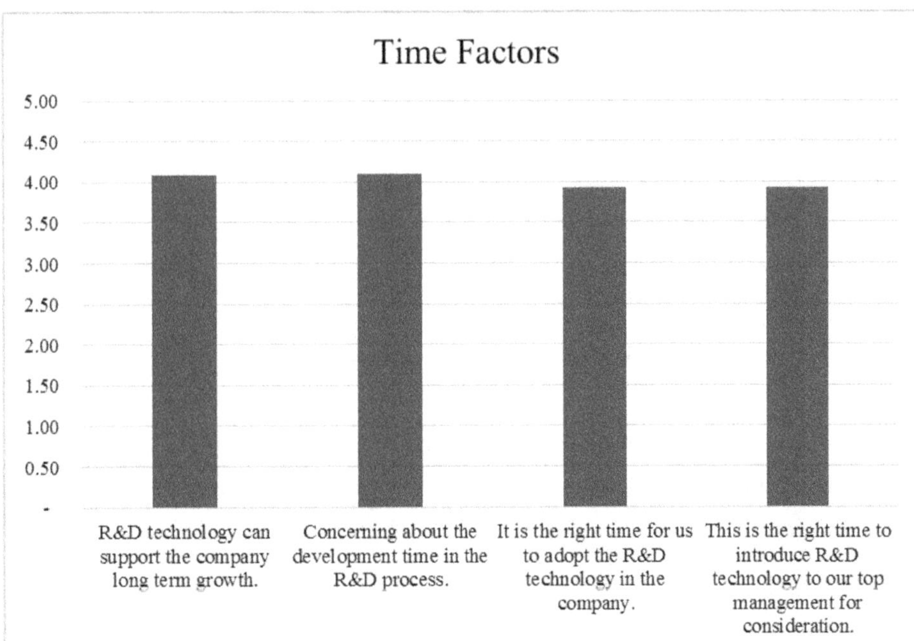

Table 5. Industry user's point of view of challenge of R&D Technology Adoption and Commercialization in the corporate

Challenge of R&D Technology Adoption and Commercialization in the Corporate	Mean Score	SD
Industry Need		
Industry requirement collection can be helped to develop the R&D projects	4.09	0.32
Industry user involve in the R&D projects in the project initial stage, it can speed up the R&D development process.	4.18	0.32
Industry user provides real market situation to help the R&D technologist explore the new idea of R&D project	4.21	0.32
Research Development		
Create a basic R&D product specification in the R&D project initial stage; it can help all stakeholders easy to know the R&D deliverables.	4.01	0.32
In the R&D project initial stage, I believe that define project risk in advance such as administrative and technical issue, it can eliminates the risk of project failure	4.00	0.32
Industry user participates in the R&D projects, it can enhance the project value of the R&D deliverables	4.12	0.32
Product Exploration		
Regular project status update of the R&D project, it can reduce the communication issue among all stakeholders	4.18	0.32
The preliminary test result and prototype sample can help the industry user easy to understand the product for future commercialization and adoption	4.17	0.32
Intellectual Property (IP) registration can protect the rights of the developer and facilitate further commercialization	4.26	0.32
Technology Implementation		
The preliminary results and prototype sample of the R&D product can help the industry user easy to know and how to use the R&D deliverables.	4.14	0.28
The promotion activity can attract the industry users to know the R&D product value for commercialization.	4.00	0.29
The R&D project trial implementation in the real industry premises, it can help the industry user to adopt the R&D deliverables	4.11	0.31
Commercialization		
Technical support of R&D technologist can help the industry user easy to adopt the R&D deliverables in the company	4.17	0.31
Clear and detail R&D product specification can speed up the industry user to adopt the R&D deliverable	4.17	0.32
Clear marketing plan can promote the R&D deliverables to the market effectively. For example, to prepare press release or to produce video clip to promote the project results	4.14	0.32

*Five-point scale: 1 = Strongly Disagree, 2 = Somewhat Disagree, 3 = Neutral, 4 = Somewhat Agree, 5 = Strongly Agree

Refer to the Table 4, the mean average weight scored for each questionnaire scored over 4. The results is to prove that all Industry User agree the proposed elements are important tasks in the R&D development process, these are quality factors, control factors, time factors, and motivating factors. For R&D deliverables commercialization, industry need, research development, product exploration, technology implementation, and commercialization are essential development stage in the R&D development pro-

Figure 8. Challenge of R&D Technology Adoption and Commercialization

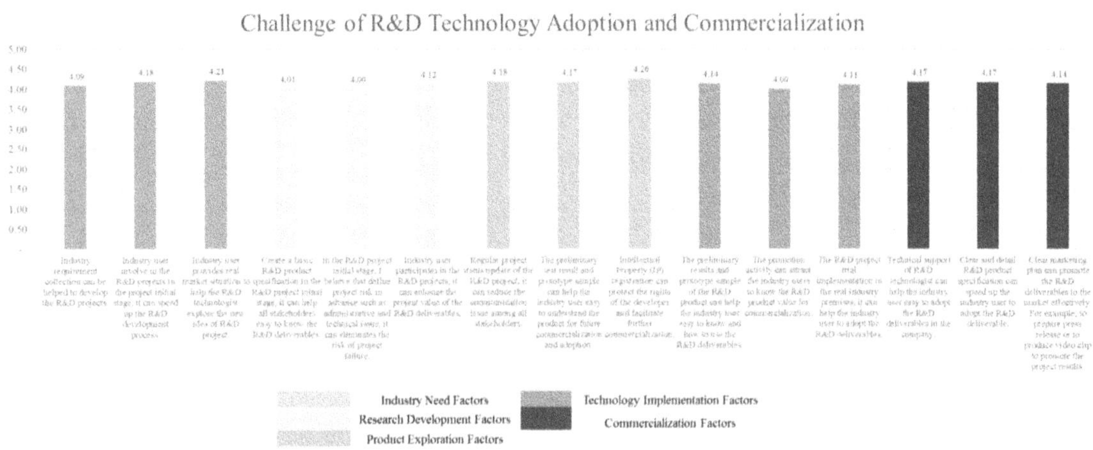

cess. It helps to enhance the rate of R&D deliverables commercialization to the market. All proposed proposition and hypothesis are supported by the research results.

Co-Relation Analysis

All proposed hypothesis scored is positive. The hypothesis test summary is showed in Table 6. The goal of a correlation analysis is to see whether two measurement variables co vary, and to quantify the strength of the relationship between the variables. A correlation matrix, which gives an overview of

Figure 9. Mean Average Weight (Industry View)

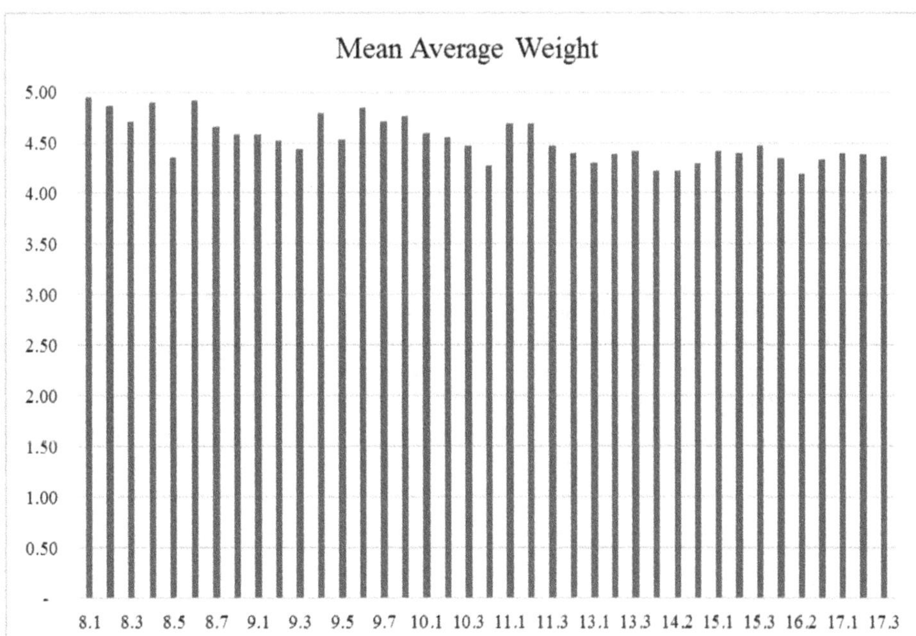

what variables tend to go up and down together and in what direction. It's good first past at relationships in the collected data from the industry user before driving into regression. The correlation is measured by a variable called Pearson's R (Pearson 2017), which ranges between -1 mean indirect relationship and 1 refer to perfect relationship. Table 5 matches each variable to all other variables. The relationship between two columns is the same, no matter which direction can be compared. For example: H1 to H1 is 1. The table show that the variables with the strongest relationship

Remark:

1. The closer the number is to positive one, the stronger the positive correction.
2. The closer the number is to negative one, the stronger the negative correlation.
3. The closer the number is to zero, the weaker the correlation.
4. Correlations greater than 0.5 are significant ($p < 0:001$), $N = 133$

The correlation coefficient functions in Microsoft excel software, a value between -1 and +1. It will tell those two variables are related to each other. The results are shown in Figure 10 is for the correlation of each hypothesis. All proposed hypotheses are positive and correlated. Most scores obtained are around 0.4 to 0.8. (See Figure 10)

Hypothesis Testing

Overall, the proposed hypotheses validation results demonstration of all 15 out of 15 hypotheses is supported. The mean scored is 4 to 4.26. It is clearly indicated that industry user believes R&D deliverables can be beneficial to the firm. The detail is summarized in Table 7.

Table 6. Co-relation Analysis for Hypothesis

	H1	H2a	H2b	H3	H4	H5a	H5b	H5c	H6	H7	H8a	H8b	H9	H10	H11
H1	1.000														
H2a	0.700	1.000													
H2b	0.656	0.742	1.000												
H3	0.589	0.557	0.535	1.000											
H4	0.601	0.613	0.683	0.684	1.000										
H5a	0.598	0.577	0.569	0.611	0.648	1.000									
H5b	0.562	0.578	0.547	0.662	0.627	0.741	1.000								
H5c	0.540	0.532	0.500	0.709	0.650	0.501	0.630	1.000							
H6	0.569	0.624	0.607	0.648	0.617	0.644	0.730	0.575	1.000						
H7	0.409	0.448	0.477	0.491	0.502	0.680	0.596	0.444	0.528	1.000					
H8a	0.549	0.525	0.463	0.511	0.545	0.483	0.569	0.559	0.626	0.441	1.000				
H8b	0.626	0.574	0.533	0.562	0.605	0.553	0.602	0.561	0.646	0.465	0.511	1.000			
H9	0.556	0.624	0.640	0.564	0.622	0.656	0.663	0.528	0.710	0.591	0.492	0.570	1.000		
H10	0.576	0.502	0.511	0.571	0.539	0.581	0.646	0.522	0.689	0.489	0.584	0.564	0.680	1.000	
H11	0.574	0.520	0.565	0.542	0.590	0.628	0.670	0.493	0.689	0.509	0.524	0.539	0.654	0.687	1.000

Figure 10. Positive Correlation

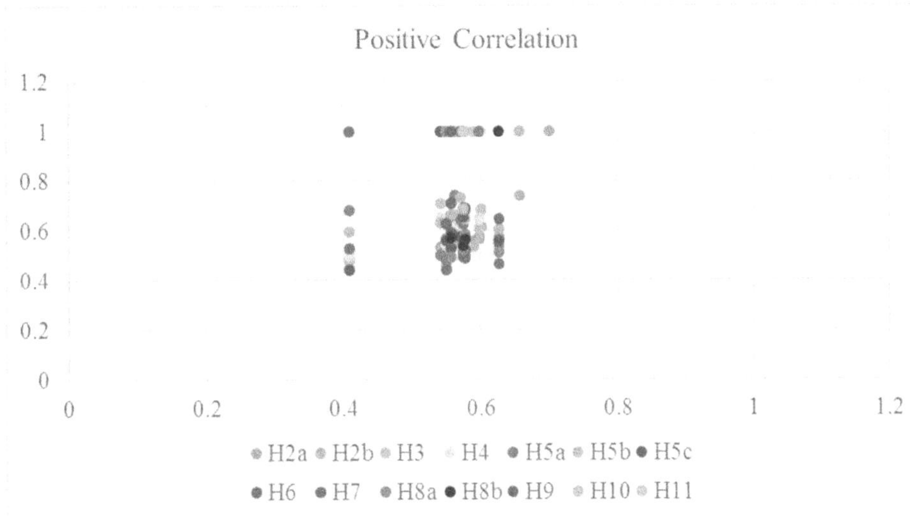

CONCLUSION AND DISCUSSION

In order to understand the market need, quantitative survey was conducted from Industry User. Their valuable comments are needed to explore the research areas and focus on the bottle neck point of ITF R&D project development. It presents the key concerns for the R&D project commercialization and adoption in the industry. The findings of Industry User is to fully understand what the user need, the time of R&D technology to market, promotion for the R&D projects, cost and benefits of the R&D deliverables, and quality of the R&D deliverables. These are the essential element in the R&D process.

Through the online questionnaire and emails to seek Industry User to provide their comment, the results found that the mean average weight scored for each questionnaire scored over 4 in part 3. It is to prove that all Industry User agree the proposed elements are important tasks in the R&D development as quality factors, control factors, time factors, and motivating factors. Overall, the proposed hypotheses validation results demonstrate of all 15 out of 15 hypothesis are supported. This chapter presents a review of literature on R&D project commercialization and adoption in Industry User (demand-side) viewpoints. Through the previous studies on new product development, new projects development, and project management to explore a novel R&D project management mechanism and model for analysing the determinants of ITF R&D projects commercialization in Hong Kong's Logistics and Supply Chain Industries.

For the industry user quantitative survey, the questionnaire was emailed to around 1100 industry user. Total 184 people filled the survey, which include 133 fully complete responses and 51 partly or incomplete responses. The results are based on 133 complete responses. The target people, who is working in the Hong Kong's Logistics and Supply Chain Industry or has an extensive working experience in this area. Refer to the research results for the Challenge of R&D Technology Adoption and Commercialization in the Corporate, the industry user totally agree those factors and 100% scored over 4 (Table 5). As such, all factors what industry user concern and understand the stated elements in Figure 8 are the critical components in the R&D project.

Table 7. Validation Results

Validation Results				
	Hypothesis	Question	Mean Scored	Validation
H1	Industry requirement collection can be helped to develop the R&D projects	Industry Need (Item 1)	4.09	Supported
H2a	Industry user provides real market situation to help the R&D technologist explore the new idea of R&D project	Industry Need (Item 2)	4.18	Supported
H2b	Industry user involve in the R&D projects in the project initial stage, it can speed up the R&D development process	Industry Need (Item 3)	4.21	Supported
H3	Create a basic R&D product specification in the R&D project initial stage, it can help all stakeholders easy to know the R&D deliverables	Research Development (Item 1)	4.01	Supported
H4	Industry user participates in the R&D projects, it can enhance the project value of the R&D deliverables	Research Development (Item 3)	4.12	Supported
H5a	Regular project status update of the R&D project, it can reduce the communication issue among all stakeholders	Product Exploration (Item 1)	4.18	Supported
H5b	The preliminary test result and prototype sample can help the industry user easy to understand the product for future commercialization and adoption	Product Exploration (Item 2)	4.17	Supported
H5c	In the R&D project initial stage, I believe that define project risk in advance such as administrative and technical issue, it can eliminates the risk of project failure	Research Development (Item 2)	4	Supported
H6	The preliminary results and prototype sample of the R&D product can help the industry user easy to know and how to use the R&D deliverables	Technology Implementation (Item 1)	4.14	Supported
H7	Intellectual Property (IP) registration can protect the rights of the developer and facilitate further commercialization	Product Exploration (Item 3)	4.26	Supported
H8a	The promotion activity can attract the industry users to know the R&D product value for commercialization	Technology Implementation (Item 2)	4	Supported
H8b	The R&D project trial implementation in the real industry premises, it can help the industry user to adopt the R&D deliverables	Technology Implementation (Item 3)	4.11	Supported
H9	Technical support of R&D technologist can help the industry user easy to adopt the R&D deliverables in the company	Commercialization (Item 1)	4.17	Supported
H10	Clear and detail R&D product specification can speed up the industry user to adopt the R&D deliverable	Commercialization (Item 2)	4.17	Supported
H11	Clear marketing plan can promote the R&D deliverables to the market effectively.	Commercialization (Item 3)	4.14	Supported

Refer to the captioned findings, industry need and R&D technologist applied research may need to be aligned for increasing the commercialization opportunity to the market. The further study will be collected the industry point of view and a real case study need to be tested to prove whether the proposed model is suitable in ITF R&D project. The author believes the survey results are a good indicator for all of us to explore how to enhance the ITF R&D project commercialization rate to the market. The analytical framework resulting from this research is proved that the elements and MSTAM are the correct direction for assisting future research of ITF R&D project commercialization.

ACKNOWLEDGMENT

I would like to give thanks to my industrial advisors, Dr. Jack So, Dr. Aldar Chan, Dr. CC Cheung, and Dr. KL Fan to provide their valuable comments to me for developing the questionnaire for my dissertation. I also would like to express their sincere thanks and warmest appreciation to those R&D Technologist Experts and Industry Experts for sharing their valuable thoughts and incredible insights.

REFERENCES

Abdullah, N. H., Shamsuddin, A., Wahab, E., & Hamid, N. A. (2012). Preliminary Qualitative Findings on Technology Adoption of Malaysian SMEs. *IEEE Conference Publications*, 15 - 20. 10.1109/CHUSER.2012.6504273

Alaoui, K. B., Naimi, Z., Benlarabi, A., & Outzourhit, A. (2015). Pilot line preindustrial reactor installation for applied research in vacuum deposition techniques for the preparation and characterization of photovoltaic cells. *IEEE Conference Publications*, 1 – 3.

Amadi-Echendu, J. E., & Rasetlola, R. T. (2011). Technology commercialization factors, frameworks and models. *IEEE Conference Publications*, 144 – 148.

Asghar, I., & Usman, M. (2013). Motivational and De-motivational Factors for Software Engineers: An Empirical Investigation. *IEEE Conference Publications*, 66 – 71. 10.1109/FIT.2013.20

Åstebro, T. (2004). Key Success Factors for Technological Entrepreneurs' R&D Projects. IEEE Transactions on Engineering Management, 51(3), 314 - 321. doi:10.1109/TEM.2004.830863

Bei, H., & Li, M. (2006). Research on the Motivation Factors for the Development of Talent in the Industry Cluster. *IEEE Conference Publications*, 1237 – 1242. 10.1109/ICMSE.2006.314221

Besrour, Bin Ab Rahim, & Dominic (2015). The Study of the Relation between Requirement Engineering Techniques and Challenges in Software Industry. *IEEE Conference Publications*, 49 – 53. 10.1109/ISMSC.2015.7594026

Blindenbach-Driessen, F. (2015). The (In) Effectiveness of Cross-Functional Innovation Teams: The Moderating Role of Organizational Context. *IEEE Transactions on Engineering Management*, *62*(1), 29–38. doi:10.1109/TEM.2014.2361623

Bruno, Ferrari, & Lopes de Oliveira e Souza (2016). A Requirements Engineering and Management Process in Concept Phase of Complex Systems. *IEEE Conference Publications*, 1 - 6.

Chen, D., Qiu, W., Min, Y., & Mei-yung, L. (2007). Activity Flow Optimization and Risk Evaluation of Complex Project. *IEEE Conference Publications*, 5196 - 5199. 10.1109/WICOM.2007.1272

Ching, T. W., & Xu, T. (2014). Performance Study of Electric Vehicles in Macau. *IEEE Conference Publications*, 1 – 5. 10.1109/CCECE.2014.6900957

Chokri, M., Aymen, F., & Lassaad, S. (2017). Prototype design of a Compact Plug-in Solar Electric Vehicle. *IEEE Conference Publications*, 1 – 4.

Chuang, Y., & Wu, L. (2007). User-Based Evaluations of Search Engines: Hygiene Factors and Motivation Factors. *IEEE Conference Publications*, 1 – 10. 10.1109/HICSS.2007.590

Chuang & Wu. (2007). User-Based Evaluation of Search Engines: Hygiene Factors and Motivation Factors. *IEEE Conference Publications*, 1 – 10.

de Souza Andrade, Soto Urbina, & de olivera N. Follador (2016). Process Proposal for the Intellectual Property Protection Management in a Technology Licensing Office from a Brazilian Scientific and Technological Institution. *IEEE Conference Publications*, 1672 – 1680.

Dong-dong, G. A. O., & Ping, L. V. (2009). A Sorting-Based Management Model to Support Early Supplier Involvement in New Product Development. *IEEE Conference Publications*, 496 - 501.

Ettlie, J. E. (1995). Early Manufacturing Involvement in New product Development. *IEEE Conference Publications*, 104 – 109. 10.1109/IEMC.1995.523917

Ferraz, A., Machado, J., & Carvalho, V. (2015). Prototype for determination of pre-transfusion tests based on image processing techniques. *IEEE Conference Publications*, 1 – 6. 10.1109/ENBENG.2015.7088849

Gordon, G., & Bush, J. B. Jr. (1997). Management of Technical Innovation. *IEEE Conference Publications*, 199 – 122.

Guzmán, L., Steinbach, S., Diebold, P., Zehler, T., Schneider, K., & Habbe, M. (2016). Evaluating the Benefits of Systematic Project Management in Large Public Sector Projects. *IEEE Conference Publications*, 32 – 35. 10.1145/2896839.2896841

Hashim & Mimos. (2002). Development of a Silicon Sensor Wafer Fab for Local Industry Requirement. *IEEE Conference Publications*, 145 – 149.

Hatcher, S., & Samuels, M. (2000). Early Involvement of Operational Test: Value Added for the CH-60S/SH-60R1. *IEEE Conference Publications*, 35 – 42. 10.1109/AERO.2000.878211

Herry. (2016). Project Management: Model Research in Success Rate of A Digital Start-Up Project. *IEEE Conference Publications*, 1 - 6.

Hirose, Y. (2012). Knowledge Transfer from Researches to Society: How to Offset the Cultural Gap? *IEEE Conference Publications*, 1107-1116.

Ho, S. C., & Chuah, K. B. (2017). *Determinants of ITF R&D Technology Commercialization in Logistics and Supply Chain Industries: R&D Technologist Perspective*. Retrieved from http://www.wikicfp.com/cfp/servlet/event.showcfp?eventid=62787

Ho, S. C., & Chuah, K. B. (2018). *Critical Success Factors for Strategic Management of ITF R&D Projects Commercialization: An Industry Expert Perspective*. Retrieved from https://www.igi-global.com/publish/call-for-papers/call-details/3018

Ho, S. C., & Chuah, K. B. (2018). *Application of MSTAM Methodology in Project Management – A Case Study of ITF Robotic Automation R&D Project*. Retrieved from https://www.igi-global.com/publish/call-for-papers/call-details/3000

Ho, S. C., & Chuah, K. B. (2018). *A Handbook for ITF R&D Project Management*. Retrieved from https://www.igi-global.com/publish/call-for-papers/call-details/2827

Holtta, Eisto, & Mahlamaki. (2009). Benefits for cast product development through early supplier involvement. *IEEE Conference Publications*, 1 – 7.

Hund, L. B., Campbell, D. L., & Newcomer, J. T. (2017). Statistical Guidance for Setting Product Specification Limits. *IEEE Conference Publications*, 1 – 6. 10.1109/RAM.2017.7889664

Iskin, I. (2011). Literature review on adoption of energy efficient technologies from a demand side management perspective: Taxonomy of adoption drivers, barriers and policy tools. *IEEE Conference Publications*, 1 – 16.

ITF IP. (2017). *Guide on Intellectual Property Arrangements for Research and Development Projects Funded Under the Innovation and Technology Support Programme and the Midstream Research Programme for Universities of the Innovation and Technology Fund*. Retrieved from https://www.itf.gov.hk/l-eng/Forms/IP_Guideline_201710.pdf

Jumie Yuventi & Weiss. (2013). Value sensitivity of Quality Function Deployment approaches in systems engineering-driven construction projects. *IEEE Conference Publications*, 847 - 852.

Keneni, B., Austin, B., Elkin, C., & Devabhaktuni, V. (2017). Educational Prototype Demonstrating Frequency Spectrum Sharing Through Channel Borrowing and Priority Assignment. *IEEE Conference Publications*, 540 – 544. 10.1109/EIT.2017.8053422

Kim, S.-K., Lee, B.-G., & Oh, K.-S. (2009). The Effect of R&D and Technology Commercialization Capabilities on the Innovation Performance of Korean IT SMEs: The Case of Direct and Indirect Recipients of Public R&D Funding. *IEEE Conference Publication*, 1531 - 1541. 10.1109/PICMET.2009.5261985

Kori, Pedaste, Altin, Tõnisson, & Palts. (2016). Factors That Influence Students' Motivation to Start and to Continue Studying Information Technology in Estonia. *IEEE Transactions on Education*, *59*(4), 255–262.

Lenina, S. V. B. (2016). Intellectual Property Protection: A Step to realize Make in India. *IEEE Conference Publications*, 2089 – 2092. 10.1109/SCOPES.2016.7955816

Lin, J., & Lin, Q. (2017). Analysis of the Key Factors of Intellectual Property Management at Art Institutions. *IEEE Conference Publications*, 206 – 208. 10.1109/ICASI.2017.7988609

Loeh, H. (2006). Project and Product Management in Collaborative Product Innovation. *IEEE Conference Publications*, 1 – 10. 10.1109/ICE.2006.7477079

Lou, Y., Fu, X., & Huang, L. (2010). Evaluation on the Commercialization Potential of Emerging Technologies Based on Structural Equation Model. *IEEE Conference Publications*, 329 – 333. 10.1109/UKSIM.2010.68

Mats & Jeffrey. (2000). Specification based Prototyping of Control Systems. *IEEE Conference Publications*, 1D3/1 - 1D3/8.

Mutingi, M., & Matope, S. (2013). Dynamics of Information Technology Adoption in a Complex Environment. *IEEE Conference Publications*, 1466 - 1471. 10.1109/ICIT.2013.6505888

NewGOV.HK. (2017). *LCQ7: Promoting development of innovation and technology*. Retrieved from: http://www.info.gov.hk/gia/general/201205/16/P201205160267.htm

Nikou, S. A., & Economides, A. A. (2014). Acceptance of Mobile-Based Assessment from the Perspective of Self-Determination Theory of Motivation. *IEEE Conference Publications*, 454 – 458. 10.1109/ICALT.2014.136

Pearson. (2017). *Pearson correlation coefficient*. Retrieved from https://en.wikipedia.org/wiki/Pearson_correlation_coefficient

Phaal, R., O'Sullivan, E., Routley, M., Ford, S., & Probert, D. (2011). A framework for mapping industrial emergence. *Technological Forecasting and Social Change*, *78*(2), 217–230. doi:10.1016/j.techfore.2010.06.018

Pirkkalainen, H., Jokinen, J. P. P., & Pawlowski, J. M. (2014). Understanding Social OER Environments—A Quantitative Study on Factors Influencing the Motivation to Share and Collaborate. *IEEE Conference Publications*, 388 – 400. 10.1109/TLT.2014.2323970

Project Network. (2015). Retrieved from https://en.wikipedia.org/wiki/Project_network

Przybysz, P. M., Duckwitz, S., Mütze-Niewöhner, S., & Schlick, C. M. (2013). Investigation of Team Composition and Task-related Conflict as Determinants of Engineering Service Productivity. *IEEE Conference Publications*, 250 – 254. 10.1109/IEEM.2013.6962412

Qing, T., Hui-yu, L., & Ming-jian, Z. (2011). Research on the Transaction Memory System in the Team of Early Supplier Involvement. *IEEE Conference Publications*, 1503 - 1507.

Regan, G., Flood, D., & McCaffery, F. (2016). Research Findings from an Industrial Trial of a Traceability Assessment and Implementation Framework. *IEEE Conference Publications*, 91 – 95. 10.1145/2904354.2904365

Roisko, V., Kämppi, P., & Luojus, S. (2013). Touch Screen Based TETRA Vehicle Radio: Preliminary Results of Multi-methodology Usability Testing Prototype. *IEEE Conference Publications*, 951 – 952. 10.1109/ICCVE.2013.6799938

Sako, M., & Kato, K. (2017). Research on the Harmonization of Intellectual Property Systems from the Point of View of Japanese Inventors. *IEEE Conference Publications*, 245 – 251. 10.1109/TEMSCON.2017.7998384

Shahzadi, R., Sadeghi, M., & Aghaz, A. (2014). Managing Conflict in Distributed Projects. *IEEE Conference Publications*, 1061 – 1065.

Shin, B. C. (2010). Preliminary test results of prototype urban maglev train. *IEEE Conference Publications*, 32 – 35.

Shtangey, S., & Tereshchenko, A. (2015). Laboratory Information Management System - Information Technology for Production Quality. *IEEE Conference Publications*, 112 – 114.

Siswanto, J., & Maulida, A. (2017). Validated ERP Modules Requirement for Micro, Small and Medium Enterprise Fashion Industry. *IEEE Conference Publications*, 1 – 6.

Smparounis, K., Mavrikios, D., Pappas, M., Xanthakis, V., Viganò, G. P., & Pentenrieder, K. (2008). A virtual and augmented reality approach to collaborative product design and demonstration. *IEEE Conference Publications*, 1 – 8.

Soh, P.-H., & Roberts, E. B. (2005, November). Technology Alliances and Networks: An External Link to Research Capability. *IEEE Transactions on Engineering Management*, *52*(4), 419–428. doi:10.1109/TEM.2005.850727

Suebsin, C., & Gerdsri, N. (2009). Key Factors Driving the Success of Technology Adoption: Case Examples of ERP Adoption. *IEEE Conference Publications*, 2638 - 2643. 10.1109/PICMET.2009.5261818

Suleiman, K. (2002). Factors Differentiating the Commercialization of Disruptive and Sustaining Technologies. *IEEE Transactions on Engineering Management*, *49*(4), 375–387.

Sultan, F., & Chan, L. (2000). The adoption of new technology: The case of object-oriented computing in software companies. *IEEE Transactions on Engineering Management*, *47*(1), 106–126. doi:10.1109/17.820730

Sun, Wu, Huang, & Lin. (2013). A Computing Resources Market Model and Supply-Demand Matching Mechanism. *IEEE Conference Publications*, 217 – 224.

Tabassum, Siddik, Shoyaib, & Khaled. (2014). Determining Interdependency Among Non-functional Requirements to Reduce Conflict. *IEEE Conference Publications*, 1 – 6.

Takada, N. (2016). Untangling the Boundaries in Technology Collaborations: The Deviation Effects of "Project Autonomy" on Innovations through Collaborations. *IEEE Conference Publications*, 399 – 409. 10.1109/PICMET.2016.7806633

Taoudi, A., Bounabat, B., & Elmir, B. (2013). Quality based Project Control using Interoperability degree as a Quality Factor. *IEEE Conference Publications*, 1 - 6. 10.1109/ISKO-Maghreb.2013.6728117

Tieng, K., Jeenanunta, C., Rittippant, N., Chongpisal, P., & Hamada, R. (2016). The influences of knowledge management on innovation within formal and non-formal R&D manufacturing firms, Thailand. *IEEE Conference Publications*, 208 – 213. 10.1109/ICMIT.2016.7605035

Walsh, Kirchhoff, & Newbert. (2002). Differentiating market strategies for disruptive technologies. IEEE Transactions on Engineering Management, 49(4), 106 – 126.

Wan, W. K., & Monsef, S. (2012). New Product Development Through Open Innovation: Role of organization structure and contextual factors. *IEEE Conference Publications*, 446 – 449.

Wang, C.-H., Huang, S.-Z., Chang, C.-H., Lin, P.-J., & Chiew, Y.-Y. (2015). Co-innovation network driven entrepreneurship in high-tech technology-evidences from China. *IEEE Conference Publications*, 1002 – 1015. 10.1109/PICMET.2015.7273122

Wang, L., & Shen, B. (2016). A Study of Design Collaboration between the Designer and Supplier in the Fashion Supply Chain. *IEEE Conference Publications*, 1 – 3.

Wu, X., Ma, T., & Tian, Y. (2009). The Empirical Impact of Internal Knowledge Structure on the Endogenous Industrial Clusters Innovation - With Wuqueqiao industrial cluster in Suzhou city as example. *IEEE Conference Publications*, 721 – 725.

Xiao, R., Wu, Z., & Sugiura, K. (2013). Slideware2.0: A prototype of presentation system by integrating web2.0 and second screen to promote education communication. *IEEE Conference Publications*, 181 - 185. 10.1109/ICICM.2013.38

Yuan, L.-S., & Dong, R.-P. (2015). A Novel Risk Assessment Algorithm for Accounting Information System Using Analytic Hierarchy Process. *IEEE Conference Publications*, 61 – 64.

Yuzuki, R., Hata, H., & Matsumoto, K. (2015). How We Resolve Conflict: An Empirical Study of Method-Level Conflict Resolution. *IEEE Conference Publications*, 21 – 24. 10.1109/SWAN.2015.7070484

Zhang, L., & Swirski, M. (2002). Managing the Specification Process in Complex Projects. *IEEE Conference Publications*, 350 – 355. 10.1109/IEMC.2002.1038456

Zhao, C. L., Ming, X. G., Wang, X. H., & Li, D. (2009). A Framework of Supplier Involved Collaborative Project Management. *IEEE Conference Publications*, 4130 - 4135. 10.1109/ICISE.2009.36

APPENDIX

Industry User Quantitative Survey Questionnaire

College of Science and Engineering
Department of Systems Engineering and Engineering Management

Dear Sir / Madam,

Re: Survey on Industry User's Expectation and Considerations Factors towards Using R&D Technology in Corporate

The City University of Hong Kong is conducting a research regarding "Determinants of R&D Projects Commercialization in Hong Kong's Logistics and Supply Chain Industries". You are cordially invited to complete this questionnaire based on your experiences and beliefs. You should be able to answer all questions in 5-10 minutes.

This questionnaire is part of an R&D survey for an *EngD Research Study* to identify the critical element and determinants factor influencing R&D results commercialization and adoption in Hong Kong's Supply Chain and Logistics Company.

Click here to participate in the online survey: https://zh.surveymonkey.com/r/CDWTZJN

You are encouraged to invite your qualified colleagues and friends to participate in this survey. This survey is completely anonymous and all information collected will be treated as *Strictly Confidential* and used for academic research purposes in aggregated forum only.

We sincerely hope that you could participate in this important research and believe that its findings will be useful for you and your company. For more information, please feel free to contact me at 9230-5672 or scho25-c@my.cityu.edu.hk. Your reply on or before 10 July 2017 will be highly appreciated.

Yours faithfully,

Ho Siu Cheung
Engineering Doctorate Candidate
Department of Systems Engineering and Engineering Management
City University of Hong Kong
(Tel: 852 9230-5672, Email: scho25-c@my.cityu.edu.hk)

For further information about this survey, please contact:

Supervisor:
Dr. Chuah, Kong Bieng
Associate Professor
EngD Deputy Programme Leader & EngD Thesis Coordinator
Department of Systems Engineering and Engineering Management
City University of Hong Kong
(Tel: 852 3442-8437, Email: mebchuah@cityu.edu.hk)

About the Innovation and Technology Fund (創新及科技基金概覽)

The Innovation and Technology Fund (ITF), administered by the Innovation and Technology Commission (http://www.itf.gov.hk/l-eng/about.asp), aims to increase the added value, productivity and competitiveness of our economic activities. The Government hopes that, through the ITF, Hong Kong companies could be encouraged and assisted to upgrade their technological level and introduce innovative ideas to their businesses.

(Reference: Project Video: http://www.lscm.hk/eng/channel.php?content_id=50076)

創新及科技基金由創新科技署管理,旨在提升本地經濟活動的增值力、生產力及競爭力。政府希望透過基金,鼓勵和協助香港的企業提升科技水平,並為其業務注入更多創新意念。(參考:項目影片: http://www.lscm.hk/chi/channel.php?content_id=50076)

Survey on Industry User's Expectation and Considerations Factors Towards Using R&D Technology in the Corporate

Part I: Participant Particulars

(Please tick the appropriate box as below)

1. Working Level:
   a. Top Level (Owner, CEO, COO, Director, GM)
   b. Senior Level (SM, M)
   c. Middle Level (AM, Supervisor, Officer)
   d. Operational Level
   e. Other
2. Educational Level:
   a. Secondary School
   b. Associate Degree / Higher Diploma
   c. Bachelor Degree
   d. Master Degree
   e. Doctoral Degree
3. Size of Company:
   a. Below 10 staffs
   b. Between 11 – 50 staffs
   c. Between 50 – 100 staffs
   d. Over 100 staffs
4. Years in the Industry:
   a. Below 1 year
   b. 1 - 5 years
   c. 6 – 10 years
   d. 11 – 15 years
   e. Over 15 years
5. Total Years of Working Experience:
   a. Below 1 year
   b. 1 - 5 years
   c. 6 – 10 years
   d. 11 – 15 years

   e. Over 15 years

6. Have you ever been participated in the R&D projects?
   a. Yes
   b. No
7. Do you believe R&D project is beneficial for company growth?
   a. Yes
   b. No

Part II: Industry User's Point of View of R&D Projects

The following questions are used to explore the possible factors that influence R&D technology development and commercialization to the Logistics and Supply Chain Industries in Hong Kong. Based on *your experiences (or what you believe),* please ticks the box below each statement that most accurately reflects the extent to which you agree or disagree with following statement (Figure 11).

Part III: Challenge of R&D Technology Adoption and Commercialization in the Corporate

The following questions are used to examine the main factors in the R&D development process which influence the R&D technology adoption and commercialization in the corporate. Based on *your experiences (or what you believe),* please ticks the box below each statement that most accurately reflects the extent to which you agree or disagree with following statement (Figure 12).

Figure 11a.

1	**Quality Factors**	**Strongly Disagree**	**Somewhat Disagree**	**Neutral**	**Somewhat Agree**	**Strongly Agree**
1.1	R&D technology can provide the new value of the company.	☐	☐	☐	☐	☐
1.2	R&D technology can improve company productivity.	☐	☐	☐	☐	☐
1.3	R&D technology can support the plurality of applications in logistics and supply chain industries.	☐	☐	☐	☐	☐
1.4	R&D technology can increase the company competitive advantage in the market.	☐	☐	☐	☐	☐
1.5	R&D technology is the critical successful factor for the company.	☐	☐	☐	☐	☐
1.6	R&D technology can support the company long term growth.	☐	☐	☐	☐	☐
1.7	R&D technology result quality is the key concern for user adoption.	☐	☐	☐	☐	☐
1.8	Concerning about the quality of R&D technology result whether suitable for the company operations.	☐	☐	☐	☐	☐

Figure 11b.

2	**Control Factors**	**Strongly Disagree**	**Somewhat Disagree**	**Neutral**	**Somewhat Agree**	**Strongly Agree**
2.1	R&D technology can guide the company to a new transformation.	☐	☐	☐	☐	☐
2.2	Concerning about technical support of R&D implementation weather enough.	☐	☐	☐	☐	☐
2.3	Concerning about risk of failure of the R&D technology adoption.	☐	☐	☐	☐	☐
2.4	Detail R&D product specification and clear guidance can help the user to use the R&D product.	☐	☐	☐	☐	☐
2.5	R&D technology process control mechanism is main factor of R&D deliverables commercialization.	☐	☐	☐	☐	☐
2.6	Trial implementation or pilot test in the company can help the user adopt the R&D deliverables.	☐	☐	☐	☐	☐
2.7	Regular review meeting with the R&D technologist can help the user easy to know the R&D technology.	☐	☐	☐	☐	☐
2.8	It would be desirable to adopt R&D technology for Hong Kong Logistics and Supply Chain Industries.	☐	☐	☐	☐	☐
3	**Motivating Factors**	**Strongly Disagree**	**Somewhat Disagree**	**Neutral**	**Somewhat Agree**	**Strongly Agree**
3.1	R&D technology can extend the business to the market.	☐	☐	☐	☐	☐
3.2	Hong Kong Government funding support for R&D technology can help the SME to build up core capability of the company.	☐	☐	☐	☐	☐
3.3	R&D technology results can enhance the interactivity between internal and external users.	☐	☐	☐	☐	☐
3.4	R&D technology promotion activities are the main factor of R&D deliverables commercialization.	☐	☐	☐	☐	☐

Figure 11c.

4	**Time Factors**	**Strongly Disagree**	**Somewhat Disagree**	**Neutral**	**Somewhat Agree**	**Strongly Agree**
4.1	R&D technology can support the company long term growth.	☐	☐	☐	☐	☐
4.2	Concerning about the development time in the R&D process.	☐	☐	☐	☐	☐
4.3	It is the right time for us to adopt the R&D technology in the company.	☐	☐	☐	☐	☐
4.4	This is the right time to introduce R&D technology to our top management for consideration.	☐	☐	☐	☐	☐
5	**Other comment:**					

Figure 12a.

1	**Industry Need**	**Strongly Disagree**	**Somewhat Disagree**	**Neutral**	**Somewhat Agree**	**Strongly Agree**
1.1	Industry requirement collection can be helped to develop the R&D projects.					
1.2	Industry user involve in the R&D projects in the project initial stage, it can speed up the R&D development process.					
1.3	Industry user provides real market situation to help the R&D technologist explore the new idea of R&D project.					
2	**Research Development**	**Strongly Disagree**	**Somewhat Disagree**	**Neutral**	**Somewhat Agree**	**Strongly Agree**
2.1	Create a basic R&D product specification in the R&D project initial stage, it can help all stakeholders easy to know the R&D deliverables.					
2.2	In the R&D project initial stage, I believe that define project risk in advance such as administrative and technical issue, it can eliminates the risk of project failure.					
2.3	Industry user participates in the R&D projects, it can enhance the project value of the R&D deliverables.					

Figure 12b.

3	**Product Exploration**	**Strongly Disagree**	**Somewhat Disagree**	**Neutral**	**Somewhat Agree**	**Strongly Agree**
3.1	Regular project status update of the R&D project, it can reduce the communication issue among all stakeholders.					
3.2	The preliminary test result and prototype sample can help the industry user easy to understand the product for future commercialization and adoption.					
3.3	Intellectual Property (IP) registration can protect the rights of the developer and facilitate further commercialization.					
4	**Technology Implementation**	**Strongly Disagree**	**Somewhat Disagree**	**Neutral**	**Somewhat Agree**	**Strongly Agree**
4.1	The preliminary results and prototype sample of the R&D product can help the industry user easy to know and how to use the R&D deliverables.					
4.2	The promotion activity can attract the industry users to know the R&D product value for commercialization.					
4.3	The R&D project trial implementation in the real industry premises, it can help the industry user to adopt the R&D deliverables.					
5	**Commercialization**	**Strongly Disagree**	**Somewhat Disagree**	**Neutral**	**Somewhat Agree**	**Strongly Agree**
5.1	Technical support of R&D technologist can help the industry user easy to adopt the R&D deliverables in the company.					
5.2	Clear and detail R&D product specification can speed up the industry user to adopt the R&D deliverable.					
5.3	Clear marketing plan can promote the R&D deliverables to the market effectively. For example, to prepare press release or to produce video clip to promote the project results					

Chapter 5
Kano-HOQ-GRA Hybrid Methodology for Customer-Driven Product Development

K. G. Durga Prasad
https://orcid.org/0000-0002-1450-159X
Gayatri Vidya Parishad School of Engineering, India

K. D. S. Sravani
Andhra University College of Engineering for Women, India

B. L. Manasa
Jawaharal Nehru Technological University Kakinada, India

ABSTRACT

In order to attain the expectations of the customers and to win their hearts and minds, manufacturing firms are looking towards customer-focused design approach for developing their products. This chapter is devoted to present a hybrid methodology for developing products with customer focus. The methodology is developed by using house of quality (HOQ) with the aid of other techniques, namely, Kano, Swing weighting method (SWM), and Grey relational analysis (GRA). HOQ serves a tool for mapping customers' perception and designers' conception during product development. The Kano helps to understand and categorize the customer needs which is required to capture the voice of customers. The mapping of customer needs and design requirements in inter-relationship matrix of HOQ is carried by using SWM. There may be a scope for ambiguity while preparing inter-relationship matrix due to insufficient data with the design team. It affects ranking of design requirements. To address this issue, GRA is employed. An illustrative example is presented in this chapter to demonstrate the methodology.

DOI: 10.4018/978-1-5225-8223-6.ch005

INTRODUCTION

The current market environment is characterized by an increasing rate of aggressive competence. In order to gain competitive advantage in the market scenario, the manufacturing industries have been changing their focus on product development from product - driven approach to customer-driven approach. They recognized the need of engaging with their customers and are searching for ways to promote better collaboration with their design team. In customer-driven approach product development begins with capturing the voice of customer and that forms the basis for the product design. This approach includes the identification of customers and their needs, and market trends for obtaining functional design targets, describing design objectives for generating product design specification, ranking design alternatives for evaluating and selecting the optimal design with customers' satisfaction (Liu et al., 2011). Therefore, customer focus and satisfaction should be viewed as driving force for developing products to ensure quality as desired by the customers. During product development, there must be appropriate mapping of customer needs and design requirements to delight the customers. House of Quality (HOQ) is the product planning matrix of Quality Function Deployment (QFD), and it is a suitable tool for addressing the issue of translating the voice of customers into the voice of designers during customer-driven product development. The successful applications of HOQ for product design and development in various manufacturing systems has been reported in the literature (Oke, 2013).

Though HOQ is widely employed in manufacturing firms, the traditional methodology for constructing HOQ has certain drawbacks. To capture the voice of customer no specific procedure has been provided. In fact identification of customer needs and their priority structure influences more on establishing the necessary product design requirements. While constructing HOQ, the prioritization of customer needs is critical, since design of products and services with QFD will be driven to fulfill these prioritized needs (Enriquez et al., 2004). In the traditional approach the priority ratings of the customer needs are obtained through assigning different importance weights for customer needs without considering the views of the customers. The weightages assigned are based on QFD team members' direct experience with the customers or on the results of customer feedback reports. If the customer preferences and the engineering capabilities are in isolation from one another, it is not possible to obtain optimal product development decisions. Therefore, there is a need to modify the methods for input to the traditional HOQ to bridge the conceptual gap between the voice of customers and voice of designers of a product (Prasad et al., 2010). In the proposed hybrid methodology Kano model analysis (KMA) is employed to understand, categorize and prioritize the customer needs. The integration of KMA with HOQ provides a scope for better understanding and obtaining more precise customer needs.

The other important aspect in constructing HOQ is the establishment of inter-relationship matrix of customer needs and design requirements. The main objective of this matrix is to establish linking between customers' expectations and product design requirements. In fact the process of translating customers' voice into designers' voice takes place in this matrix. This matrix evaluates the strength of the linkages between the customer requirements and the design characteristics of a product. Each cell of this matrix expresses a relationship, which represents the strength of the impact of a design characteristic on a corresponding customer need. In the usual practice, the strength of a relationship is assessed with a rating scale, which uses 0, 1, 3, and 5 to represent no, weak, medium, and strong relationships, respectively (Lin et al., 2006). Occasionally there may be the existence of improper relationships between customer needs and design requirements due to insufficient and vague information and this leads to mismatching between them. The incorrect mapping influences on fulfillment of product design requirements to satisfy the customer needs. To tackle this issue grey relational analysis (GRA) has been proposed in the product development methodology as it addresses the issue of uncertainty of experts' opinions, lack of adequate information, inaccuracy of the available data and difficulty in quantification of certain factors (Malekpoor et al., 2017). Before adopting GRA procedure, swing weighting method (SWM) has been used to establish preliminary data on strength of the relationships among the customer needs and design requirements. Later GRA procedure is applied which includes various steps such as generation of the data set of referential series, calculation of grey relational coefficient for each attribute and then determining the grey relational grades of the parameters which are to be prioritized (Prasad et al., 2017). The grey relational grades provide the basis for ranking the product design requirements. An illustrative example is presented in this chapter with a view to demonstrate the proposed hybrid methodology.

BACKGROUND

To understand the hybrid methodology for product development presented in this chapter, it is necessary to know the techniques employed for constructing the road map of the hybrid methodology. A brief discussion on the techniques incorporated in the methodology are presented in the following sections.

House of Quality

House of Quality (HOQ) is the basic design tool of Quality Function Deployment (QFD), a customer-oriented product development technique (Prasad et al., 2014) developed in Japan in the late 1960's by Prof. Yoji Akao and Prof. Shigeru Mizuno with a view to incorporating customers' needs into the design of a product before it was manufactured (Islam et al., 2007). HOQ is a kind of conceptual map that provides the means for inter-functional planning and communications (Hauser and Clausing, 1988). It is mainly built on the belief that products or services should be designed to reflect the customer needs. The HOQ appears like a normal house with foundation, walls and roof. HOQ has a number of components, which are namely customer needs (WHATs), product design requirements (HOWs), prioritized customer needs, inter-relationship between customer needs and product design requirements (WHATs vs HOWs), relationship among product design requirements (HOWs vs HOWs) and prioritized design requirements. The prioritized product design requirements form the foundation of the house. Customer needs and prioritized customer needs form the walls while relationships between customer needs and design requirements form the main body of the house. Relationships among product design requirements form the roof of the house and product design requirements form the ceiling of the house (Islam et al., 2007). However, the HOQ matrix has two principal parts. The horizontal portion of the HOQ matrix called 'customer portion' that contains information related to the customer and a list of customer needs and their priorities occupy in this customer portion. The vertical portion of the matrix is named as 'technical portion' as it contains technical information that responds to the customer desires. The intersection of the customer portion and technical portion forms center portion of the matrix called inter-relationship matrix in which customer desires are integrated with technical measures. The exact translation between customer needs and design requirements takes place in this inter-relationship matrix. The values in the cells of the inter-relationship matrix indicate the strength of the relationships among customer needs and design requirements. The strengths of the inter-relations are usually expressed by the product design team with the help of a rating scale. The design team can also use symbols such as circle, solid circle, square and triangle etc. to express the strength of the relationships. The inter-relationship values in HOQ matrix are further reflected in the computation of relative importance values of the design requirements. The prioritized design requirements are the outcome of HOQ and the customer perceptions are reflected in the priorities of the design requirements.

The HOQ has been attracted by many number of researchers for its application in various fields. Prasad (1998) and Chan (2002) discussed the historical developments of QFD, definitions, contents and applications of QFD from its inception. Carnevalli and Miguel (2008) presented a review, analysis, classification and codification of the literature on QFD produced between 2002 and 2006. Sharma et al. (2008) presented a literature review and a reference bank of more than 400 QFD publications (Subbaiah et al., 2011). It is observed from the literature that even if the QFD process contains four phases, most of the organizations and researchers were confined to apply their customized version of the HOQ. In the recent times there is an increased focus on the development of HOQ-based hybrid methodologies with the support of multi-criteria decision making techniques, intelligent quantitative techniques, various optimization techniques, value engineering techniques, marketing research techniques, cost management techniques etc. to address various short comings in the traditional methodology of HOQ, thereby improving the reliability of decisions during product development. In this chapter hybrid methodology is proposed by embedding KMA and GRA in to HOQ with a view to better understanding of the customers and dealing with the ambiguity due to data insufficiency.

Kano Model Analysis

Kano model analysis (KMA) is a customer preferential ranking technique, which was proposed by Prof. Noriaki Kano and his colleagues in the 1980s. Though a number of methods and tools have been developed accordingly to help companies for better understanding of customers' requirements, Kano's model is a widely used tool for understanding the voice of customers and their impact on customer satisfaction (Wang and Ji, 2010). The KMA classifies the customer needs into the following categories (Chaudha et al., 2011; Ghorbani et al., 2013).

- **Attractive (A):** The sufficiency of this category needs will cause customers to feel excited and their absence will not dissatisfy customers.
- **Must-be (M):** These needs are considered by the customers as basic requirements and their presence will not increase the satisfaction significantly. However, the absence of these needs will cause extreme dissatisfaction.
- **One-dimensional (O):** The sufficiency of these needs gives satisfaction to the customers and insufficiency causes dissatisfaction. These needs are linear and symmetric, because they are usually connected to customer's explicit needs and expectations.
- **Indifferent (I):** The sufficiency or insufficiency of this category needs will not result in either customer satisfaction or customer dissatisfaction.
- **Reverse (R):** These are the needs that if they are provided, customers will be dissatisfied and vice-versa.
- **Questionable (Q):** This rating indicates that either the question was phrased incorrectly, or the customer misunderstood the question, or an illogical response was given.

The following step by step procedure is usually employed for execution of KMA (Matzler and Hinterhuber, 1998).

Step 1: Identification of customers' needs

The customer's needs are usually identified by conducting personal interviews with the customers or focus group discussions with the members who already know the product or service. The focus group discussions are useful in identifying attractive needs whereas individual interviews are helpful in identifying one-dimensional needs.

Step 2: Developing and administering Kano questionnaire

Kano questionnaire is provided with a pair of questions in functional and dysfunctional forms for each customer need. It helps to categorize customer's needs into six Kano categories. There are five possible answers for each pair of questions: like, must-be, one-dimensional, neutral, live with and dislike. The Table 1 shows the traditional form of Kano questionnaire which is to be administered to customers for obtaining their responses on each customer need. Customers are used to express their response with a tick mark. Prior to the distribution of the Kano questionnaires, it is advised to conduct group interviews instead of simply distributing and collecting the questionnaires. Since the Kano questionnaire may be new to the respondents (customers), it is required to explain the purposes of the survey and give necessary directions about filling the questionnaires.

On the basis of customer responses to both the forms of questions, the customer need is classified as one of six Kano categories for that customer by checking the Kano evaluation table (Berger et al., 1993) which is shown in Table 2.

Step 3: Evaluation and interpretation of the results

The expectation levels of customers' needs can be evaluated and interpreted on the basis of the frequency of responses to the Kano questionnaire survey. In order to estimate the average impact on customer satisfaction, Berger et al., (1993) proposed customer satisfaction index (CS). It is calculated by dividing the sum of frequencies of attractive (f_A) and one-dimensional needs (f_O) with the sum of the frequencies of attractive (f_A), one-dimensional (f_O), must-be (f_M) and indifferent needs (f_I).

The mathematical expression for CS is given below

Table 1. Kano questionnaire pattern

Question No.	Kano question	Customer's response				
		I like it that way	**I must be that way**	**I am neutral**	**I can live with that way**	**I dislike it that way**
1	Functional form of the question					
	Dysfunctional form of the question					

Table 2. Kano evaluation table

Customer requirements		**Dysfunctional (negative) question**				
		Like	**Must - be**	**Neutral**	**Live with**	**Dislike**
Functional (positive) question	Like	Q	A	A	A	O
	Must-be	R	I	I	I	M
	Neutral	R	I	I	I	M
	Live with	R	I	I	I	M
	Dislike	R	R	R	R	Q
Note: A: attractive; M: must-be; R: reverse; O: one dimensional; Q: questionable; I: indifferent						

Source: (Matzler et al., 1996)

$$CS_i = \frac{f(A)+f(O)}{f(A)+f(O)+f(M)+f(I)} \tag{1}$$

To estimate the average impact on dissatisfaction, Berger et al., (1993) proposed customer dissatisfaction index (DS). It is calculated by dividing the sum of frequencies of must-be and one-dimensional needs with the sum of the frequencies of attractive, one-dimensional, must-be and indifferent needs. The mathematical expression for DS is given below

$$DS_i = \frac{f(M)+f(O)}{f(A)+f(O)+f(M)+f(I)} \tag{2}$$

The values of customer satisfaction index lies between 0 and 1. The values close to 1 indicate greater satisfaction whereas the values close to 0 indicate lower satisfaction. The customer dissatisfaction index takes the values between – 1 and 0. The values close to – 1 indicate great dissatisfaction and values close to 0 indicate low dissatisfaction. The strategic implications of Kano model analysis are: fulfill all must-be needs, be competitive regarding one - dimensional needs, standout with regard to attractive needs, do not spend time or money for developing requirement that is perceived as indifferent, and avoid reverse requirements (Dominici et al., 2013).

Swing Weighting Method

The decision making problems often involve tradeoffs between multiple attributes of different options. Therefore, multi-attribute decision making is an important field in decision analysis. Practically all approaches to multi-attribute decision making explicitly or implicitly make use of the concept of the relative importance or weights of attributes (Weber and Borcherding, 1993). There are different attribute weight elicitation methods. Swing weighting method is a simple and popular method for eliciting the weights of an additive multi-attribute utility function (Troffaes and Sahlin, 2017). It is the modified version of Simple Multi-Attribute Rating Technique (SMART). Explicitly Swing Weighting Method (SWM) considers the range of consequences on the given attributes. In SWM, the decision maker is asked to consider a situation where he is stuck with a hypothetical alternative that has all the attributes at their worst levels. First the decision maker is asked to move one attribute to its best level: "Select an attribute that you first would like to change to the best level and assign 100 points to this most important attribute". Next the decision maker is asked to choose an attribute change from the worst to the best level which he considers to be the second most desirable improvement and to assign points less than 100 to that attribute change. It is continued with all the remaining attributes. Finally the given points are normalized to sum up to one to get the attribute weights (Poyhonen and Hamalainen, 2001). The following step by step procedure of SWM (Park and Kim, 1998).can be carried to elicit the attribute weights through an interaction with the decision maker.

Step 1: The decision makers have to show two alternatives. One leads to the worst consequence on each attribute, and the other one leads to the best consequence.

Step 2: Ask the decision maker to imagine that he or she is under a condition of the worst alternative (i.e., the one leading to the worst consequence on each attribute) and to rank the attributes, one at a time, by indicating which attribute will improve an objective utility function (e.g., customer satisfaction) most of its level swings from the worst to the best consequence of the attribute.

Step 3: Assign a relative value of 100 to the most significant attribute which was ranked first in step 2. Rate swings of all other attributes on a 0 – 100 scale. Therefore, an attribute whose swing leads to half the improvement compared to the most significant swing would get a weight of 50, and a completely irrelevant attribute would get a weight of 0.

Step 4: Normalize the weights so that they add up to one.

The SWM is much useful in multi-attribute decision making and the use of judgments by interval in this method is an important tool to address imprecision. As SWM is an appropriate method to establish the relationship coefficients between customer requirements and design requirements, it is employed in the proposed hybrid methodology for establishing inter-relationship matrix of HOQ.

Grey Relational Analysis

Prof. Deng Julong founded the grey system theory in the year 1982. Since then, it has been widely used in many fields such as agriculture, electric power, IT, transportation, economics, management, etc. The Grey system theory works on unascertained problems which have little research data and poor information (Wei & Jun-fu, 2011). Various kinds of systems that exist in human society with different nature and all of them are classified into three types (white system, black system, and grey system). A system whose information is completely clear is called a white system and a system whose information is not clear at all is called a black system (a black box). Grey system has the information which might be incomplete or unknown (Wang, 2014). The meaning of the term 'grey' indicates the characteristic between black and white. The application of 'grey' system helps to bridge the gap between 'black' and 'white'. Indeed, the incomplete information is the basic characteristic of the problems considered in grey systems theory (Madhuri & Chandulal, 2010). The statistical analysis requires much data and a typical model for the data distribution. But GRA requires less data and can analyze many factors, which can overcome the disadvantages of the statistics method. It is an important approach of grey system theory in the application of determining a set of alternatives in terms of decision attributes. Grey relational analysis (GRA) is the part of grey system theory and the computations in GRA are simple and straight forward, calculations are based on original data and also it is considered to be one of the best methods to make decisions in the business environment (Samvedi et al., 2012; Prasad et al., 2017). GRA is based on geometrical mathematics, which comply with the principles of normality, symmetry, entirety, and proximity (Fu et al., 2001). There are three major steps involved in GRA. The first step is data pre-processing. It is usually required when the range or unit in one data sequence is different from others or the sequence scatter range is too large. Here, the data must be normalized, scaled and polarized into a comparable sequence before proceeding with the other steps. The process is called the generation of grey relation or standard processing. There are two processes involved in pre-processing stage; data representative and data normalization. Initially the original data series (X) is represented as reference (x_0) and comparative series (x_i). Then the data is normalized and the normalization is based on the characteristics (expectancy) of data sequence. Usually treating the series data set for normalizing through considering the expectancies such as 'larger-is-better', 'smaller-is-better' and 'nominal-is-best'.

- If the expectancy is 'larger-the-better', the sequences after data pre-processing can be specified as

$$x_i^*(j) = \frac{x_i(j) - \min x_i(j)}{\max x_i(j) - \min x_i(j)} \tag{3}$$

- If the expectancy is 'smaller-the-better', the sequences after data pre-processing can be specified as

$$x_i^*(j) = \frac{\max x_i(j) - x_i(j)}{\max x_i(j) - \min x_i(j)} \tag{4}$$

where $i = 1, 2, \ldots, m$ and $j = 1, 2, \ldots, n$. The m and n represents number of data items and number of parameters respectively. The $\max x_i(j)$ and $\min x_i(j)$ indicate the maximum and minimum values of entity j respectively.

The second step in GRA is the calculation of grey relational coefficients using the following equation.

$$\gamma_{0i}(j) = \frac{\Delta\min + \rho\Delta\max}{\Delta_{0i}(j) + \rho\Delta\max} \tag{5}$$

where $\Delta\min = \min\min \Delta_{0i}(j)$ and $\Delta\max = \max\max \Delta_{0i}(j)$

$\Delta_{0i}(j)$ = absolute value of the difference between x_0* and x_i*at the point of j and
ρ = distinguished coefficient or identification coefficient ($\rho \in [0,1]$).

The value may be adjusted based on the actual system requirements. The value of ρ indicates the distinguished ability. If the value of ρ is the smaller and the distinguished ability is the larger. In general the value of ρ is taken as 0.5 (Tosun, 2006). In GRA procedure, the final step is the computation of grey relational grades for all the attributes. The grey relational grade (GRG) is the weighting-sum of the grey relational coefficients. The GRG can be computed by using the following equation.

$$\text{Grey relational grade}\left(\Gamma_{oi}\right)=\sum_{j=1}^{n}\left(w_i\left(j\right)\times\gamma_{oi}\left(j\right)\right) \tag{6}$$

where $w_i\left(j\right)$ = priority ratings or weightages of the attributes

The grey relational grade (GRG) represents the level of correlation which exists between the reference sequence and the comparability sequence. If they are identical, then the value of GRG is equal to 1. The GRG also indicates the extent of influence that the comparability sequence could exert over the reference sequence. Grey relational analysis is actually a measurement of absolute value of data difference between sequences, and it could be used to measure approximation correlation between sequences (Fung, 2003). The major advantage of grey theory is that it can handle both incomplete information and unclear problems very precisely. It serves as an analysis tool especially in cases when there is insufficient data (Wen, 2004). As GRA is suitable for solving problems with complicated interrelationships between multiple factors and variables (Moran et al., 2006), it is employed in the proposed hybrid methodology to obtain the priority structure of design requirements.

KANO-HOQ-GRA HYBRID METHODOLOGY

The hybrid methodology proposed in this chapter provides a customer - focused product development approach in which customer expectations are of central concern within the design process. The outline of the proposed hybrid methodology is shown in Figure 1. The major steps involved in the procedure to carry out the proposed methodology are discussed below:

Step 1: Identification of customer needs

The identification of customer needs is a critical aspect in constructing HOQ for product development. Customer needs are usually identified by interacting directly with the customers by asking what they are looking for in a product. Market survey reports are also useful means to the product design team for understanding the needs of the customers.

Step 2: Prioritization of customer needs using KMA

The priority structure of customer needs gives a scope on understanding about the intensity of desires about the product which is required for the design team while establishing design requirements. Therefore it is necessary to measure the preferences of customers while prioritizing customer needs. To this end Kano model analysis helps to determine the priorities of the customer needs. The procedure to carry out KMA involves the preparation and distribution of Kano questionnaire to the customers, evaluation of their responses, categorization of their needs and computation of customer satisfaction indices. The normalized values of satisfaction indices give the weightages for the customer needs. The KMA procedure is discussed in detail in the earlier section of this chapter which provides a clear understand-

Figure 1. Outline of the Kano-HOQ-GRA hybrid methodology for product development

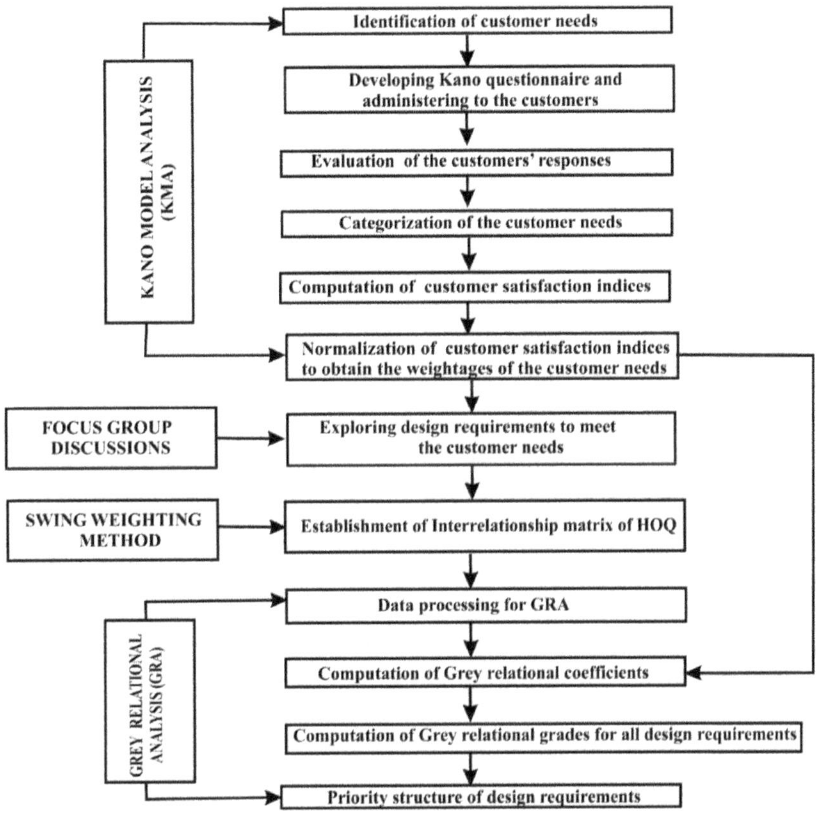

ing on determining the priority ratings of customer needs. These customer priority ratings (weightages) are to be used in the later stage of the methodology while computing the grey relational coefficients for all the customer needs using GRA. Finally the preferences of the customers are reflected in the priority structure of the product design requirements.

Step 3: Identification of design requirements

The technical portion of HOQ starts with the establishment of design requirements and completed with the determination of priority structure of the design requirements. The customers are usually expressing their wants and needs in their own language. As the customers may not have sufficient knowledge of the individual functions of the product, they are unable to describe the features clearly in terms of technical specifications that are to be incorporated in the final product. But the core philosophy of QFD is the exact translation of customer needs into design requirements to produce products to meet those needs. Therefore, the each customer need may be translated into one or more design requirements. Therefore, it is essential for a design team to establish necessary design requirements to incorporate customer views in the final product. The design requirements are explored through brainstorming sessions conducted among the focus group members. The outcome of the focus group discussions provide the list of design requirements.

Step 4: Establishment of inter-relationship matrix of HOQ using SWM

The technical portion of HOQ starts with the establishment of design requirements and completed with the determination of priority structure of the design requirements. As the translation of customer needs into product design requirements takes place in this matrix, it is necessary to present strength of the relationships among customer needs and design requirements in the inter-relationship matrix. In this methodology Swing weighting method has been proposed which provides the relationship coefficients. The relationship coefficients indicate the strengths of the relationships among customer needs and design requirements. The detailed procedure of SWM discussed in the background section of the chapter helps to apply SWM in preparing inter-relationship matrix of HOQ.

Step 5: Prioritization of design requirements using GRA

The HOQ starts with the voice of customer and ends with the establishment of priority structure of the design requirements. It is most possible to happen and inevitable for certain extent of inadequacy and lack of precision of data on establishing relationship coefficients among customer needs and design requirements which may lead to develop incorrect priority structure of design requirements. In order to tackle this issue, grey relational analysis is employed in the product development methodology. The procedure to carryout GRA is mentioned in detail in the previous section of the chapter which guides the team to apply for determining the priorities of the design requirements.

ILLUSTRATIVE EXAMPLE

In the growth and attainment of the present day standard of living, the role played by refrigeration is more significant. In India, refrigerators have the highest aspiration value of all consumer durables, with the exception of televisions (Durga Prasad et al., 2014). As the large number of brands have entered the market and the customers have extensive choice, the refrigerator industry has become highly competitive. This accounts for the high growth rate of the refrigerator market. To demonstrate the proposed hybrid methodology, a case study of designing a domestic refrigerator is considered as an illustrative example. The study is carried out at Visakhapatnam, in the state of Andhra Pradesh, India. In order to identify the customer needs and their categorization, Kano questionnaire has been developed through personal interviews with the customers, market surveys, and brain storming sessions with the targeted customers. The summary of Kano model questionnaire is shown in Table 3.

The questionnaire has been administered to 128 customers under different demographic characteristics. The six basic customer needs such as good refrigeration effect (GRE), more preservation (MP), high storage volume (HSV), less energy consumption (LEC), less price (LP) and service reliability (SR) were identified. The Kano categorization of the needs and their weightages are determined using KMA methodology as discussed earlier in this chapter. The customer satisfaction indices (CSI) for all the customer needs and weightages are presented in the Table 4.

The design requirements for domestic refrigerator mentioned in Table 5 (Prasad et al., 2014) are the outcome of brainstorming sessions held with the technical experts in the field of refrigerator manufacturing and maintenance.

Table 3. Kano questionnaire on domestic refrigerator administered to customers

Sl. No	Questions (Functional and dysfunctional)	Customer response to questions				
		I like it that way	It must be that way	I am neutral	I can live with it that way	I dislike it that way
1(a)	If the energy consumption is less, how do you feel?					
1(b)	If the energy consumption is more, how do you feel?					
2(a)	If the preservation for more varieties of items is provided, how do you feel?					
2(b)	If the preservation for less varieties of items is provided, how do you feel?					
3(a)	If the refrigeration effect is more, how do you feel?					
3(b)	If the refrigeration effect is less, how do you feel?					
4(a)	If more storage volume is provided, how do you feel?					
4(b)	If less storage volume is provided, how do you feel?					
5(a)	If the price of the refrigerator is more, how do you feel?					
5(b)	If the price of the refrigerator is less, how do you feel?					
6(a)	If the after sales service is most reliable, how do you feel?					
6(b)	If the after sales service is least reliable, how do you feel?					

Table 4. Kano categorization and weightages of customer needs

Sl. No	Customer needs (CNs)	Kano categorization of customer needs						Total	Grade	CSI	Weightage (w)
		A	O	M	I	R	Q				
1	GRE	09	60	67	0	0	0	128	M	0.539	0.12
2	MP	18	72	38	1	0	0	128	O	0.703	0.16
3	HSV	30	79	14	0	0	0	128	O	0.852	0.19
4	LEC	57	59	12	0	0	0	128	O	0.906	0.20
5	LP	71	20	37	0	0	0	128	A	0.711	0.16
6	SR	80	23	28	0	0	0	128	A	0.805	0.18

Table 5. List of design requirements for domestic refrigerator

Sl. No	Design Requirements
1	DR1: Enhancing Compressor Performance (ECOMP)
2	DR2: Enhancing Condenser Performance (ECONP)
3	DR3: Enhancing Evaporator Performance (EEP)
4	DR4: Use Good Thermal Insulation Material (UGTIM)
5	DR5: Quick Response to Trouble Shooting (QRTS)
6	DR6: Effective Refrigerator Controls (ERC)
7	DR7: Optimum Design of Refrigerator Compartments (ODRC)

After identifying the design requirements, the next step is the establishment of inter-relationship matrix for constructing HOQ. It is necessary to obtain the strength of relationships among the customer needs and design requirements for preparing inter-relationship matrix. The assessment of strength of the relationships and the computation of relationship coefficients between customer needs and design requirements have been carried out by using SWM as discussed previously in the background section. The Table 6 shows the strength of relationships among customer needs and design requirements.

The normalization of the ratings presented in the Table 6 gives the relations ship coefficients which are to be incorporated in the inter-relationship matrix of HOQ. The relationship coefficients between customer needs and design requirements with regard to each customer need are determined. For instance, the calculation of relationship coefficients for the customer need i.e., GRE and all the design requirements is presented below.

Relationship coefficient between GRE and ECOMP:

$$f_{11} = \frac{100}{100 + 80 + 90 + 70 + 30 + 60 + 40} = 0.215$$

Table 6. Strength of relationships among customer needs and design requirements

		Design requirements							Total
		ECOMP	ECONP	EEP	UGTIM	QRTS	ERC	ODRC	
Customer needs	GRE	100	80	90	70	30	60	40	530
	MP	85	65	74	90	40	50	100	504
	HSV	80	70	80	90	50	60	100	530
	LEC	80	70	60	60	100	90	40	500
	LP	100	90	80	70	50	60	70	520
	SR	90	80	70	100	50	70	53	513

Relationship coefficient between GRE and ECONP:

$$f_{12} = \frac{80}{100 + 80 + 90 + 70 + 30 + 60 + 40} = 0.167$$

Relationship coefficient between GRE and EEP:

$$f_{13} = \frac{90}{100 + 80 + 90 + 70 + 30 + 60 + 40} = 0.193$$

Relationship coefficient between GRE and UGTIM:

$$f_{14} = \frac{70}{100 + 80 + 90 + 70 + 30 + 60 + 40} = 0.144$$

Relationship coefficient between GRE and QRTS:

$$f_{15} = \frac{30}{100 + 80 + 90 + 70 + 30 + 60 + 40} = 0.064$$

Relationship coefficient between GRE and ERC:

$$f_{16} = \frac{60}{100 + 80 + 90 + 70 + 30 + 60 + 40} = 0.129$$

Relationship coefficient between GRE and ODRC:

$$f_{17} = \frac{40}{100 + 80 + 90 + 70 + 30 + 60 + 40} = 0.086$$

In the same way all the relationship coefficients between each remaining customer need and all the design requirements are computed and then presented in Table 7, which will be the inter-relationship matrix of HOQ.

The subsequent step in the implementation of hybrid methodology is the application of GRA. The GRA procedure includes processing the data presented in the inter-relationship matrix of HOQ, calculation of grey relationship coefficients (GRCs) and finally the computation of grey relation grades (GRGs) of all the design requirements. While processing the data normalization has been carried by considering expectancies such as 'larger the better' and 'smaller the better' and using the procedure discussed earlier in the discussion of grey relational analysis. For instance, in the case of good refrigeration effect (GRE) the criteria larger the better can be adopted. The normalized data for all the design requirements with respect to each customer need has been prepared and shown in Table 7. The GRC values for all the cus-

Table 6. Inter-relationship matrix of HOQ

		Design requirements						
		ECOMP	**ECONP**	**EEP**	**UGTIM**	**QRTS**	**ERC**	**ODRC**
Customer needs	GRE	0.215	0.167	0.193	0.144	0.064	0.129	0.086
	MP	0.168	0.128	0.146	0.178	0.079	0.099	0.198
	HSV	0.161	0.124	0.137	0.173	0.010	0.113	0.188
	LEC	0.179	0.165	0.142	0.192	0.086	0.131	0.102
	LP	0.190	0.174	0.157	0.125	0.095	0.112	0.144
	SR	0.167	0.142	0.121	0.101	0.203	0.182	0.081

tomer needs is also computed. The procedure carried to determine GRC of all customer need with reference to a design requirement is presented below by considering all customer needs against the ECOMP.

For GRE: $\gamma_0\left(1\right) = \dfrac{0 + 0.5\left(1\right)}{\left(1 - 1\right) + 0.5\left(1\right)} = 1$

For MP: $\gamma_0(2) = \dfrac{0 + 0.5\left(1\right)}{\left(1 - 0.747\right) + 0.5\left(1\right)} = 0.664$

For HSV: $\gamma_0(3) = \dfrac{0 + 0.5\left(1\right)}{\left(1 - 0.848\right) + 0.5\left(1\right)} = 0.766$

For LEC: $\gamma_0(4) = \dfrac{0 + 0.5\left(1\right)}{\left(0 - 0.122\right) + 0.5\left(1\right)} = 0.803$

For LP: $\gamma_0(5) = \dfrac{0 + 0.5\left(1\right)}{\left(0 - 0\right) + 0.5\left(1\right)} = 1.0$

For SR: $\gamma_0(6) = \dfrac{0 + 0.5\left(1\right)}{\left(0 - 0.702\right) + 0.5\left(1\right)} = 0.626$

Similarly GRC values for all the customer needs with reference to each design requirement are computed and are presented in Table 8.

Table 7. Normalized data in inter-relationship matrix of HOQ

		Design requirements						
		ECOMP	**ECONP**	**EEP**	**UGTIM**	**QRTS**	**ERC**	**ODRC**
Customer needs	GRE	1.00	0.682	0.854	0.529	0.000	0.430	0.145
	MP	0.747	0.411	0.563	0.831	0.000	0.168	1.000
	HSV	0.848	0.640	0.713	0.915	0.000	0.573	1.000
	LEC	0.122	0.254	0.471	0.000	1.000	0.575	0.849
	LP	0.000	0.168	0.347	0.684	1.000	0.821	0.484
	SR	0.702	0.500	0.327	0.163	1.000	0.827	0.000

Table 8. Grey relational coefficients for all customer needs with respect to each design requirement

		Design requirements						
		ECOMP	**ECONP**	**EEP**	**UGTIM**	**QRTS**	**ERC**	**ODRC**
Customer needs	GRE	1.000	0.611	0.773	0.514	0.333	0.467	0.369
	MP	0.664	0.459	0.533	0.747	0.333	0.375	1.000
	HSV	0.766	0.581	0.635	0.854	0.333	0.539	1.000
	LEC	0.803	0.663	0.514	1.000	0.333	0.465	0.370
	LP	1.000	0.748	0.590	0.422	0.333	0.371	0.508
	SR	0.626	0.500	0.426	0.373	1.000	0.742	0.333

The GRC values for all the customer needs and the weightages of the customer needs obtained through KMO are utilized to determine the grey relational grades (GRG) of the design requirements. The weighted sum of the GRC values gives the GRG of the design requirements which are presented in the Table 9.

RESULTS AND DISCUSSION

The proposed hybrid methodology is developed by integrating Kano model analysis, Swing weighting technique and Grey relational analysis in the HOQ appropriately to obtain the rank order of design requirements. In the customer portion of HOQ, the priority structure of customer needs is obtained by using KMA. From the Table 4, it is observed that the priority value for less energy consumption is 0.2, where the highest is compared to that of the other customer needs. It indicates that most of the custom-

Table 9. Grey relational grades for all the design requirements

	Design Requirements						
	ECOMP	**ECONP**	**EEP**	**UGTIM**	**QRTS**	**ERC**	**ODRC**
Grey Relational Grade (GRG)	0.805	0.600	0.573	0.678	0.456	0.504	0.609
Ranks of design requirements	1	4	5	2	7	6	3

ers prefer to purchase a domestic refrigerator that has to be suitable for less energy consumption. The high storage volume has the second priority and is equal to 0.19. This means that, the customers are looking forward to domestic refrigerator with more storage space. The needs such as service reliability, more preservation, less price and good refrigeration effect has the next level priorities of the customers in order. These priority values of the customer needs form a basis for designing a domestic refrigerator. The SWM is employed to construct inter-relationship matrix of HOQ and the relationship coefficients obtained through SWM are presented in Table 6. The relationship coefficients presented in the Table 6 depicts the strength of relationships among the customer needs and design requirements. According to grey relational analysis, the alternative which possesses highest grey relational grade is the most prioritized one among all the alternatives. From the Table 9, it is observed that the priority scores (GRG) of enhancing compressor performance, use of good thermal insulation material, optimum design of refrigerator compartments and enhancing condenser performance are 0.805, 0.678, 0.609 and 0.6 respectively. Primarily the design team has to focus orderly on these design requirements to improve the customer satisfaction to the maximum extent. In the next level the team should pay attention orderly on enhancing evaporator performance, effective refrigerator controls and response to trouble shooting as they have next level priority values. These priority ratings of the design requirements will play a key role in the establishment of parts deployment matrix of QFD. In fact, while designing a product at the conceptual stage, the design team is required to know trade-off among the design requirements. The prioritized list of design requirements is the optimum order to ensure the deployment of customer needs into the product design.

CONCLUSION

The Kano-HOQ-GRA hybrid methodology for product development discussed in this chapter paves way for product development team to develop a product with customer focus. It provides an opportunity for management of a firm to establish value added relationship with their customers as the methodology predominantly considers the priorities of the customers. If the customer preferences and the engineering capabilities are in isolation from one another, it is not possible to obtain optimal product development decisions. The Kano model analysis modifies the methods for input to the traditional HOQ and helps to bridge the conceptual gap between the voice of customers and voice of designers of a product. The Swing weighting method serves the purpose of systematic mapping of customer needs and design requirements while constructing HOQ. The methodology enables design and manufacturing engineers to make appropriate decisions in selecting design requirements to deploy the customer needs in to a product. The illustrative example discussed in this chapter provides a guidance and understanding of the hybrid methodology for product development in any manufacturing firm.

REFERENCES

Berger, C., Blauth, R., Boger, D., Bolster, C., Burchill, G., DuMouchel, W., ... Walden, D. (1993). Kano's method for understanding customer-defined quality. *Center for Quality of Management Journal*, *2*(4), 3–35.

Carnevalli, J. A., & Miguel, P. C. (2008). Review, analysis and classification of the literature on QFD-Types of research, difficulties and benefits. *International Journal of Production Economics*, *114*(2), 737–754. doi:10.1016/j.ijpe.2008.03.006

Chan, L. K., & Wu, M. L. (2002). Quality function deployment: A comprehensive review of its concepts and methods. *Quality Engineering*, *15*(1), 23–35. doi:10.1081/QEN-120006708

Chaudha, A., Jain, R., Singh, A. R., & Mishra, P. K. (2011). Integration of Kano's model into quality function deployment (QFD). *International Journal of Advanced Manufacturing Technology*, *53*(5), 689–698. doi:10.100700170-010-2867-0

Dominici, G., & Palumbo, F. (2013). The drivers of customer satisfaction in the hospitality industry: Applying the Kano model to Sicilian hotels. *International Journal of Leisure and Tourism Marketing*, *3*(3), 215–236. doi:10.1504/IJLTM.2013.052623

Durga Prasad, K. G., Venkata Subbaiah, K., & Narayana Rao, K. (2014). Supply chain design through QFD-based optimization. *Journal of Manufacturing Technology Management*, *25*(5), 712–733. doi:10.1108/JMTM-03-2012-0030

Enriquez, F. T., Osuna, A. J., & Bosch, V. G. (2004). Prioritizing customer needs at spectator events: Obtaining accuracy at a difficult QFD arena. *International Journal of Quality & Reliability Management*, *21*(9), 984–990. doi:10.1108/02656710410561790

Fu, C., Zheng, J., Zhao, J., & Xu, W. (2001). Application of grey relational analysis for corrosion failure of oil tubes. *Corrosion Science*, *43*(5), 881–889. doi:10.1016/S0010-938X(00)00089-5

Fung, C. P. (2003). Manufacturing process optimization for wear property of fiber-reinforced polybutylene terephthalate composites with grey relational analysis. *Wear*, *254*(3-4), 298–306. doi:10.1016/S0043-1648(03)00013-9

Ghorbani, M., Mohammad Arabzad, S., & Shahin, A. (2013). A novel approach for supplier selection based on the Kano model and fuzzy MCDM. *International Journal of Production Research*, *51*(18), 5469–5484. doi:10.1080/00207543.2013.784403

Hauser, J. R., & Clausing, D. (1988). The House of Quality. *Harvard Business Review*, *66*(3), 63–73.

Islam, R., Ahmed, M., & Alias, M. H. (2007). Application of Quality Function Deployment in redesigning website: A case study on TV3. *International Journal of Business Information Systems*, *2*(2), 195–216. doi:10.1504/IJBIS.2007.011619

Lin, M. C., Wang, C. C., & Chen, T. C. (2006). A strategy for managing customer-oriented product design. *Concurrent Engineering*, *14*(3), 231–244. doi:10.1177/1063293X06068390

Liu, C., Ramirez-Serrano, A., & Yin, G. (2011). Customer-driven product design and evaluation method for collaborative design environments. *Journal of Intelligent Manufacturing*, *22*(5), 751–764. doi:10.100710845-009-0334-2

Madhuri, Ch. B., & Chandulal, J. A. (2010). Evaluating websites using COPRAS-GRA combined with grey clustering. *International Journal of Engineering Science and Technology*, *2*(10), 5280–5294.

Malekpoor, H., Chalvatzis, K., Mishra, N., Mehlawat, M. K., Zafirakis, D., & Song, M. (2017). Integrated grey relational analysis and multi objective grey linear programming for sustainable electricity generation planning. *Annals of Operations Research*, 1–29.

Matzler, K., & Hinterhuber, H. H. (1998). How to make product development projects more successful by integrating Kano's model of customer satisfaction into quality function deployment. *Technovation*, *18*(1), 25–38. doi:10.1016/S0166-4972(97)00072-2

Matzler, K., Hinterhuber, H. H., Bailom, F., & Sauerwein, E. (1996). How to delight your customers. *Journal of Product and Brand Management*, *5*(2), 6–18. doi:10.1108/10610429610119469

Moran, J., Granada, E., Míguez, J. L., & Porteiro, J. (2006). Use of grey relational analysis to assess and optimize small biomass boilers. *Fuel Processing Technology*, *87*(2), 123–127. doi:10.1016/j.fuproc.2005.08.008

Oke, S. A. (2013). Manufacturing quality function deployment: Literature review and future trends. *Engineering Journal (New York)*, *17*(3), 79–103.

Park, T., & Kim, K. J. (1998). Determination of an optimal set of design requirements using house of quality. *Journal of Operations Management*, *16*(5), 569–581. doi:10.1016/S0272-6963(97)00029-6

Poyhonen, M., & Hamalainen, R. P. (2001). On the convergence of multiattribute weighting methods. *European Journal of Operational Research*, *129*(3), 569–585. doi:10.1016/S0377-2217(99)00467-1

Prasad, B. (1998). Review of QFD and related deployment techniques. *Journal of Manufacturing Systems*, *17*(3), 221–234. doi:10.1016/S0278-6125(98)80063-0

Prasad, K., Subbaiah, K., & Prasad, M. (2017). Supplier evaluation and selection through DEA-AHP-GRA Integrated approach-A case study. *Uncertain Supply Chain Management*, *5*(4), 369–382. doi:10.5267/j.uscm.2017.4.001

Prasad, K. D., Subbaiah, K. V., & Rao, K. N. (2014). Multi-objective optimization approach for cost management during product design at the conceptual phase. *Journal of Industrial Engineering International*, *10*(1), 48. doi:10.100740092-014-0048-8

Prasad, K. G. D., Subbaiah, K. V., Rao, K. N., & Sastry, C. V. R. (2010). Prioritization of Customer Needs in House of Quality Using Conjoint Analysis. *International Journal of Qualitative Research*, *4*(2), 145–154.

Samvedi, A., Jain, V., & Chan, F. T. (2012). An integrated approach for machine tool selection using fuzzy analytical hierarchy process and grey relational analysis. *International Journal of Production Research*, *50*(12), 3211–3221. doi:10.1080/00207543.2011.560906

Sharma, J. R., Rawani, A. M., & Barahate, M. (2008). Quality function deployment: A comprehensive literature review. *International Journal of Data Analysis Techniques and Strategies*, *1*(1), 78–103. doi:10.1504/IJDATS.2008.020024

Subbaiah, K. V., Durga Prasad, K. G., & Narayana Rao, K. (2011). Customer-driven product planning using conjoint analysis and QFD-ANP methodology. *International Journal of Productivity and Quality Management*, *7*(3), 374–394. doi:10.1504/IJPQM.2011.039353

Tosun, N. (2006). Determination of optimum parameters for multi-performance characteristics in drilling by using grey relational analysis. *International Journal of Advanced Manufacturing Technology*, *28*(5-6), 450–455. doi:10.100700170-004-2386-y

Troffaes, M., & Sahlin, U. (2017). *Imprecise swing weighting for multi-attribute utility elicitation based on partial preferences*. PMLR.

Wang, C. H. (2014). Integrating correspondence analysis with Grey relational model to implement a user-driven STP product strategy for smart glasses. *Journal of Intelligent Manufacturing*, *27*(5), 1007–1016. doi:10.100710845-014-0931-6

Wang, T., & Ji, P. (2010). Understanding customer needs through quantitative analysis of Kano's model. *International Journal of Quality & Reliability Management*, *27*(2), 173–184. doi:10.1108/02656711011014294

Weber, M., & Borcherding, K. (1993). Behavioral influences on weight judgments in multiattribute decision making. *European Journal of Operational Research*, *67*(1), 1–12. doi:10.1016/0377-2217(93)90318-H

Wei, C., & Jun-fu, C. (2011). Theoretical discussion of applying grey system theory in neuropsychological studies. *Grey Systems. Theory and Application*, *1*(3), 268–273.

Wen, K. L. (2004). The grey system analysis and its application in gas breakdown and var compensator finding. *International Journal of Computational Cognition*, *2*(1), 21–44.

Chapter 6
Role of R&D Practices for Effective Product Development Process in NPD

Sudeshna Roy
Jadavpur University, India

Pranab K. Dan
Indian Institute of Technology Kharagpur, India

Nipu Modak
Jadavpur University, India

ABSTRACT

The necessity of new product development (NPD) in the global competition is a well-established fact. Imperativeness of research and development (R&D) practices and product development process (PDP) in NPD are inevitable. In case of R&D practices, fuzzy-front-end (FFE) activities and improvisation are the two sub-factors which are not directly related to R&D but motivate it indirectly with their actions. For PDP, modular product development (MPD) and market analysis are recognized as the factors directly influencing the PDP of the firm for NPD success. This chapter considers product quality and technological developments as performance attributes for development of comprehensive framework by structural equation modeling (SEM) approach. Primary data from 263 experts of Indian manufacturing industries has been collected for analysis purpose. This empirical research portrays the role of R&D practices along with its indirectly related success factors for effectively controlling PDP along with its sub-factors for developing high quality products with technological developments.

DOI: 10.4018/978-1-5225-8223-6.ch006

INTRODUCTION

New product development (NPD) is an essential activity each firm be involved to guarantee success and survival in competitive market environment (Sholeh et al., 2018). Globalization brings the world in a single platform and widens the scope of selecting the products as per the requirements and choice of end-users. This enforces the firms to develop new products by optimal utilization of available resources (Roy et al., 2018). SME sector is the backbone of Indian economy by contributing almost 6.11% of manufacturing GDP (www.cii.in). It provides resilience to economic adversities and protract in global competition. This evinces the necessity of NPD activities in SMEs for sustaining in dynamic manufacturing environment by producing innovative concepts and strategies for future growth (Roy et al., 2018). NPD success is a very rare target to be attained due to associated high costs, requirement of updated knowledgebase about customers' demand, adoption of state-of-the-art technologies and sound experience on NPD activities for fulfilment of NPD objectives (Largosen, 2005).Though involvement in NPD activities and setting policies for success increases in all types of industries, NPD effort somehow fails at the final stage due to lack of systematic approach from the early stage of development (Florén et al., 2018). NPD is a series of activities every firm involves for developing new products (Bhuiyan, 2011). It consists of seven stages from strategy development followed by generation of initial ideas, screening and evaluation of those developed ideas, design and development evolves structuring of the developed idea into a product that is producible, testing and finally the commercialization (Booz et al., 1982). Control of various parameters on NPD activities during the development process is well-noticed. Role of these parameters are critically vital and must be taken care of for successful completion of NPD process. These parameters or constituents are famously known as critical success factors (CSFs) (Ernst, 2002). Previous researchers have signified the vitality of various CSFs for successful NPD. Identification of these factors is not enough, but realization of their beneficial role and their optimal utilization in practical field is equally matter of high concern to avoid the organizational failure (Rockart, 1979). Success of the firm is not elusive; it can be defined by various dimensions like financial and economic performance, customer acceptance, technological developments, quality aspects and time based measures (Huang et al., 2004). These success measures are crucial for measuring firm's NPD performance in turn firm's success.

Research and development (R&D) is concerned as one of the imperative factors influencing NPD success. It offers continuous improvement of the development activities by adopting updated technologies and optimal usage of available resources (Nicholas et al., 2015). There are various sub-factors like technology (Wang et al., 2014), learning (Roy and The´rin, 2008) and intellectual capital (Hsu and Fang, 2009) which directly impacted on R&D practice for its successful implementation in practical field. These sub-factors are well-discussed in previous literature. There are also various sub-factors which are not directly linked with R&D practices but encourage the R&D practices in indirect manner. Fuzzy-front-end (FFE) activities is one of those indirect sub-factors of R&D which is the very much prior phase of ideation for development of new products to structure innovative ideas with optimal utilization of technical and marketing knowledge (de SousaMendes and Ganga, 2013). Like FFE activities, improvization is another indirect constituent of R&D practices. It is the degree of innovativeness introduced to the newly developed products. This can be achieved through continuous research activities and adoption of these research findings for promoting new ideas for firm's better performance (Akgun et al., 2007). The successful implementation of R&D practice with the realization of the essentiality of these sub-factors helps in technological developments of the firm in turn the NPD success.

Like R&D, product development process (PDP) is another constituent needs to be taken care of for successful NPD. Modification and upgradation of PDP is always welcomed as nature of customer demand is highly volatile in nature. They are always looking for the innovative products rather than existing ones. PDP is closely associated with two sub-factors namely – modular product design (MPD) and market analysis. MPD is the attempt of escalating the range of new products considering the development time and cost as well (Lau, 2011). The main objective of MPD is introduction of varieties of new products through assimilating simple modules for developing complex products (Robertson and Ulrich, 1998). Besides, market analysis is one of the vital sub-factors of PDP to accumulate the customers' needs. Based on these demands, organizations set their direction of NPD. Data survey through direct as well as telephonic interviews, e-mail sharing, skype calling are often performed to gather the vital information regarding the customers' demand (Acur et al., 2012). PDP and its allied sub-factors develop high quality new products as per customers' requirements to achieve firm's success (Chaudhuri and Boer, 2016).

A single comprehensive framework accommodating both R&D practices and PDP for establishment of outcome considering technological developments and product quality is practically unavailable in existing literature. The drivers along with their indicators and the outcome effects with sub-factors of R&D and PDP for developing technologically updated high quality products are rarely been addressed. This study addresses on these research gaps such as: (1) what are the indirectly influencing sub-factors of R&D practices and their indicators and how critical they are for NPD success?; (2) what are the sub-factors of PDP, their antecedents and their impact on NPD success?; (3) what are the main performance attributes corresponding to R&D and PDP respectively? How these attributes are impacted by the collective influence of R&D practice and PDP for NPD success? These research gaps needs to be bridged for drawing valuable insights which helps in smooth continuation of NPD activities for firm's success. This research is a novel approach for realizing the importance in implementation of R&D practice and PDP collectively to build the suitable ambience escalating the technological developments and product quality leading to NPD success.

The objective of this research has been to build a comprehensive structural model portraying the interrelationship about the critical linkages of R&D and PDP for NPD success. First, this research examines the indirectly related sub-factors of R&D practices namely – FFE activities and improvization. It involves identification of their indicators for measuring the constructs from the primary data accumulated from experts of Indian manufacturing industries. Second, it embraces the sub-factors of PDP such as MPD and market analysis along with the recognition of their indicators. Third, this study analyzes the combined impact of R&D practice and PDP on NPD success in terms of technological developments and product quality by structural equation modelling (SEM) approach using IBM SPSS AMOS 21.0 software packages. Finally, this empirical research considers the practical data collected from Indian manufacturing companies developing engineering products. The development of structural model considering Indian scenario is still needs to be explored. This study justifies efforts to bridge this gap which is another novelty of this research. The developed comprehensive model helps in managerial actions through gaining the purposeful insights.

The structure of this paper is as follows: The very next section incorporates theoretical background which describes the theory of formulation of R&D practices and PDP for NPD success in terms of technological developments and product quality. Based on these backgrounds the hypotheses interconnecting the constructs are developed. Considering these developed hypotheses a path model incorporating the hypotheses among the constructs depicting the combined impact of R&D and PDP on NPD success is structured. The research methodology section is segmented in three sub-sections discussing data analysis techniques, design of questionnaire and data collection. It also represents the profiles of the respondents contributed in this research. The forth section namely – results and discussions incorporates the analysis of measurement validity along with the results including the discussions of measurement model as well as structural model as attained from IBM SPSS AMOS 21.0. Finally, the conclusion section explains the theoretical contribution and managerial implications to express the essentiality of this novel approach in the available literature. Limitations and scope of future work for further enhancement of this study is also incorporated in this section.

THEORETICAL FORMULATION

Modifications and up-gradation in manufacturing offers a tough challenge to firms for successful development of new products to survive in the global platform (Eid, 2009). Successful development of new products is difficult task to be performed and requires various constituents to be taken care of. R&D is one of those very common practices which play a vital role in NPD by offering innovative ideas and controlling the development process effectively for providing a complete shape to those innovative ideas in a form of new products. Necessity of useable theoretical background to draw the hypotheses interconnecting the constituents of R&D practices and PDP is essential to be taken care of which in turn offers an accomplishment in NPD.

Research and Development (R&D) Practices

R&D has grown as an inseparable activity of the firm assuring immense success through developing new products (Tripathy et al., 2012). R&D requires a sufficient fund allotment for developing own design unit. This can produce new products with additional features offering competitive advantages to their respective competitors (Sun and Wing, 2005). A well-organized R&D unit demands enough investments in infrastructure, human resources, and intellectual resources for unprompted NPD (Haverila, 2012). These features improve the manufacturability of the firm for efficient development of new products (Kim and Kim, 2009). This shows that:

H2a: R&D practices of the firm ensure NPD success

R&D practices starts with the idea generation and conceptualization for structuring the innovative and creative ideas for NPD. This phase is actually the pre-planning phase termed as FFE activities (Ayag, 2005). The compactness and effectiveness of FFE activities is the decisive parameter of the future of new products aimed to be developed as per the innovative ideas of the R&D experts. FFE activities reduce the chances of uncertainty in the newly developed products through focusing on market uncertainty based on the end-users demand (Verworn, 2009). This prior planning considering all the uncertainties helps in developing high-tech new products within the estimated development time and cost (Ayag, 2005). This signifies that effective FFE activities structured indirectly from R&D practices escalates the NPD success of the firm which is expressed as:

H1b: Effective FFE activities motivate the NPD success of the firm

Improvization is another indirect consequence of R&D practices which offers instant planning and implantation of those plans for shaping these ideas in practice. It offers the innovativeness to the new products to be developed through learning and unlearning process (Akgun et al., 2007). Improvization is a result of team learning to adopt novel concepts, approaches and techniques for introducing new features to the products as per the customers' requirements. This risk-taking mentality of the firm strengthens the improvisation practice abandoning the chance of failure. It encourages the eccentric concepts and procedures to offer new products better than the existing ones (Miner et al., 2001). This discussion evinces the fact that:

H1c: Improvization intensifies the NPD success offering an improved firm performance

Development of new products requires a structured new product development process. This is the result of continuous R&D activities of the firm for developing new products as per market demand (Lynn et al., 1999). Development of the PDP necessitates a common understanding of the process through R&D (Lester, 1998). Research activities within the organization categorize the decisive factors of PDP and set the process as per the required decision points as recognized by the process manager for successful NPD (Copper, 1995). PDP introduces uniqueness to the newly developed products which is the result of shaping the innovative ideas in a tangible form with the help R&D teams (Gatignon et al., 2002; Song and Montoya-Weiss, 1998). This infers that:

H1d: R&D assists to set the PDP and stimulates the process for NPD success

Product Development Process (PDP)

PDP is a series of phases of developing final product usable to the consumers (Graner, 2016). This process of development of is very much crucial for successful completion of new products as well. For optimal utilization of firm's resources a detailed idea of PDP started from idea generation to commercialization is essential (Tzokas et al., 2004). It enhances the NPD performance through technological developments of new products. PDP significantly improves NPD success of the firm assuring industrial sustainability for long-run (Ernst, 2002). This infers the fact that:

H1a: PDP significantly influences NPD success of the firm

Modular product design (MPD) is a process of developing a product which is the assimilation of various small products termed as 'modules' to perform as a single unit (Bladwin and Clark, 2000). These modules can be treated as an independent product. In case of MPD, the interdependencies of these modules are aimed to be minimized for functioning in more operative fashion (Ulrich and Eppinger, 2000). Development of MPD introduces flexibility to the newly developed complex products as the formation of small modules is easier than development of large products with complex features (Bladwin and Clark, 2000). Again, MPD introduces the ease of replacing the faults within the final product just by substituting the faulty module without changing the complete product. Moreover, the combinations of various modules may produce a wide range of new products as per customers' requirements (Pine, 1997). This facilitates the NPD success of the firm through high-end customization with quality assurance. From this, it can be demonstrated that:

H2b: MPD ensures the NPD success of the firm with quality assurance

PDP remains an incomplete task without adoption of market analysis practice for establishing a clear idea of the target market and consumption pattern of the customer (Shinno and Hashizume, 2002). It is a very foremost step of NPD incorporating market-plan generation to market testing before final product launch. It accumulates primary information from the market to offer the new products with the desired features (Graner, 2016; Heirati and Aron O'Cass, 2016). Market analysis also helps to analyse the current situation of the prospective competitors and identifies the similarities as well as differences of the same type of products with their own products (Acur et al., 2012). A thorough market research approaches to gather huge information on customer demand influences NPD success of the firm. This can be inferred as:

H2c: High degree of market analysis conveys NPD success to the firm

This theoretical background helps to structure a research model representing the role of R&D practices to control PDP exploring the above mentioned hypotheses for examining the linkages of these constituents and their sub-factors for achieving NPD success. Figure 1 demonstrates the framed path model showing the combined effect of R&D practice and PDP on NPD success quantified in terms of technological developments with quality assurance.

RESEARCH METHODOLOGY

This research studies the role of R&D practice for effective PDP in developing high quality new products with technological developments. Accumulation of primary data from manufacturing experts of Indian companies is performed to comprehend the linkages of the factors and their sub-factors for realization

Figure 1. Path model of input and output constructs along with the developed hypotheses for NPD success

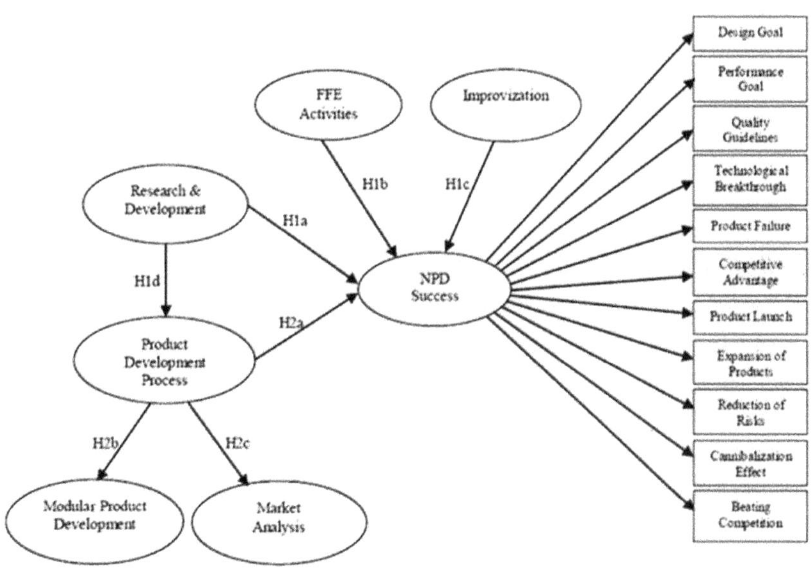

of their criticality in practical field. Exploratory factor analysis (EFA) has been conducted for extracting the variables with higher loading values used for further framework development. Structural equation modeling (SEM) approach is employed for forming this structural framework interlinking the success factors and measures along with their indicators based on the experts' opinion.

Methodology

SEM is a multivariate data analysis technique for representing, estimating, and testing relationships between latent constructs and manifest variables. These manifest variables are used to measure the respective latent construct (Rigdon, 1998). SEM analyzes the hypotheses developed from theoretical background by using the primary data collected from experts of Indian manufacturing sector. First, exploratory factor analysis (EFA) is performed for identify the indicators with higher loading values. After that, those indicators high higher loadings are utilized for structural model formation. SEM comprises of both measurement model and structural model. Measurement model calculates the standardized regression weights (SRWs) of the manifest variables whereas structural model develops the interrelationships among the latent constructs to identify their impacts. Multiple numbers of model fitness tests are performed for validation of the developed model (Joreskog, 1973).

Design of Questionnaire

A semi-structure questionnaire is developed for assimilating the responses of industry experts for both input and output constructs. In this study R&D practices and PDP are considered as input variables having sub-factors namely – FFE activities, improvization and MPD and market analysis respectively. NPD success is the output construct measured in terms of product quality and technological developments. The questionnaire is segmented in three sections –first section covers the detailed information about the profiles of the respondents, second and section accumulate the answers regarding input and output constructs respectively. 7 point likert scale has been used to rate the responses where '1' denotes strongly disagree and '7' denotes strongly agree regarding the importance in implementation of the factors in practical field for input constructs in second section. In case of output construct, '1' represents very low and '7' denotes very high for measuring performance attributes. Moreover, in the second section there is a provision of sharing the own suggestions of the respondents about the additional measures for each construct. This structure of the questionnaire is itself a novel approach for data collection. For construct validity of the questionnaire a pilot study is been constructed using 50 responses from Kolkata and Howrah. This offers validation of the questionnaire and the further modification for better data collection.

Sample and Data Collection

This research explores the present scenario of the Indian manufacturing enterprises mainly SMEs developing new engineering products. This involves primary data collection from SMEs in India considering the design and development experts as the targeted samples. List of manufacturing companies are gathered from Capitaline Plus database. Few respondents are contacted based on the references from other companies using snowball sampling method. Presidents, Vice-Presidents, Managers are mainly targeted for the data survey. Maximum data are collected by direct interviewing and remaining is through e-mail sharing and telephonic interviews. Initially, 380 experts have been approached for the interview.

Among them, 263 responses are found useful for data analysis. The demographic profile of respondents is enlisted in Table 1.

Results and Discussions

Analysis of Measurement Validity

Principal component based exploratory factor analysis is performed for extracting measures with higher loading values. In this study, 34 among 47 manifests are identified and extracted having factor loadings (FLs) greater than 0.6 as per the conventional practice. The values of FLs respective to each measure are listed in Table 2. The composite reliability (CR) and Cronbach's alpha (α) reliability testing is executed to check the reliability of the accumulated data. Average variance extracted (AVE) is performed for measuring discriminant validity. These values of reliability (CR, α) and validity (AVE) indices are enlisted in Table 2 as found in IBM SPSS 21.0 software. Threshold value for CR is 0.5, above which the data is considered as highly reliable. Values between 0.3 to 0.5 are also taken as reliable one (Holmes and Smith, 2001). In case of α, values greater than 0.8 is considered as reliable (Ong et al., 2004). For discriminant validity the desired values of validity indices greater than 0.05 (Homles and Smith, 2001). After checking theses indices, the accumulated primary data is used for model formation.

Table 1. Demographic profile of respondents

Sample Characteristic	Classifications	Approached	Responded	Percentage
Geographical location	Indian manufacturing SMEs	380	263	69.21
Type of organization	Fabrication	63	46	73.02
	Electrical equipment	51	33	64.71
	Industrial valves	42	32	76.19
	Textile Machineries	41	27	65.85
	Fire fighting equipment	36	26	72.22
	Hydraulics & pneumatic	32	25	78.13
	Burners and heaters	28	22	78.57
	Material handling equipment	25	21	84
	Cell and battery	23	14	60.87
	R&D sectors	20	9	45
	Air ventilators	19	8	42.11
Respondent's Profile	Executive	75	69	92
	Manager	90	83	92.22
	Senior Manager	70	48	68.57
	Vice President	76	36	43.9
	President	69	27	39.13

Structural Equation Modeling (SEM)

In this study, first, SEM is employed to develop a framework comprising of all the factors and their sub-factors including the indicators as well. It forms the measurement model and structural model performing the factor analysis and estimation of path values respectively. Various fit indices are calculated for validation of models.

Measurement Model

Factor loadings of the measurement variables to their respective constructs are measured by exploratory factor analysis. The values of loadings range from 0.484 to 0.820 as mentioned in Table 2. Among these loadings 34 manifest variables have values greater than 0.6. So, these variables are used for further model development. After dimension reduction, confirmatory factor analysis (CFA) is used for estimating the model fit. The standardized regression weights (SRWs) of the manifest variables to their constructs are calculated and the values range from 0.31 to 0.62 as listed in Table 2. The measurement model also have good model to data fit as the values obtained: χ^2 =366.415, df= 209, χ^2 /df= 1.753, RMSEA= 0.057, GFI=0.878, AGFI=0.836. The standard values of fit indices are represented in Table 3 (Byrne, 2010).

In case of measurement model, the confirmatory factor analysis validates the occurrence of measurement variables on their respective latent. The manifest variables are segmented in two sections such as recognized from literature review and from experts' opinion. For R&D, number of R&D persons in research activities, years of experience of R&D persons, investments for better R&D infrastructure, qualification of R&D experts, numbers of patents and R&D culture in the firm are the indicators identified from the available literature. R&D management vision and direction and adoption of cleaner technology research are the manifests documented from the experts' opinion. EFA is performed by using IBM SPSS 21.0 to calculate the value of FLs for dimension reduction by discarding the manifest variables with lower loading values (<0.6). After executing EFA, number of R&D persons in research activities (m1), years of experience of R&D persons (m2), investments for better R&D infrastructure (m3), qualification of R&D experts (m4) and R&D management vision and direction (m5) are selected for their higher loading values which are further employed for structural model formation. From expert opinion, hands-on experience on PDP (m5) is selected for its higher loading value. Similarly, for each construct the factor analysis is performed to analyse the standardized regression weights (SRWs) linking the predictor to their dependent variables as listed in Table 2.

Structural Model

Analysis of measurement model is followed by structural model formation. It tests the hypotheses developed between the constructs by using IBM SPSS AMOS 21.0 as shown in Figure 2. The model to data fitness is tested and approved by the parameters obtained mentioned as χ^2 =463.034, df= 286, χ^2 / df= 1.619, RMSEA= 0.043, GFI=0.926, AGFI=0.897 (Hair et al., 1995). As, these indices are within desired range, the model is treated as acceptable. The path estimates in between the constructs range from 0.35 to 0.97 as listed in Table 4.

Table 2. Latent constructs and their indicators including factor loadings, standardized regression weights, validity and reliability indices

Latent constructs with their manifest variables along with validity and reliability indices	FL	SRWs
Research and Development (R&D)[AVE=.64; CR=0.76; α=0.829]	-	-
From literature review:	-	-
1. Number of R&D persons in research activities (m1)	.762	.58
2. Years of experience of R&D persons (m2)	.680	.52
3. Investments for better R&D infrastructure (m3)	.621	.52
4. Qualification of R&D experts (m4)	.609	.51
5. Number of patents	.558	
6. R&D culture in the firm	.492	
From experts' opinion:	-	-
7. R&D management vision and direction (m6)	.618	.40
8. Adoption of cleaner technology research	.523	
FFE Activities (FFEA)[AVE=.49; CR=0.58; α=0.707]	-	-
From literature review:	-	-
1. Interdisciplinary idea generation and screening (m7)	.665	.50
2. Intensity of initial planning (m8)	.641	.36
3. Effort to reduce market uncertainty (m9)	.636	.51
4. Level of communication in early phases of product development (m10)	.627	.46
5. Understanding of target market and users need	.597	
Improvization (I)[AVE=0.52; CR=0.61; α=0.722]	-	-
From literature review:	-	-
1. Difference of NPD process as in practical case vs. desired process (m11)	.820	.57
2. Improvisation of team in developing this product as per the plan (m12)	.798	.60
3. Improvisation of team in commercializing this product as per plan (m13)	.725	.62
Product Development Process (PDP) [AVE=0.62; CR=0.73; α=0.817]	-	-
From literature review:	-	-
1. Fund allocation for updated development activities (m14)	.793	.51
2. Advanced product development methods (m15)	.782	.51
3. Developmental time and cost (m16)	.761	.45
4. Training for NPD Management (m17)	.653	.52
From experts' opinion:	-	-
7. Hands-on experience on PDP of new products (m18)	.636	.53
8. Reviewing of the product	.598	
Modular Product Design (MPD) [AVE=.56; CR=0.64; α=0.745]	-	-
From literature review:	-	-
1. Adoption of Modular product design	.781	.57
2. Increase in product variety	.699	.57
3. System reliability improvement	.656	.50
4. Product component commonality	.624	.49
From experts' opinion:	-	-

continued on following page

Table 2. Continued

5. Product component recycling	.610	.47
Market Analysis (MA) [AVE=0.59; CR=0.69; α=0.778]	-	-
From literature review:	-	-
1.Development of market plan	.809	.61
2. Consumption pattern of the product	.786	.51
Latent constructs with their manifest variables along with validity and reliability indices	FL	SRWs
Market Analysis (MA) [AVE=0.59; CR=0.69; α=0.778]	-	-
From literature review:	-	-
3. Target market and growth pattern	.699	.54
4. Focus on customer satisfaction	654	.47
5. Market testing	.601	.45
From experts' opinion:	-	-
6. Advertisement and promotion	.593	
7. Competitor monitoring	.520	
NPD Success [AVE=0.67; CR=0.82; α=0.897]	-	-
From literature review:	-	-
1. Achievement of design goal (m28)	.784	.40
2. Achieved product performance goal (m29)	.766	.31
3. Meeting quality guidelines (m30)	.731	.37
4. Technological breakthrough (m31)	.685	.33
5. Product failure rate (m34)	.632	.44
6. Competitive advantages through technological performance (m35)	.621	.38
7. Rate of product launching (m32)	.617	.43
From experts' opinion:	-	-
8. Expansion product family with varieties of products (m33)	.588	
9. Reduction of risks (m36)	.537	
10. Cannibalization effect	.503	
11. Beating competition to market	.484	

Table 3. Fit Indices (Byrne, 2010)

Fit Indices	Desired Range
χ^2 /degrees of freedom	≤ 2.00
RMSEA(Root Mean Square Error of Approximation)	Values less than 0.05 show good fit Values as high as 0.08 represent reasonable fit Values > 1.0 show poor fit
Goodness-of-fit index (GFI)	≥ .90
Average Goodness-of-fit index (AGFI)	≥ .90

Table 4. Statistics of Path Estimates

Path Descriptions	Hypotheses	Standardized Estimates	t-Values
R&D→NPD Success	H1a	0.50 (***)	7.807
FFEA→NPD Success	H1b	0.73 (***)	9.362
Imp→NPD Success	H1c	0.42 (***)	7.309
R&D→ PDP	H1d	0.97 (***)	12.359
PDP→NPD Success	H2a	0.80 (***)	10.107
PDP→MPD	H2b	0.63 (***)	8.885
PDP→MA	H2c	0.35 (***)	7.161

Notes:
- *[*** indicate the significance at p value <0.01*
- *Model fit indices:* $\chi 2$ *= 463.034, degrees of freedom = 286,* $\chi 2$ */degrees of freedom = 1.619, RMSEA = 0.043, GFI = 0.926, AGFI = 0.897]*

Figure 2 represents the complete framework comprising of both measurement and structural model with all the latent constructs and their indicators as well. Measurement model shows the regression weights of manifest variables and path estimates among latent constructs are covered by structural model. The arrowheads also have the significance in the model. It shows the impact of a construct on another construct which is depended on the same. The structural model here expresses the impact of R&D practices on PDP for encouraging NPD success of the firm. R&D itself comprises of two indirect constituents namely – FFE activities and improvization. Similarly PDP consists of MPD and market analysis. The linkages of factors with their sub-factors and their combined impact on NPD success are well-represented in the developed model.

CONCLUSION

Theoretical Contribution

This research proposes a clear realization of the inevitability of R&D practices for controlling PDP to enhance the NPD success in Indian manufacturing industries. First, the identification of the indicators of R&D practices is executed. It also includes the recognition of sub-factors of R&D practices which are not directly linked to R&D but influenced by R&D practices of the firm. Second, the impact of R&D practices on PDP is also being examined. For this, the sub-factors of PDP along with their indicators are located. Third, the combined impact of these constituents on NPD success influenced by the optimal embracement of their sub-factors is also investigated. Finally, the NPD success of the firm is quantified in terms of technological developments and quality assurance of the newly developed products as the performance attributes.

Figure 2. SEM model after execution comprising of all the latent constructs and their indicators representing the path estimates

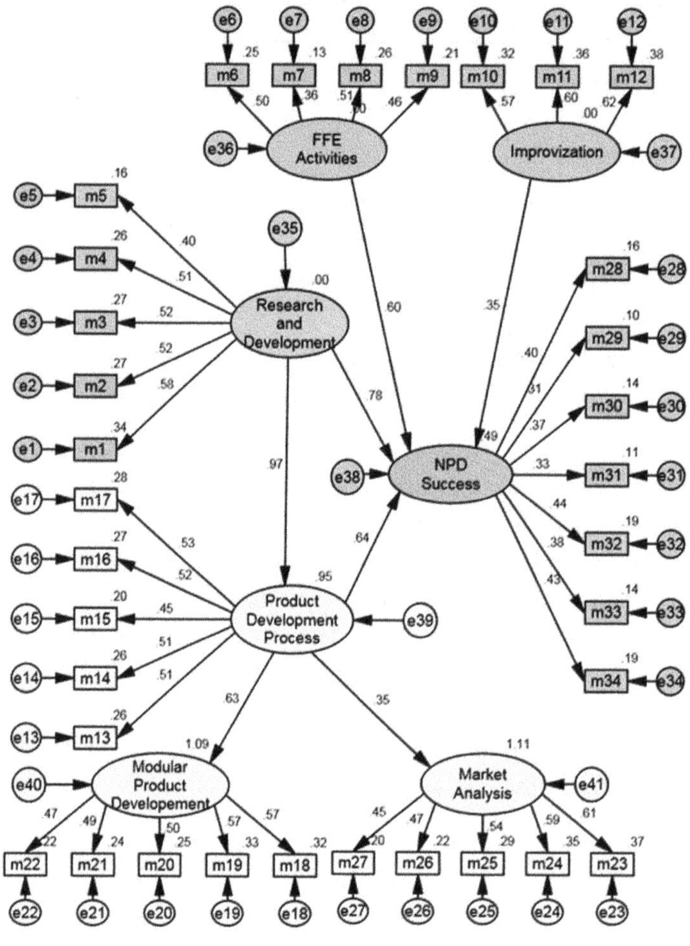

Managerial Implications

The interpretations of this empirical study clarifies the requirement of adopting R&D practices and their sub-factors allied indirectly for influencing PDP and its sub-factors in turn the NPD success of the firm. The analysis produces the results help to draw the significant managerial implications.

First, the realization of R&D practices and PDP R& is attained for successful NPD through utilizing the available resources in the firm. This evinces that FFE activities and improvization are two constituents of R&D practices which are not directly connected but have an effective impact on NPD success of the firm. As per experts' opinion, R&D management vision and direction embraces R&D practice of the firm leading towards better NPD performance. In case of FFE activities, elevated level of communication in early phases of product development enriches the FFE activities of the firm as well as the NPD success.This analysis computes that number of R&D persons in research activities, effort to reduce market uncertainty and improvisation of team in commercializing this product as per plan are the most significant attributes of R&D practices, FFE activities and improvisation respectively.

Second, in case of PDP, the hands-on working experience on PDP of new products is one of the crucial characteristics the firm need to be adopted for better completion of the development process of the new products. MPD and market analysis are the identified sub-factors of PDP among which product component commonality is the suggested feature for intensifying the MPD of the firm. In case of market analysis, though advertisement & promotion and competitor monitoring are suggested as the potential indicators, but the analysis shows their insufficient contribution for market analysis in realistic meadow. For PDP, hands-on experience is identified as the most significant attribute whereas, for market analysis development of the market plan is the vital most attribute among others. In case of MPD both adoption of modular product design and increase in product variety are prioritized as having the highest contribution for implementing MPD in the firm.

Finally, recommendations of experts' for NPD success comprises – expansion product family with varieties of products, reduction of risks, cannibalization effect and beating competition to market. But these indicators are recognized as less imperative than the available ones in the existing literature. Achieved product performance goal is acknowledged as the most vital attribute of NPD success succeeded by achievement of design goal, competitive advantages through technological performance, meeting quality guidelines, technological breakthrough, product failure rate and pace of product launch. This result signifies that quality assurance of the newly developed products is more imperative in respect to the technological developments of the new products.

Limitations and Future Scope

Limitation is one of the imperative phenomenons associated with each research. This is not the exceptional as well. This empirical research requires large number of samples accumulated as the responses of the industry experts. Data accumulation of such huge number of samples is a difficult task to be performed. Moreover, the reliability and validity of the responses are needed to be tested for avoiding misleading results. Proper techniques needed to be adopted for discarding the erroneous responses. The data have been collected from some strategic locations where industry density is quite high excluding many firms which are out of the major cities.

This research can be expanded to the other sectors than the engineering product development firms. There is also a scope of analyzing the study in sub-sectorial level considering a single type of product. Similarly, the comparison among the developed frameworks for different sub-sectors may well be attempted.

ACKNOWLEDGMENT

Department of Science and Technology (DST) of Government of India was substantially supported this empirical research as a grant titled as INSPIRE Fellowship for completing the PhD research of the corresponding author. Experts of various manufacturing companies of India are also being thanked for their kind support and participation.

REFERENCES

Acur, N., Kandemir, D., & Boer, H. (2012). Strategic alignment and new product development: Drivers and performance effects. *Journal of Product Innovation Management*, *29*(2), 304–318. doi:10.1111/j.1540-5885.2011.00897.x

Akgün, A. E., Byrne, J. C., Lynn, G. S., & Keskin, H. (2007). New product development in turbulent environments: Impact of improvisation and unlearning on new product performance. *Journal of Engineering and Technology Management*, *24*(3), 203–230. doi:10.1016/j.jengtecman.2007.05.008

Ayag, Z. (2005). An integrated approach to evaluating conceptual design alternatives in a new product development environment. *International Journal of Production Research, 43*(4), 687-713.

Baldwin, C. Y., & Clark, K. B. (2000). *Design Rules*. Cambridge, MA: MIT Press.

Bhuiyan, N. (2011). A framework for successful new product development. *Journal of Industrial Engineering and Management, 4*(4), 746-770.

Booz, A. (1982). *New product management for the 1980's. Boo*, Allen & Hamilton, Inc.

Byrne, B. M. (2010). *Structural Equation Modeling with AMOS: Basic Concepts, Applications, and Programming*. New York: Taylor and Francis Group LLC.

Chaudhuri, A., & Boer, H. (2016). The impact of product-process complexity and new product development order winners on new product development performance: The mediating role of collaborative competence. *Journal of Engineering and Technology Management*, *42*, 65–80. doi:10.1016/j.jengtecman.2016.10.002

Cooper, R. G., & Kleinschmidt, E. J. (1995). Benchmarking the firm's critical success factors in new product development. *Journal of Product Innovation Management*, *12*(5), 374–391. doi:10.1016/0737-6782(95)00059-3

de Sousa Mendes, G. H., & Miller Devós Ganga, G. (2013). Predicting success in product development: The application of principal component analysis to categorical data and binomial logistic regression. *Journal of Technology Management & Innovation*, *8*(3), 83–97.

Ernst, H. (2002). Success factors of new product development: A review of the empirical literature. *International Journal of Management Reviews*, *4*(1), 1–40. doi:10.1111/1468-2370.00075

Florén, H., Frishammar, J., Parida, V., & Wincent, J. (2018). Critical success factors in early new product development: A review and a conceptual model. *The International Entrepreneurship and Management Journal*, *14*(2), 411–427. doi:10.100711365-017-0458-3

Gatignon, H., Tushman, M. L., Smith, W., & Anderson, P. (2002). A structural approach to assessing innovation: Construct development of innovation locus, type, and characteristics. *Management Science*, *48*(9), 1103–1122. doi:10.1287/mnsc.48.9.1103.174

Graner, M. (2016). Are methods the key to product development success? An empirical analysis of method application in new product development. In *Impact of Design Research on Industrial Practice* (pp. 23–43). Springer International Publishing. doi:10.1007/978-3-319-19449-3_2

Hair, J. F., Anderson, R. E., Tatham, R. L., & Black, W. C. (1995). *Multivariate Data Analysis with Readings*. Englewood Cliffs, NJ: Prentice Hall.

Haverila, M. J. (2012). Product–firm compatibility in new product development in technology companies. *The Journal of High Technology Management Research*, *23*(2), 130–141. doi:10.1016/j.hitech.2012.06.008

Heirati, N., & O'Cass, A. (2016). Supporting new product commercialization through managerial social ties and market knowledge development in an emerging economy. *Asia Pacific Journal of Management*, *33*(2), 411–433. doi:10.100710490-015-9437-9

Holmes, P., & Smith, P. (2001). *Introduction to structural equation modelling using LISREAL*. Perth: ACSPRI-Winter training Program.

Hsu, Y. H., & Fang, W. (2009). Intellectual capital and new product development performance: The mediating role of organizational learning capability. *Technological Forecasting and Social Change*, *76*(5), 664–677. doi:10.1016/j.techfore.2008.03.012

Huang, X., Soutar, G. N., & Brown, A. (2004). Measuring new product success: An empirical investigation of Australian SMEs. *Industrial Marketing Management*, *33*(2), 117–123. doi:10.1016/S0019-8501(03)00034-8

Joreskog, K. G. (1973). A general method for estimating a linear structural equation system. In A. S. Goldberger & O. D. Duncan (Eds.), *Structural Equation Models in the Social Sciences* (pp. 85–112). New York: Academic Press.

Kim, B., & Kim, J. (2009). Structural factors of NPD (new product development) team for manufacturability. *International Journal of Project Management*, *27*(7), 690–702. doi:10.1016/j.ijproman.2008.11.003

Lagrosen, S. (2005). Customer involvement in new product development: A relationship marketing perspective. *European Journal of Innovation Management*, *8*(4), 424–436. doi:10.1108/14601060510627803

Lau, A. K. (2011). Critical success factors in managing modular production design: Six company case studies in Hong Kong, China, and Singapore. *Journal of Engineering and Technology Management*, *28*(3), 168–183. doi:10.1016/j.jengtecman.2011.03.004

Lester, D. H. (1998). Critical success factors for new product development. *Research Technology Management*, *41*(1), 36–43. doi:10.1080/08956308.1998.11671182

Lynn, G. S., Abel, K. D., Valentine, W. S., & Wright, R. C. (1999). Key factors in increasing speed to market and improving new product success rates. *Industrial Marketing Management*, *28*(4), 319–326. doi:10.1016/S0019-8501(98)00008-X

Micro, Medium & Small Scale Industry. (2018, August 28). Retrieved from https://www.cii.in/Sectors.aspx

Miner, A. S., Bassof, P., & Moorman, C. (2001). Organizational improvisation and learning: A field study. *Administrative Science Quarterly*, *46*(2), 304–337. doi:10.2307/2667089

Nicholas, J., Ledwith, A., Aloini, D., Martini, A., & Nosella, A. (2015). Searching for radical new product ideas: Exploratory and confirmatory factor analysis for construct validation. *International Journal of Technology Management*, *68*(1-2), 70–98. doi:10.1504/IJTM.2015.068779

Ong, C. S., Lai, J. Y., & Wang, Y. S. (2004). Factors affecting engineers' acceptance of asynchronous e-learning systems in high-tech companies. *Information & Management, 41*(6), 795–804. doi:10.1016/j.im.2003.08.012

Pine, J. B. II. (1997). *Mass Customization: The New Frontier in Business Competition*. Cambridge, MA: Harvard Business School Press.

Rigdon, E. E. (1998). Structural Equation Modelling. In G. A. Marcoulides (Ed.), *Mahwah, Modern methods for business research* (pp. 251–294). Lawrence Erlbaum Associates Publishers.

Robertson, D., & Ulrich, K. (1998). Planning for product platforms. *Sloan Management Review, 39*(4), 19.

Rockart, J. F. (1979). Chief executives define their own data needs. *Harvard Business Review, 57*(2), 81–93. PMID:10297607

Roy, M. J., & Thérin, F. (2008). Knowledge acquisition and environmental commitment in SMEs. *Corporate Social Responsibility and Environmental Management, 15*(5), 249–259. doi:10.1002/csr.145

Roy, S., Dan, P., & Modak, N. (2018). Effect of teamwork culture on NPD team's capability in Indian engineering manufacturing sector. *Management Science Letters, 8*(7), 767–784. doi:10.5267/j.msl.2018.5.009

Roy, S., Dan, P. K., & Modak, N. (2018). Cascading effects of management actions on NPD in the manufacturing sector: The Indian context. *Journal of Manufacturing Technology Management, 29*(7), 1115–1137. doi:10.1108/JMTM-11-2017-0231

Shinno, H., & Hashizume, H. (2002). Structured method for identifying success factors in new product development of machine tools. *CIRP Annals-Manufacturing Technology, 51*(1), 281–284. doi:10.1016/S0007-8506(07)61517-0

Sholeh, M., Ghasemi, A., & Shahbazi, M. (2018). A new systematic approach in new product development through an integration of general morphological analysis and IPA. *Decision Science Letters, 7*(2), 181–196. doi:10.5267/j.dsl.2017.5.004

Song, X. M., & Montoya-Weiss, M. M. (1998). Critical development activities for really new versus incremental products. *Journal of Product Innovation Management, 15*(2), 124–135. doi:10.1016/S0737-6782(97)00077-5

Sun, H., & Wing, W. C. (2005). Critical success factors for new product development in the Hong Kong toy industry. *Technovation, 25*(3), 293–303. doi:10.1016/S0166-4972(03)00097-X

Tripathy, S., Kumar Ray, P., & Sahu, S. (2012). Factors governing R&D practices in Indian manufacturing firms: Structural equation modelling. *International Journal of Modelling in Operations Management, 2*(1), 45–68. doi:10.1504/IJMOM.2012.043960

Tzokas, N., Hultink, E. J., & Hart, S. (2004). Navigating the new product development process. *Industrial Marketing Management, 33*(7), 619–626. doi:10.1016/j.indmarman.2003.09.004

Ulrich, K. T., & Eppinger, S. D. (2000). *Product Design and Development* (2nd ed.). Boston: Irwin/McGraw-Hill.

Verworn, B. (2009). A structural equation model of the impact of the "fuzzy front end" on the success of new product development. *Research Policy*, *38*(10), 1571–1581. doi:10.1016/j.respol.2009.09.006

Wang, J., Liang, Z., & Xue, L. (2014). Multinational R&D in China: Differentiation and integration of global R&D networks. *International Journal of Technology Management*, *65*(1-4), 96–124. doi:10.1504/IJTM.2014.060959

APPENDIX

- Interview Protocol with description of latent variables

1. R&D practices

- Number of R&D persons: Number of R&D persons in research activities.
- Years of experience of R&D persons: Experience of the R&D team members which help in the new product development of the firm.
- Investments for better R&D infrastructure: Investment in R&D infrastructure and methods for sustainable product development.
- Qualification of R&D experts: Qualification of the R&D team members which affects the continuous flow of the product development.
- Number of patents: Number of patents of the firm (if any).
- R&D culture in the firm: R&D oriented culture in the firm.

    a. Fuzzy front end (FFE) activities:

- Interdisciplinary idea generation and screening: Interdisciplinary idea generation and screening ideas by historical analogy.
- Intensity of initial planning: Intensity of initial planning for interdisciplinary idea selection.
- Effort to reduce market uncertainty: Effort to reduce market uncertainty during development of new products.

    b. Improvization:

- Difference of NPD process as in practical case versus desired process: Figuring out of new product-development process as it went along versus following a rigid well-defined plan.
- Improvisation of team in developing this product as per the plan: Improvisation of team in developing this product versus strictly following the plan.
- Improvisation of team in commercializing this product as per plan: Improvisation of team in commercializing this product versus strictly following the plan.

2. Product Development Process (PDP):

- Fund allocation for updated development activities: Fund investment for completion of the process of new product development.
- Advanced product development methods:Eagerness for technological innovations by applying updated tools and techniques.
- Developmental time and cost: Time required for developing the product and the cost associated for development.

- Training for NPD Management: Arrangement of systematic training programs for successful management of new product development.
- Market research activities: Involvement of activities for gathering information about the customers' demand and preferences.
- Adoption of TQM: Managerial approach for achieving long-term success through improving processes and products in the field respective field they work.

    a. Modular product design (MPD):

- Adoption of Modular product design:Idea generation of developing a product through assimilating a set of small products designed independently functioning together as a whole.
- Increase in product variety: Varieties of products are developed by assembling the individual modular parts.
- System reliability improvement: Improvement of system reliability through developing the final product by assembling the modular parts.
- Product component commonality: Reusing common components in a range of products.

    b. Market analysis:

- Development of market plan: Well-established market plan generation involves activities in accomplishing the specific marketing objectives within a set of time frame.
- Consumption pattern of the product: Identification of need of the product and the rate of consumption.
- Target market and growth pattern: Identification of particular group of consumers at which a product is aimed and the growth pattern as well.
- Focus on customer satisfaction: Identification of customers' needs for providing high-end customer satisfaction.
- Market testing: Experiments conducted before commercializing the new products to the open market to test its scope of the success.

6. New product development (NPD) Success:

- Achievement of design goal: Attainment of design specifications.
- Achieved product performance goal: Attainment of product specifications as per customer demand.
- Meeting quality guidelines: Achievement of quality of the new products as previously specified.
- Technological breakthrough: Technological breakthrough attained during NPD activities.
- Product failure rate: Rate of failure of newly developed products in the firm.
- Competitive advantages through technological performance: Technological superiority of the newly developed products with respect to the competitors.
- Rate of product launching: Frequency of newly launched products in the firm.

Chapter 7
Multi-Objective Territory Design for Sales Managers of a Direct Sales Company

Elias Olivares-Benitez
Universidad Panamericana, Mexico

Pilar Novo Ibarra
Universidad Panamericana, Mexico

Samuel Nucamendi-Guillén
Universidad Panamericana, Mexico

Omar G. Rojas
https://orcid.org/0000-0002-0681-3833
Universidad Panamericana, Mexico

ABSTRACT

This chapter presents a case study to organize the sales territories for a company with 11 sales managers to be assigned to 111 sales coverage units in Mexico. The assignment problem is modeled as a mathematical program with two objective functions. One objective minimizes the maximum distance traveled by the manager, and the other objective minimizes the variation of the sales growth goals with respect to the national average. To solve the bi-objective non-linear mixed-integer program, a weights method is selected. Some instances are solved using commercial software with long computational times. Also, a heuristic and a metaheuristic based on simulated annealing were developed. The design of the heuristic generates good solutions for the distance objective. The metaheuristic produces better results than the heuristic, with a better balance between the objectives. The heuristic and the metaheuristic are capable of providing good results with short computational times.

DOI: 10.4018/978-1-5225-8223-6.ch007

INTRODUCTION

The problem addressed in this chapter is based on the case of a direct sales company. This is an assignment problem that results in a configuration of territories for the sales managers of the company. The problem is modeled as an optimization problem and is described below.

In the highest level of the sales structure, 11 managers must serve a market consisting of 111 sales coverage units. A sales coverage unit (SCU) is a geographic area with dispersed sellers and a supervisor with an office located in a specific city or town into the area. Managers have regional offices with fixed locations, from where they visit the different supervisors of the sales coverage units. These visits will depend on the sales coverage units assigned to each manager. In the most basic structure, this is an assignment problem, where sales coverage units are assigned to managers. Figure 1 shows the structure of the sales organization.

The visits of managers to the cities of the supervisors are scheduled at different moments during the year, with a long separation in time between visits. The visits are done to check the results of sales of the SCU, to define strategies for the short-term future, and to make training of the supervisors and the sellers. Because of the policies of the company, the manager must spend some time in the regional office between visits for administrative activities. Therefore, there is no opportunity to sequence visits in single or multiple routes. The visits must consider the movement from the city of origin of the manager to the city of the supervisor, and back to the city of origin. Transportation costs have a positive correlation with the distance between cities.

Currently, the company has an inherited structure that was determined with unknown objectives. The basic structure was of SCUs already assigned to managers. The basic structure of assignment is updated each year when new SCUs are added or some SCUs are deleted because of lack of activity. The new SCUs added are assigned to the closest manager. At this point, the company wanted to reorganize its sales structure based on two objectives: to minimize the distance each manager traveled and to balance the territories with respect to the expected average sales growth. The motivation of the study was to solve this problem for the company while some operational constraints were observed.

Figure 1. Structure of the sales organization

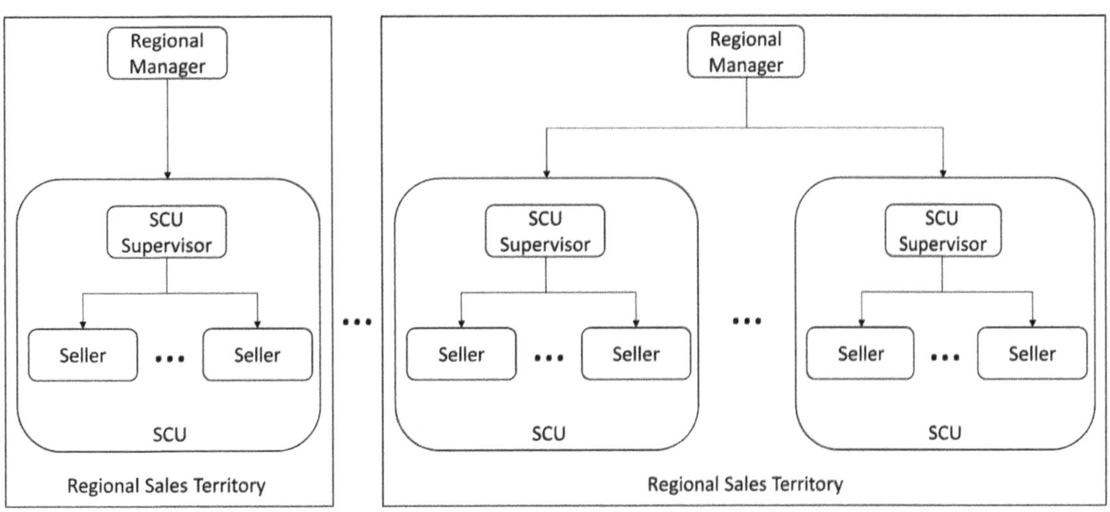

To avoid spending too much time on transportation, one objective of the problem is to minimize the maximum distance that any manager must travel. This also helps to balance the size of each territory, regarding the number of customers. In the other side, one of the constraints of the problem is that the sum of all the transportation costs should not exceed an annual budget.

For each one of the sales coverage units, a goal of sales growth is determined for the year. Together, the growth goals of each SCU result in a national growth average. An individual growth average can be calculated for each one of the managers according to the sales coverage units assigned to them. One of the objectives of the problem is to make the assignment such that the differences of the individual growth averages with respect to the national growth average are minimized. In this way, the managers have a feeling of equity in the distribution of responsibilities.

The problem must consider that each one of the managers must have assigned at least one sales coverage unit and that each sales coverage unit is assigned to a single manager. The idea is to keep the current number and location of managers. In a feasible solution without this consideration, some managers might be fired, but the company wants to keep the salesforce because of social responsibility concerns. Another issue to be observed is to avoid duplication of assignments, to elude administrative confusion of the supervisors reporting to multiple managers, and to keep order in the computation of the individual growth averages.

The situation described above is modeled as a mathematical program with two objectives, to be optimized. Because of the mathematical function used to calculate the averages of sales growth, the model results being non-linear. This non-linearity in the model makes impossible to solve the full instance of the case with the mathematical program in a reasonable time. Then, a partition of the instance was selected to test the correctness of the model and to gain insight in the interpretation of the solution. To cope with the complexity of the problem, a heuristic procedure was developed. Also, a simulated annealing metaheuristic is developed to solve the problem. The results of the model solved with a commercial optimizer, the heuristic, and the metaheuristic are compared.

This work makes two significant contributions: the most important is the modeling of an optimization problem to solve a real situation for the sales department of a large company, and the second is the development of a heuristic and a metaheuristic to solve this problem.

BACKGROUND

It is possible to define the problem described before as a multi-objective assignment problem. An assignment problem is a special type of linear programming, in which the realization of "tasks" is assigned to the "servers." The servers can be people, machines, vehicles, plants, etc. to which one or more tasks are assigned. In order to fit the definition of an assignment problem, five assumptions must be fulfilled (Hillier & Lieberman, 2010):

1. The number of servers is equal to the number of tasks.
2. Each server is assigned only to one task.
3. Each task must be done by only one server.
4. There is a cost associated with the server that performs the task.
5. The goal is to determine how all assignments should be made to minimize total costs.

Although the last three assumptions are met for this problem, the first two contradict it, since in this work there are more SCUs to assign than the number of sales managers, and although each task is performed by only one server, more than one task is assigned to each. Due to this reason, the proposed model does not obey the typical model for an allocation problem but rather presents more complexity. Neither should it be ignored that, being a problem with two objectives, there is not a single purpose of minimizing total costs; and other aspects are to be taken into account.

In multi-objective optimization, unlike optimization with a single objective, the solution corresponds to a family of points instead of a single point, known as the Pareto front. All the solutions in the front are optimal in the sense that there are no better solutions that do not degrade at least one of the objectives of the problem (non-dominated solutions). In a multi-objective problem, a solution is not obtained that is optimal and satisfies all the objectives, but rather seeks to satisfy in the best way the objective functions proposed. These objectives are sometimes prioritized according to the importance of each one (Fonseca & Fleming, 1993).

Tavares-Pereira, Figueira et al. (2007) raised the problem of dividing one territory into homogeneous zones as a problem of multi-objective districting. They proposed an evolutionary algorithm to obtain an approximation of the Pareto front. Their method works with multiple objectives, allows solving large instances in reasonable computational time, and generates good quality solutions. The problems of districting can be handled with approaches of division or agglomeration. The first aims to obtain territories that create a unit while the second obtains territories composed of separate units but with a connection between them.

Lei et al. (2014) presented a problem of the traveling salesperson and districting with multiple periods. The problem consists of designing districts and sub-districts for a multiple traveler salesperson problem that has dynamic customers in several periods. Each salesperson serves all customers in their district but performs a single route during each period. Their work is interesting as an example that, in real life, problems include different criteria and aspects to take into account to obtain a satisfactory solution.

Haase & Müller (2014) find an approach to organize the sales force in a company, which seeks to maximize the total benefit. Their problem has a broad scope, looking for a solution that delivers the size of the sales force, the locations of sales representatives, the division of territories, and the allocation of resources to these territories.

Rios-Mercado & Fernández (2009) used the GRASP (Greedy-Randomized Adaptive Search Procedure) method for territorial commercial design, to apply it in the case of a beverage distribution company. What they sought was to minimize the dispersion of the territory, balance the activity in the territories, and the contiguity of these territories.

Gliesch et al. (2018) address a commercial districting problem for the distribution of products. The problem requires to balance the districts and to optimize compactness. To solve the problem, they developed a multi-start algorithm that uses two alternating tabu search metaheuristics to improve each objective each time.

Territory design problems are also known as districting problems and can be applied to many different situations. For example, Lin et al. (2018) studied a districting problem to deliver care services to elderly people. They propose a multi-objective mixed-integer non-linear programming model to serve objectives as the balancing of the workload, maximization of geographic compactness, and minimization of workforce size. They used a non-dominated sorting genetic algorithm, modified, to solve the problem.

Kong et al. (2019) analyze a bi-objective districting problem. They used a weighted function that combined the objectives of district balance and compactness. They developed a center-based mixed-integer linear programming model to solve the problem. To improve the speed and quality of the solutions, they use district centers obtained from a multi-start weighted K-medoids algorithm.

There are other metaheuristics used to solve multi-objective problems. The algorithm of Ant Colony was introduced around 1990 and is inspired by the actual behavior of ants in their habitat, which is determined by the paths they take when going in search of food (Blum, 2005). Doerner et al. (2004) used this method for the problem of selecting a multi-objective portfolio and compared it with other heuristics using random instances. Alves & Climaco (2000), Coello (2000), and Dorigo & Gambardella (1997) also proposed metaheuristics or heuristics comparable to the Ant Colony algorithm for multi-objective problems.

In many situations, it is useful to transform the multi-objective problem into a single objective problem in order to solve it with traditional methods. For these cases, there will be a subject who will be responsible for making decisions regarding the approach of this objective and the solution that best meets the needs of the same problem. Deb (2001, 2014) presented the differences between solving problems using more than one objective or reducing them to only one.

Brockhoff & Zitzler (2006) studied problems that required more than three objectives, and when seeing how complicated it was to find a Pareto front in terms of computational solution time, as well as in the decision-making process, they presented the question of whether all the objectives were necessary to solve the problem satisfactorily. It is a fact that, in real life, problems do not present a single objective, but it is worthwhile to carefully analyze each of the objectives and to decide which of those directly affect what you want to achieve, or if it is possible, to aggregate some of them in one. Authors such as Deb & Saxena (2005), Zitzler et al. (2003), and Zitzler & Kunzli. (2004) questioned the same within the field of large-scale multi-objective problems, and were concerned with reducing the objectives to simplify all possible problems.

In recent decades, heuristic procedures have been developed to help solve optimization problems that were previously very complicated or impossible to solve. These heuristic techniques can be combined with each other and being developed to adapt to different problems, without strict rules when using them (Lee & EI-Sharkawee, 2008).

Segura-Ramiro et al. (2007) developed a heuristic for a problem of territory design in a beverage distribution company. They aimed to minimize the dispersion of these territories in two phases, beginning by locating the centers of the territories and then assigning customers to each of the centers.

Other authors have used heuristics to solve similar problems. For instance, Resende & Wernecks (2004) used a combination of heuristics for the k-means problem. Rosing & Revelle (1997) explain the construction of a solution using a two-phase heuristic.

These applications and research show that the need to solve a problem has been presented in different situations and that, because of so many differences, it is necessary to adapt and search for new methods that satisfy what is required in each case.

This work proposes a comparison between three different methods of solution for the assignment of territories to the sales managers. The first method uses a commercial software that obtains a global optimum for the multi-objective problem, after reducing it to a single objective model by assigning a weight to each objective. For the second method, a heuristic was developed to obtain a solution, and it is expected that it will obtain solutions of good quality in comparison with the global optimum obtained for each instance. The third method uses a simulated annealing framework to obtain a solution to the weighted single objective model.

MATHEMATICAL MODEL

The problem described in the introduction is modeled as a mathematical program. The notation is introduced to understand the model.

Notation

Sets and Indices

P Set of sales managers,
L Set of cities or towns with offices for supervisors of the Sales Coverage Units (SCU),
i Index for set P, $i = 1, \ldots, |P|$,
j Index for set L, $j = 1, \ldots, |L|$.

Variables

$a_{ij} \in \{0,1\}$ binary variable with a value of 1 if manager i is assigned to city j and of 0, otherwise,
$g_i \in R^+$ growth goal in percentage calculated for manager i according to the cities assigned,
$d_{imax} \in R^+$ longest distance covered by any manager in one year.

Parameters

$M_j \in R^+$ Growth goal for city j for the next year,
$V_j \in R^+$ Sales in city j in the previous year,
$T \in R^+$ Growth goal in the overall percentage of the country (national average) for the next year,
$B \in R^+$ Total budget for the sales managers for the next year.

Model

$$min \quad d_{imax} \tag{1}$$

$$min \sum_{i \in P} \left| \frac{T - g_i}{T} \right| \tag{2}$$

Subject to:

$$T = \frac{\sum_{j \in L} M_j}{\sum_{j \in L} V_j} - 1 \tag{3}$$

$$g_i = \frac{\sum_{j \in L} M_j a_{ij}}{\sum_{j \in L} V_j a_{ij}} - 1$$

$$\forall i \in P \quad (4)$$

$$\sum_{j \in L} a_{ij} \geq 1$$

$$\forall i \in P \quad (5)$$

$$\sum_{i \in P} a_{ij} = 1$$

$$\forall j \in L \quad (6)$$

$$\sum_{j \in L} c_{ij} a_{ij} \leq B$$

$$\forall i \in P \quad (7)$$

$$a_{ij} \in \{0,1\} \; \forall i \in P, j \in L \quad (8)$$

The first objective (1) seeks to minimize the longest distance traveled by any manager *I*, while the second objective (2) minimizes the absolute difference between the percentage of growth of the country and the percentage assigned to each manager, to balance the goals.

Equation (3) shows how the percentage of the country's growth is calculated, dividing the sum of the goals of each SCU in the next year between the sum of the sales of each SCU in the previous year, and subtracting one. This variation is known, and represents 2.5% in the case study, but it is presented to know how to calculate it in case there are changes in the parameters, or to choose different instances for the same problem. Equations (4) follow the same line as (3), but multiplying by the assignment variable, in order to obtain the goal that each manager will have once the municipalities are assigned. Constraints (5) ensure that each manager is assigned at least to one SCU, and equations (6) limit the assignment of only one manager to each city. There are also constraints (7) to attain the total budget of the next year, preventing from exceeding it. Finally, constraints (8) ensure that the binary variable results in a value of 1 if an SCU is assigned to a manager and 0 otherwise.

METHODOLOGY

Because of the non-linearity in the model, it was impossible to solve the full real instance of the case with the mathematical program in a reasonable time. Then, a partition of 10 sales coverage units and 5 managers was selected to test the correctness of the model and to gain insight in the interpretation of the solution.

When having two objective functions, it is important to pay attention to how both behave, if when one improves the other worsens or vice versa, or even if they do not relate to each other. For the solution in the software, the two objective functions were presented in a single one, and being both minimization problems it was possible to add them. A fraction (α) was determined to weigh the objective functions. The first objective was multiplied by α, while the second was multiplied by (1-α). Different values were used to observe the effect of giving more importance to each of the objective functions. The instance with 10 sales coverage units and 5 managers was solved with this approach.

Later, for a fixed weight of (α), two partitions of the full instance were solved optimally. One partition, named 6x40, has 6 managers and 40 sales coverage units. Another partition, named 6x45, has 6 managers and 45 sales coverage units.

To cope with the complexity of the problem, a heuristic procedure was developed. Since more importance is given to the objective related to the distance, the heuristic was designed to favor this criterion. The performance of the heuristic is calculated by the comparison of results with those obtained with the mathematical model, solved by a commercial optimization software for instances 6x40 and 6x45.

Also, a metaheuristic based on Simulated Annealing was developed to solve the model with the weighted objective with a fixed weight of (α). The metaheuristic was used to solve the same instances 6x40 and 6x45 to compare the results with the heuristic and the optimal solutions.

Optimal Solutions

For the solution in the software, the two objective functions were presented in a single one. The objective function that combines both objectives is:

$$\min\ [(\alpha)(f_1)(d_{i\max}) + (1-\alpha)(f_2)(\sum_{i\in P} | \frac{T-g_i}{T} |)] \tag{9}$$

where:

$f1 = 1$
$f2$ = factor to normalize values of the objective functions

$0 \leq \alpha \leq 1$

Since the first objective function calculates a distance and the second one a very small number, it was necessary to multiply the second objective function by a factor that normalizes both. Otherwise, one of the objectives would be much more important than the other. Different values were given for α to solve a small partition of the original instance with 10 sales coverage units and 5 managers.

After obtaining results of the first partition of the instance, a value of 0.8 (80%) was assigned to the objective corresponding to the maximum distance covered by the managers. The complement of 0.2 (20%) was assigned to the weight of the objective related to the balance of sales growth averages. Then, partitions named 6x40 and 6x45 were solved with this approach.

Heuristic

The heuristic makes two basic steps: first, it assigns the closest SCU to each manager in order to meet the constraint of assignments to all the managers; in a second step, the remaining sales coverage units are assigned to their closest manager.

The proposed heuristic arose from a preference of the decision maker in which the importance of giving more weight to minimizing the maximum distance traveled is preferred. The process to obtain a solution with the help of the heuristic is shown in Algorithm 1.

Algorithm 1: Heuristic to Solve the Territory Design Problem

```
Input: Cities to be assigned to Managers (L), Set of Managers (P), Distance
between them and Cost
Sort cities by distance smallest to largest
A = Cities already assigned
For i = 1 To L
If City i not in A Then
Assign City i to P (Closest)
Add i to A
End if
Next i
Output: Best Solution Found
```

Metaheuristic

The general procedure for the Simulated Annealing metaheuristic is described in many works in the literature (Delahaye et al., 2019), as shown in Algorithm 2 for the minimization of an objective function.

Algorithm 2: General Procedure for Simulated Annealing

```
Input: Structure of variables and data of the instance, Cooling Schedule,
Neighborhood structure
Generate an initial solution s0
Solution s = s0
Initialize Temperature T = T0
Repeat
```

```
For n iterations
Generate a neighbor s'= N(s)
If Objective function (s') ≤ Objective function (s)
Accept s'
s = s'
Else
Calculate probability p = e^(k*(Objective function (s) - Objective function
(s'))/T)
Obtain a random probability p_r
If p_r ≤ p
Accept s'
s = s'
End if
End if
End for
Update T = T - delta
Until T < T_min
Output: Best solution found
```

For the problem presented here, the distinctive elements are the generation of the initial solution and the neighborhood structure. The generation of the initial solution followed the procedure described above for the heuristic, assigning the SCUs to the closest manager. The neighborhood structure followed the next rule: one SCU is selected randomly and reassigned to a different manager. Figure 2 shows the scheme of the neighborhood, where an insertion operation is done. In this case, the objective function corresponds to the weighted combined function (9). To handle feasibility, only solutions that meet the budget constraint were allowed; otherwise, the solution was discarded and a new neighbor was generated.

The number of iterations for each temperature was fixed as the number of SCUs of the instance. The number of changes in the temperature was set as the number of managers of the instance, adapting delta according to this. The temperature was initialized to 2000, and the last temperature was established as 200. The value of parameter k helps to normalize the difference of the objective functions in order to calculate reasonable values for the acceptance probabilities. The first time a worse solution appears the

Figure 2. Insertion operation in the neighborhood structure of the simulated annealing procedure

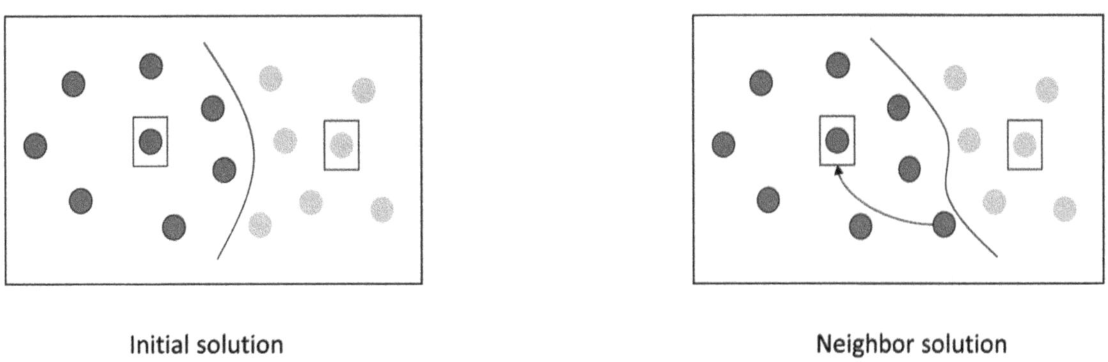

value of k is calculated to produce a probability $p = 0.8$, and this value is fixed for all the future computations in the algorithm.

SOLUTIONS AND RECOMMENDATIONS

Optimal Solutions

To deal with the bi-objective nature of the problem, a weights method is used to solve the mixed-integer non-linear problem. The model was solved using LINGO 17.0 in a 64 Gb RAM, 3.4 GHz Workstation. The weights method for multi-objective problems allows constructing a Pareto front for the instance changing the weights iteratively. Because of the non-linearity in the model, it was impossible to solve the full instance of the case with the mathematical program in a reasonable time. Then, a partition of the instance was selected to test the correctness of the model and to gain insight in the interpretation of the solution. A partition of 10 sales coverage units and 5 managers was solved, and the Pareto front obtained showed a compromise between the results of the objective functions, i.e. when one decreases, the other grows. Table 1 shows the results of both objectives, as well as the value of the combined objective function (9). "OF Maximum Distance" corresponds to the objective (1) and it is the longest distance that managers will have to travel according to the assigned cities, while "OF Goal Variation" results from the sum of all the variations of the goal percentage of the managers with respect to the percentage of the total goal in objective (2).

The first thing that stands out when observing the results shown in Table 1 is that when $\alpha = 0$, that is, when all the weight is charged to the objective of balancing the growth goals and none to minimize the distances, the maximum distance shows a very high value in comparison with the later results. This does not happen when loading the entire weight towards minimizing the maximum distance. In this case, the variation of the sales goals increases, but not to the same extent as in the first case. Additionally, a total variation of the goals of 1.025 does not represent a problem here, but a travel distance of 450.33 Km. would not be a viable option and the solution obtained would be discarded.

Figure 3 shows the behavior of the maximum distance and the variation of the goals. When changing the value of α, it can be seen that the line changes level. The line corresponding to the maximum distance allows us to visualize the sharp drop that was mentioned when analyzing Table 1, compared to the line of the goal variation, which shows a more homogeneous behavior and the distance in the y-axis between one point and another.

Table 1. Results modifying α

α	$1 - \alpha$	OF Maximum Distance	OF Goal Variation	OF Weighted
0.0	1.0	450.33	0.2463420	10.8390
0.2	0.8	102.00	0.2542593	29.3499
0.4	0.6	67.29	0.4205125	38.0175
0.6	0.4	55.02	0.5890438	43.3792
0.8	0.2	53.25	0.6563833	48.3762
1.0	0.0	51.84	1.0257350	51.8400

Figure 4 shows the Pareto front for the instance. It can be seen how the smaller the maximum distance, the sum of the variations increases in value. It is concluded based on these results, that in order to obtain a good result from one of the objective functions, the other one is compromised and negatively affected.

For the nature of the problem, in which there are managers who must travel by land during the year to reach their assigned SCU departing from a base, it is more logical to observe the distances they travel and be inclined to minimize them. In this case, it is also important that the managers' goals be balanced to avoid the workload, but to a lesser extent than to avoid them traveling very long distances. Once the tables are observed, the numbers show what was assumed at the beginning, that it is more convenient to

Figure 3. Values of the objective functions varying the value of α

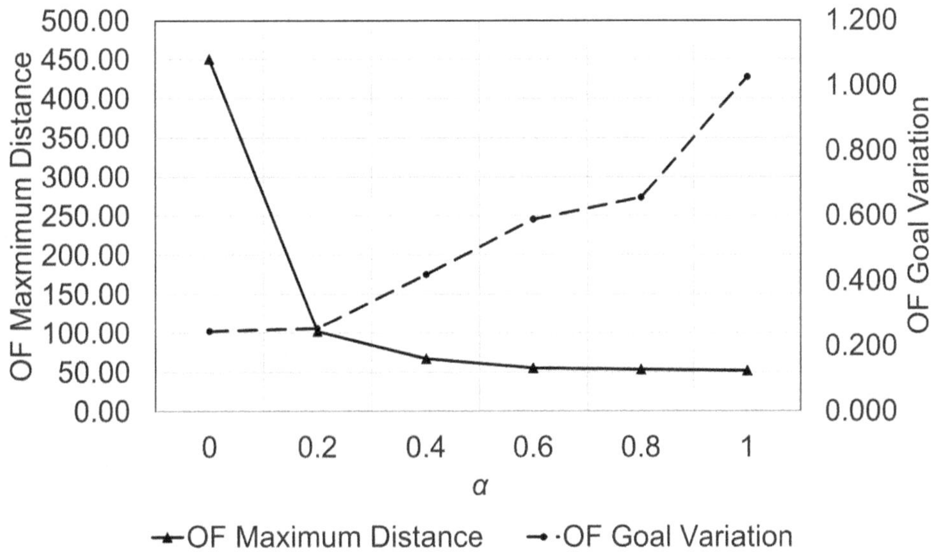

Figure 4. Pareto front for the instance with 10 sales coverage units and 5 managers

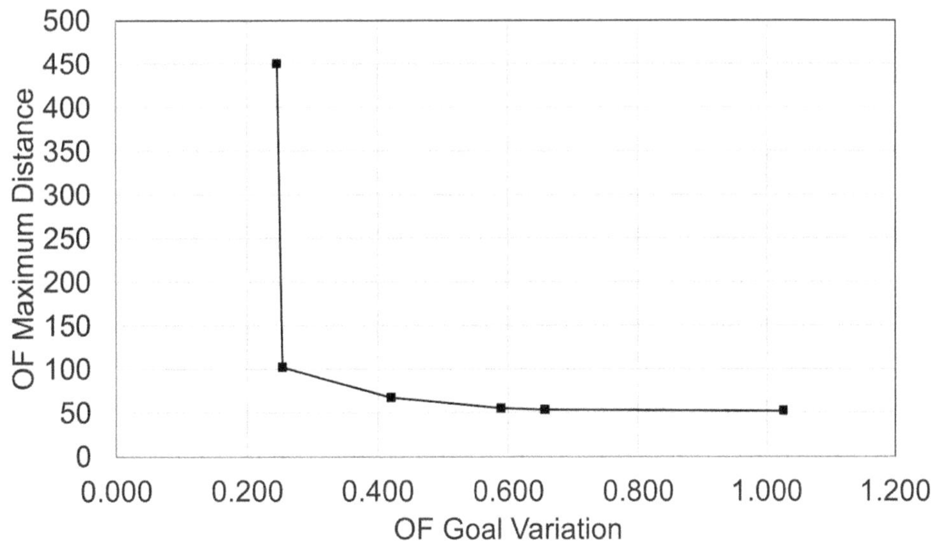

give greater weight to the objective function that minimizes the maximum distance. For later instances a value of 0.8 will be used, leaving only 20% of importance to the objective of balancing the goals.

Heuristic and Metaheuristic vs. Optimal Solutions

Using fixed weights for the objective functions, instance 6x40 took near 4 hours (3:56 hr) to be solved in the commercial optimization software, while instance 6x45 took near 72 hours to be solved, both to find the global optimum. These same instances were solved with both methods, the proposed heuristics and non-linear programming (NLP) so that it was possible to compare both results. The runtime for the heuristic and the metaheuristic were less than one minute for both instances. The metaheuristic was executed 10 times, and the best value is reported for the weighted objective function. Table 2 shows the results for the instance 6x40 while Table 3 shows the results for instance 6x45.

It is relevant to discuss what Tables 2 and 3 show. In both instances, the total objective function results in a lower value when using the global optimization software with the NLP. If the breakdown of both objectives is observed, it can be concluded that the heuristic satisfies to a great extent the first function, which minimizes the maximum distance, especially in the small instance. It can be said that the heuristic is not interested in minimizing the variation of the goals, but it shows a good quality of response for the other objective. The metaheuristic balances better the objectives with a good result for the objective of the distance, but a worse value for the variation of the goals. However, for both instances, the weighted objective function is closer to the optimum than the heuristic.

Figures 5, 6 and 7 show the results for the three instances mentioned for objectives in equations (1), (2), and (9) respectively.

Figure 5 shows that the heuristic yields better results for the objective function of the maximum distance, behaving in the same way as the NLP, but with lower values. In contrast to this, Figure 6 shows that the heuristic is not related in terms of the size of the instance and the behavior of the second objective function, while the NLP remains less variable regardless of the size of the instance. For the total objective function, Figure 7 shows similar behavior for both methods, but the heuristic yields higher

Table 2. Results for instance 6x40

Objective Function	NLP	Heuristic	Metaheuristic	Heuristic vs. NLP	Metaheuristic vs. NLP
OF Maximum Distance	151.120	125.160	148.312	-17.18%	-1.86%
OF Goal Variation	0.7185623	10.3254	1.46287	1336.95%	103.58%
OF Weighted	127.2193482	190.99152	131.522856	50.13%	3.38%

Table 3. Results for instance 6x45

Objective Function	NLP	Heuristic	Metaheuristic	Heuristic vs. NLP	Metaheuristic vs. NLP
OF Maximum Distance	96.460	87.010	94.327	-9.80%	-2.21%
OF Goal Variation	0.5132239	6.4859	1.35214	1163.76%	163.46%
OF Weighted	81.68437032	126.68392	87.360432	55.09%	6.95%

Figure 5. Maximum distance for instances 5x10, 6x40 and 6x45 with the three methods

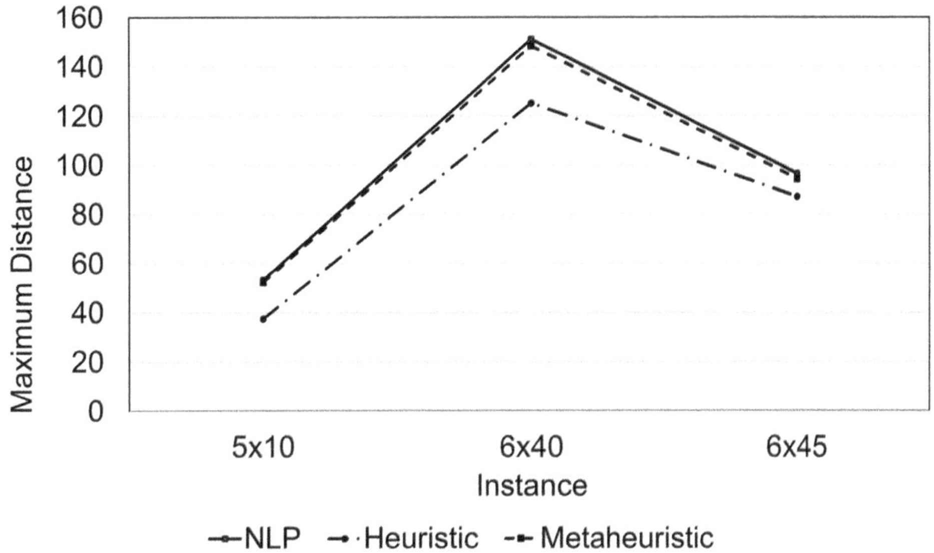

Figure 6. Goal variation for instances 5x10, 6x40 and 6x45 with the three methods

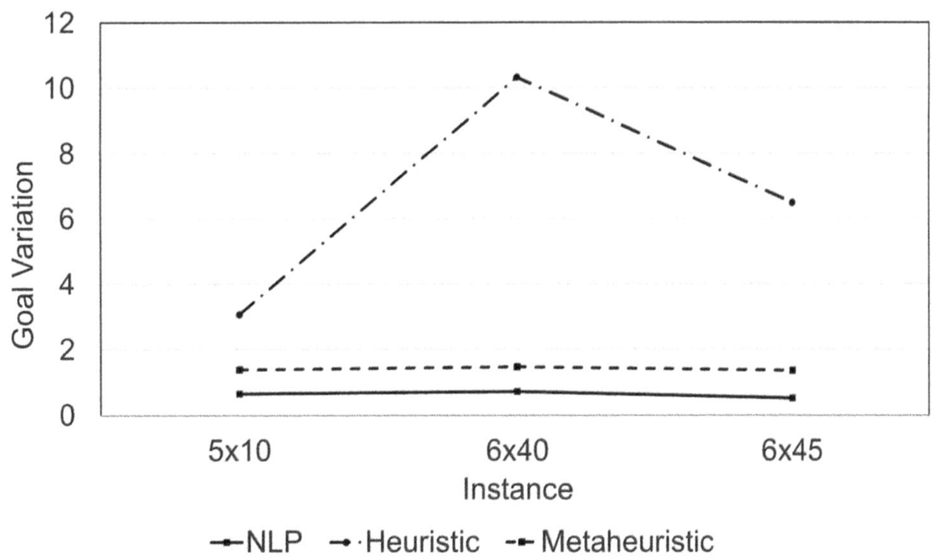

results in all cases. The metaheuristic behaves better than the NLP but worse than the heuristic for the maximum distance. For the case of the goal variation the metaheuristic shows worse results than the NLP, but better than the heuristic. In the global comparison with the weighted function, the metaheuristic is close to the NLP and better than the heuristic.

For a better understanding of the results, Figure 8 shows the comparison of the heuristic and the metaheuristic vs. the global optimum obtained with the NLP method. No relationship is observed regarding the size of the instance and the percentage variation of the results of the heuristic and the metaheuristic

Figure 7. Weighted objective function for instances 5x10, 6x40 and 6x45 with the three methods

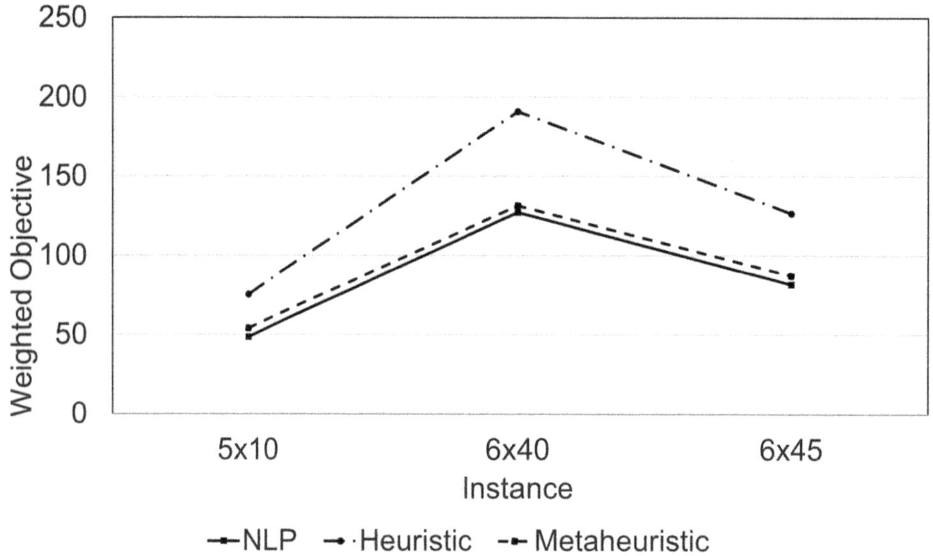

with respect to the overall solution. It can only be concluded that the heuristic and the metaheuristic will always obtain a value higher than the global optimum for the objective function in the conditions that were established.

Initially, it was established that there were 111 SCU to be assigned to 11 managers. During this study, the largest instance that was possible to solve with NLP was 45 SCU to assign to six managers due to the infeasibility of computational time required for larger instances. The real instance can be solved with the heuristic, having the knowledge that it will benefit the minimization of distances and obtaining a good quality result for this objective. The metaheuristic was also applied to the original instance. The result is shown in Table 4. The data used for the analysis corresponds to the planning of the year 2019. Thus, it is not possible to compare exactly with the assignment that the company would have done, because of the change in sales goals. However, the last assignment of the year 2018 was considered to simulate what would be the numbers if this structure would be conserved for the planning of the year 2019. This is included in the last column of Table 4. Good behavior of the metaheuristic is observed with a better result than the heuristic for the weighted function and compared to the simulation of the current assignment.

FUTURE RESEARCH DIRECTIONS

For future work it would be interesting to perform a deep analysis to reduce the objective functions to a single one and, for example, to transform the objective of minimizing the variations, adapting it into a restriction, to use the method of epsilon-constraint. For this, it would be necessary to establish a value that should not be exceeded to ensure that there are no large variations between the goals of the managers.

Once the sales team has been structured, a new problem could arise; when adding a new SCU, it is desired to assign it to the best possible manager. You can consider a new mathematical model that assigns the municipality to a manager without the need to restructure the entire country.

Figure 8. Weighted objective with respect to NLP for instances 5x10, 6x40 and 6x45 with the three methods

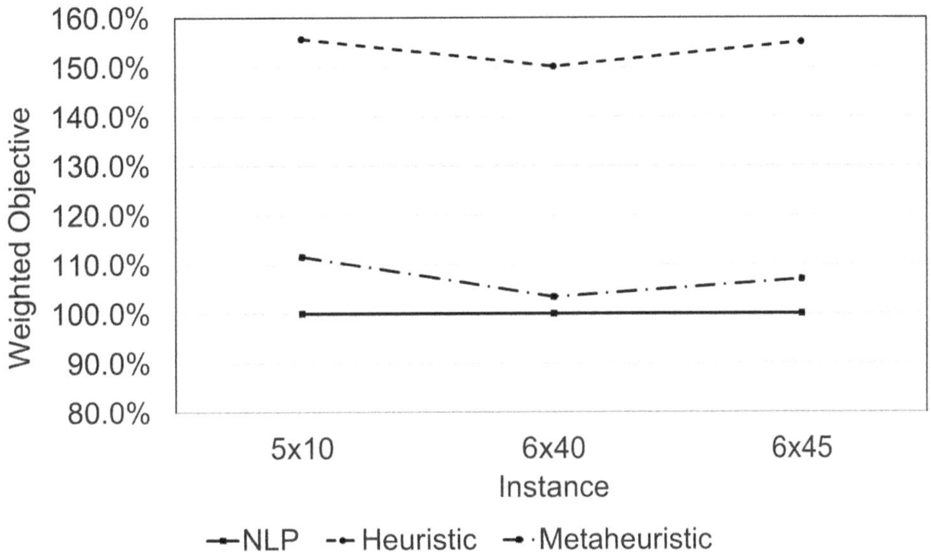

Table 4. Results for instance 11x111

Objective function	Heuristic	Metaheuristic	Simulated Current Assignment
OF Maximum Distance	140.52	152.26	178.33
OF Goal Variation	32.6	1.762713	42.3
OF Weighted	399.30	137.32	514.90

CONCLUSION

A problem was stated for a company that seeks to structure its sales force, which is established around the country so that none of its managers must travel a disproportionate distance or deals with a much higher or lower sales goal than other managers.

A mathematical model was presented that includes all the aspects to take into account to obtain a satisfactory result. Having two objective functions opens up different possibilities to solve the problem, and in this case three of these options were compared to determine if a heuristic method and a meta-heuristic procedure, which require less computational cost, can replace a global solution obtained with commercial software.

It is possible, having observed the results, to conclude that the used methods worked adequately. The variations of the goals are more complicated to be presented as a minimization objective; thus, the heuristics did not yield satisfactory results for this, but the metaheuristic achieves a more balanced result. The global optimization software method balances both objectives adequately, but these results could generate a higher monetary cost, since the concepts of distance traveled, and the money spent on these trips are linked. The application of the heuristic to the instances results in good results for the objective related to the distance, but poor results for the objective related to the differences in growth goals. The heuristic and the metaheuristic are fast enough to solve the full instance in reasonable time.

For larger instances, which cannot be solved in the optimization software or require a long solution time, there is the option to use the heuristic method knowing that it will greatly benefit to minimize the distances, but not to balance the goals. The metaheuristic achieves a better balance between the objectives with reasonable computational time. When compared to the numbers that would result in conserving the current assignment of the company, the metaheuristic results better, although the heuristic has a good behavior also. Now, these developed solutions respond to planned objectives of the company, instead of having a sales structure inherited from ancient executives that designed the system with unknown aims.

The contributions of this work are two: the most important is the modeling of an optimization problem to solve a real situation for the sales department of a large company, and the second is the development of a heuristic and a metaheuristic to solve this problem. The results obtained were presented to the company for future implementation.

ACKNOWLEDGMENT

This research was supported by Universidad Panamericana [grant number UP-CI-2018-ING-GDL-06].

REFERENCES

Alves, M., & Climaco, J. (2000). An Interactive Method for 0—1 Multiobjective Problems Using Simulated Annealing and Tabu Search. *Journal of Heuristics*, *6*(3), 385–403. doi:10.1023/A:1009686616612

Blum, C. (2005). Ant colony optimization: Introduction and recent trends. *Physics of Life Reviews*, *2*(4), 353–373. doi:10.1016/j.plrev.2005.10.001

Brockhoff, D., & Zitzler, E. (2006). Are all objectives necessary? On dimensionality reduction in evolutionary multiobjective optimization. In PPSN IX, Reykjavik. LNCS 4193 (pp. 533–542). Springer. doi:10.1007/11844297_54

Coello, C. (2000). An Updated Survey of GA-Based Multiobjective Optimization Techniques. *ACM Computing Surveys*, *32*(2), 109–143. doi:10.1145/358923.358929

Deb, K. (2001). *Multi-Objective Optimization Using Evolutionary Algorithms*. Chichester, UK: Wiley.

Deb, K. (2014) Multi-objective Optimization. In Search Methodologies (pp. 403-449). New York, NY: Springer Science+Business Media. doi:10.1007/978-1-4614-6940-7_15

Deb, K., & Saxena, D. K. (2005). On finding pareto-optimal solutions through dimensionality reduction for certain large-dimensional multi-objective optimization problems. Kangal report no. 2005011, Kanpur Genetic Algorithms Laboratory (KanGAL).

Delahaye, D., Chaimatanan, S., & Mongeau, M. (2019). Simulated Annealing: From Basics to Applications. In M. Gendreau & J.-Y. Potvin (Eds.), *Handbook of Metaheuristics* (pp. 1–35). Cham, Switzerland: Springer International Publishing AG. doi:10.1007/978-3-319-91086-4_1

Doerner, K., Gutjahr, W., Hartl, R., Strauss, C., & Stummer, C. (2004). Pareto Ant Colony Optimization: A Metaheuristic Approach to Multiobjective Portfolio Selection. *Annals of Operations Research, 131*(1-4), 79–99. doi:10.1023/B:ANOR.0000039513.99038.c6

Dorigo, M., & Gambardella, L. (1997). Ant Colony System: A Cooperative Learning Approach to the Travelling Salesman Problem. *IEEE Transactions on Evolutionary Computation, 1*(1), 53–66. doi:10.1109/4235.585892

Fonseca, C., & Fleming, P. (1993). *Genetic Algorithms for Multiobjective Optimization: Formulation, Discussion and Generalization.* Retrieved from http://citeseerx.ist.psu.edu/viewdoc/download?doi=10.1.1.48.9077&rep=rep1&type=pdf

Gliesch, A., Ritt, M., & Moreira, M. C. O. (2018). A Multistart Alternating Tabu Search for Commercial Districting. *Lecture Notes in Computer Science, 10782*, 158-173. 10.1007/978-3-319-77449-7_glieas11

Hillier, F. S., & Lieberman, G. J. (2010). *Introduction to operations research.* Dubuque, IA: McGraw-Hill.

Kong, Y., Zhu, Y., & Wang, Y. (2019). A center-based modeling approach to solve the districting problem. *International Journal of Geographical Information Science, 33*(2), 368–384. doi:10.1080/13658816.2018.1474472

Lee, K., & El-Sharkawee, M. (2008). *Modern Heuristic Optimization Techniques.* IEEE Press.

Lei, H., Laporte, G., Liu, Y., & Zhang, T. (2014). Dynamic design of sales territories. *Computers & Operations Research, 56*, 84–92. doi:10.1016/j.cor.2014.11.008

Lin, M., Chin, K. S., Ma, L., & Tsui, K. L. (2018). (in press). A comprehensive multi-objective mixed integer nonlinear programming model for an integrated elderly care service districting problem. *Annals of Operations Research.* doi:10.100710479-018-3078-6

Resende, M., & Werneckz, R. (2004). A hybrid heuristic for the p-median problem. *Journal of Heuristics, 10*(1), 59–88. doi:10.1023/B:HEUR.0000019986.96257.50

Rios Mercado, R., & Fernández, E. (2009). A reactive GRASP for a commercial territory design problem with multiple balancing requirements. *Computers & Operations Research, 36*(3), 755–776. doi:10.1016/j.cor.2007.10.024

Rosing, K., & Revelle, C. (1997). Heuristic concentration: Two stage solution construction. *European Journal of Operational Research, 97*(1), 75–86. doi:10.1016/S0377-2217(96)00100-2

Segura-Ramiro, A., Ríos-Mercado, R., Alvarez, A., & De Alba Romenus, K. (2007). *A location-allocation heuristic for a territory design problem in a beverage distribution firm. International Conference of Industrial Engineering*, Cancun, México.

Tavares-Pereira, F., Figueira, J., Mousseau, V., & Roy, B. (2007). Multiple criteria districting problems: The public transportation network pricing system of the Paris region. *Annals of Operations Research, 154*(1), 69–92. doi:10.100710479-007-0181-5

Zitzler, E., & Kunzli, S. (2004). *Indicator-based selection in multiobjective search. In PPSN VIII Proceedings* (pp. 832–842). Springer.

Zitzler, E., Thiele, L., Laumanns, M., Fonseca, C. M., & Grunert da Fonseca, V. (2003). Performance assessment of multiobjective optimizers: An analysis and review. *IEEE Transactions on Evolutionary Computation, 7*(2), 117–132. doi:10.1109/TEVC.2003.810758

Chapter 8
Semantic Web Service for Global Apparel Business

Kamalendu Pal
City, University of London, UK

ABSTRACT

Nowadays a substantial share of the production processes of the world's apparel business is taking place in developing countries. In the apparel business, supply chain coordination needs resource and information sharing between business partners. Semantic web service computing (SWSC) provides numerous opportunities and value-added service capabilities that global apparel business requires to exchange information between distributed business partners. The ability to dynamically discover and invoke a web service is an important aspect of semantic web service-based architectures. This chapter describes the main features of an ontology-based web service framework, known as CSIA (collaborative service integration architecture) for integrating distributed business information systems in a global supply chain. The CSIA framework uses a hybrid knowledge-based system, which consists of structural case-based reasoning (S-CBR), rule-based reasoning (RBR), and an ontology concept similarity assessment algorithm.

INTRODUCTION

In recent decades, global apparel businesses have an inclination to be worldwide activity due to the economic advantage of globalization of product design and development. As a consequence, apparel businesses are operating increasingly globalized multi-tier supply chains and deliver products and services to customers all over the world. Optimizing design and manufacturing cost, an increase of outsourcing activities and globalization of markets have led to integrated supply chain planning and management processes. In this way, apparel supply chains heavily depend on their collaborative corporate partners; and they are depend on each other for resources and information sharing. Right information, at the right time makes apparel supply chain operation much more agile. It has been acknowledged by academics and practitioners that the autonomy of supply chain partners information sharing (Pal, 2018) need to keep in mind at the time of supply chain information sharing infrastructure design. There exist different approaches for sharing information within garment manufacturing supply chain partners.

DOI: 10.4018/978-1-5225-8223-6.ch008

How to establish an open, flexible integrated business information environment is the key to solve this problem. Traditional stereo-type information systems integration techniques are very tightly coupled, and these systems are not very useful in heterogeneous system integration purpose. When data exchange format or business unit process logic is modified, adaptability adjusted of system integrated to both sides is necessary. In other words, interoperability plays an important role when heterogeneous information system needs to be integrated. Service Oriented Computing (SOC) using web services and Service Oriented Architecture (SOA) offer a promising solution to this kind of system integration dilemma. It provides a framework to represent business processes as independent modules (services) with clear and accessible interfaces. This service interactions take place using a standard description language (e.g. XML) and it makes easy for integration of different services to build a supply chain business application.

These applications are using the Internet, Intranet, or any other forms of computer network forms to connect with customers, suppliers and other associated business partners. Web service has generally been pioneered by information technology (IT) companies, where demand is constantly changing, and products have very short product life cycles and short order-to-delivery times as a result. Apparel businesses successfully engaging in service-oriented applications can convert data from their back-end systems into a common readable format and thus are able to share information and conduct online transactions with their business partners via the data communication network. It also encompasses the use of innovative business concepts, such as dynamic pricing through software agent-based negotiation mechanism (Pal & Karakostas, 2016), competition via purchasing consortia and direct online sales to customers.

In a distributed web service environment, service discovery is one of the main functionalities to locate the desired services. Service discovery is a process of finding the desired service(s) by matching service descriptions against service requests. A service description provides service-related information which can be advertised by a service provider and search during service discovery process. Such information usually includes functional properties and non-functional properties.

Ontologies are the basis for adding semantic expressiveness to service descriptions and requirements. Ontology is an explicit and formal specification of a shared conceptualization. A service ontology is accordingly an explicit and formal specification of core concepts of the functional and non-functional properties of service. A domain ontology (or domain-specific ontology) models a specific domain and represents the meanings of terms as they apply to that domain. Ontological relations such as "is-subclass-of" or "part-of" are used for ontological inference.

Semantic service discovery is a service discovery process based on ontology concepts. By using ontology concepts defined in a service ontology expressively in a service description, semantics of the service description can be defined. These service descriptions are therefore expressive semantic descriptions. At the same time, by having both ontology-based descriptions and requirements, an ontology-enhanced reasoning engine (i.e. capable of ontological inference) can be used to locate services automatically and accurately.

Hence, the problem requires a methodical approach which has specific knowledge for capturing the web service execution experiences and appropriate reasoning mechanisms based on the enhanced service descriptions. Semantic web services empower web services with semantics. Moreover, the popularity of semantic web service-based computing (Berners-Lee et al, 2001) has attracted attention to the area of service modeling. For example, one of the main research projects including the US-based initiative – Ontology Web Language Service (OWL-S) (Martin et al, 2004). European projects include DIP (Data, Information, and Process Integration with Semantic Web Services), SUPER (is security-focused research project using social media in emergency management), and SOA4All (a project that provides a comprehensive global service delivery platform) (Roman et al, 2015). In these and other projects, researchers have proposed several frameworks for semantic web services, especially WSD-S (Web Service Description Language – Semantics) (McIlraith et al, 2003) (Martin et al, 2004); and WSMO (Web Service Modelling Ontology) (Romana et al, 2005). Despite these efforts, web service discovery is still a complex task and is difficult to implement manually. Hence, semi-automated or fully dynamic web service discovery presents a real research challenge. To address the problem, this paper describes the functionalities of a hybrid knowledge-based service concept matching framework, CSIA, which uses Structured Case-Based Reasoning (S-CBR) and Rule-Based Reasoning (RBR). The remainder of the chapter is organized as follows. Section 2 outlines supply chain management business activities description. Section 3 introduces briefly web service concept and motivation. Section 3 describes the system architecture of CSIA and its service concept similarity assessment algorithm. This includes a retail supply chain management business scenario and the service concept similarity assessment algorithm; including its evaluation. Section 4 presents a review of relevant research approach for web service discovery. Section 5 provides the future research directions of this current project. Section 6 ends with concluding remarks.

SUPPLY CHAIN MANAGEMENT BUSINESS ACTIVITIES

A global apparel supply chain consists of interconnected business facilities and their associated activities work together to provide value-added service to its customers. Customer-engaged supply chains always need different stakeholders' information for their supply chains. An entire network of manufacturers and distributors, transportation and logistics agencies, financial institutions, warehouses and freight-forwarders work together to make sure that the right goods and services are available at the right price, where and when the customers want them. Having supplied value-added services (e.g. products and associated customer services), the supply chain does not terminate. The global apparel supply chain is comprised of several steps from the front end, through the customer request, supply chain order processing initiation, quality assurance assessment for products and services, customer support facilities, to maintenance and repair facilities. Textile and garment manufacturing companies are investing state-of-the-art practices to optimize both the cost and operational efficiency of their supply chain.

In a typical apparel supply chain, raw materials (e.g. cotton, yarn, fabric, interlining, buttons, garments, packaging, tagging) are purchased from suppliers and products are manufactured at one or more manufacturing plants. Then they are transported to intermediate storage (e.g. warehouse, distribution centre) for packing and shipping to retailers or customers. The path from supplier to customer can include a number of intermediaries such as wholesalers, warehouse, and retailers, depending on the products and markets. In this way, supply chain management relates to business activities such as inbound and outbound transportation, warehousing, and inventory control. Importantly, it also embodies the information system necessary to monitor all of these business activities. Figure 1 shows a simple diagrammatic representation of a textile and garment manufacturing supply chain, which highlights some of the main internal business activities.

Increased internalization of apparel industries is changing the potential practices of global manufacturing and distribution supply chains, and many retail companies have adopted new business models, either by outsourcing or by establishing business-alliances in other countries. Globalization has also led to changes in operational practices, where products are manufactured in one part of the world and sold in another. The garment manufacturing supply chain has become more global in its geographical scope;

Figure 1. A diagrammatic representation of a supply chain

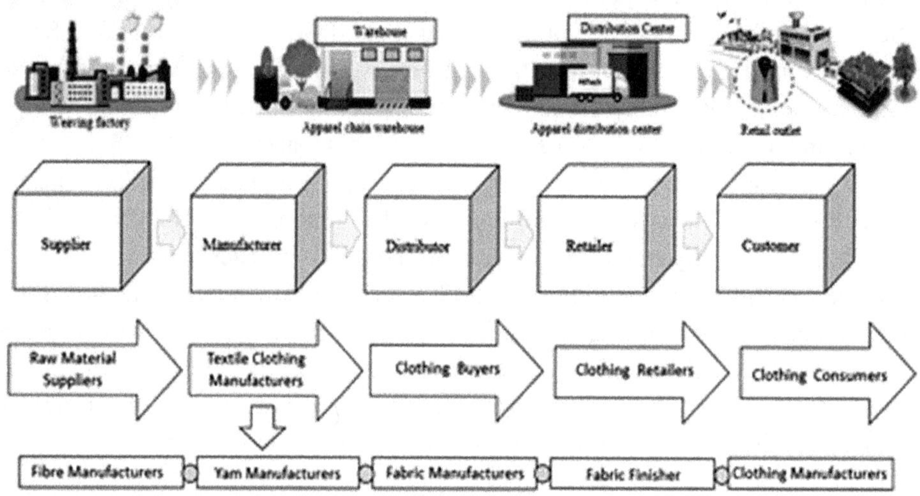

the internal market it getting more competitive and customer demand oriented. Customers are looking for more variety as well as better quality assured products and services. Effective garment manufacturing supply chain management is therefore increasingly appreciated by business today.

OVERVIEW OF WEB SERVICE AND MOTIVATION

Web services have become the popular choice for the implementation of service delivery systems which are distributed and interoperable. These services are built by a set of core technologies that cater the functionalities for communication, description, and discovery of services. The standards that provide these functionalities are Simple Object Access Protocol (SOAP), Web Services Description Language (WSDL), and Universal Description, Discovery, and Integration (UDDI) (OASIS, 2004). These XML-based standards use common Internet Protocols for the exchange of service requests and responses. (Extensible Markup Language, XML, is a common platform-independent data format across the enterprise) Figure 2A shows the relationship of these technologies as a standard stack for web services, and Figure 2B describes briefly service publishing, service requesting and service finding mechanisms using a simple diagrammatic representation.

When a service provider creates a new service, it describes the service using standard WSDL, which defines the service in terms of the messages to be exchanged between services and how they can be found by specifying the location of the service with an appropriate Universal Resource Locater (URL). To make the service available to consumers, the provider registers the service in a UDDI registry by supplying the details of the service provider, the category of the service, and technical details on how to bind to the service. The UDDI registry will then maintain pointers to the WSDL description and to the service. When a consumer wants to use a service, it queries the UDDI registry to find the service that matches its needs and obtains the WSDL description of that service, as well as the access point of the service. The consumer uses the WSDL description to construct a SOAP message to be transported over HTTP (Hyper Text Transmission Protocol) with which to communicate with the service.

Web services are loosely coupled software components that are published, located and invoked across a network computing infrastructure. Software-based web services are the building blocks for Service Oriented Computing (SOC), and they can be composed to provide a coarse-grained functionality and to automate business processes. In addition, technological improvements are providing more advanced communication facilities (e.g. online vendor managed inventory replenishment, payment using mobile hand-held devices). Business service facilities are providing more flexibility to its end-users and at the same time managing these business processes are becoming more complex. In SCM, many applications can be built by calling different web services available on the web or corporate intranets. These applications are highly dependent on discovering of correct web services. In particular, the description of web service consists of the technical parameters, constraints and policies that define the terms to invoke the web service. A web service definition needs four important things – *name, description, input* and *output*. *Name* provides business service name and it is used as a unique identifier; *description* represents the brief outline of the service; the *input* consists of number of parameters; and the *output* is also represented by a set of service parameters. SOAP based protocol provides the mechanism to exchange structured information in a decentralized and distributed information system.

In this way, web services aim to use the Web as a world-wide infrastructure for distributed computation purposes in order to carry out seamless integration of business processes. However, as the set of available web services increases, it becomes crucial to have automated service discovery mechanisms to help in finding services that match a requester's requirement. Finding appropriate web services depends on the facilities available for service providers to describe the capabilities of their services, and for service requesters to describe their needs in an unambiguous form that is ideally machine-readable. In order to achieve this objective, ordinary web service descriptions need to be enriched using domain ontology (or semantic mark-up). The next section introduces the concept of semantic annotation mechanisms of web services.

Semantic Web Service and Ontology

Semantic web service is an emerging information technology (IT) paradigm in which the main goal is to realize the development of distributed applications in a heterogeneous business environment. It is built on top of the Web Services technology that provides means for software development that enable

Figure 2. Diagrammatic representation of web service technologies

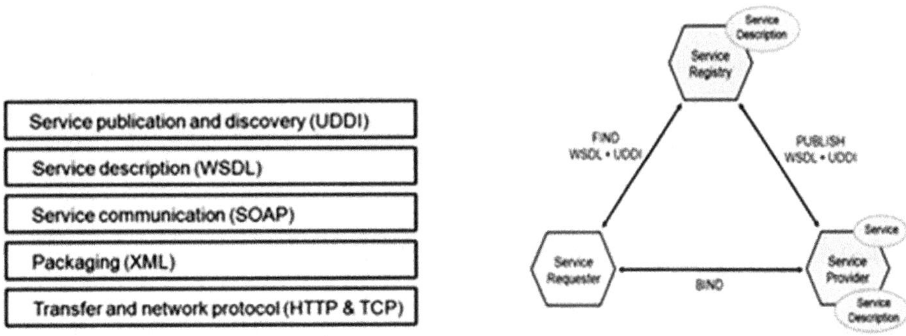

Figure 2A Web service standards stack

Figure 2B Web service relationships diagram

dynamic, execution-time discovery, composition, and invocation of Web Services. The underlying technology is extended with rich semantic representations developed in the area of the Semantic Web and with capabilities for automatic reasoning developed in the field of artificial intelligence.

Semantic Web is the new vision of the Web whose main goal is to make Web contents not only human readable but also machine readable and processable. The Semantic Web is not a separate Web but an extension of the current one, in which information is given well-defined meaning, better enabling computers and people to work in cooperation. The essential property of the World Wide Web is its universality. The power of a hypertext link is that "anything can link to anything". Web technology, therefore, must not discriminate among different types of information. On the contrary, information varies along many axes. One of these is the difference between information produced primarily for human consumption and that produced mainly for machines.

The Semantic Web brings structure to the meaningful contents of Web pages, creating an environment where software agents, roaming from page to page, can readily carry out sophisticated tasks for users. So, for example, such an agent coming to the Web page of the Department of Logistic Science will know not only that the page has keywords such as "Department, people, research activities, publications" but also that Professor Kulkarni works at this Department on Tuesdays, Thursdays, and Fridays and that the script takes a date range in dd-mm-yyyy format and returns his appointment times. These semantics were encoded into the Web page using off-the-shelf software for writing Semantic Web pages along with resources listed on the Company's site.

Like the Internet, the Semantic Web must be as decentralized as possible. Decentralization requires compromises: The Web had to throw away the idea of the total consistency of all its interconnections, ushering in the information message "Error 1001: Not Found" but allowing unchecked exponential growth. Traditional knowledge-representation systems typically have been centralized, requiring everyone to share the same definition of common concepts such as "parent" or "vehicle". But central control is stifling and increasing the size and scope of such a system rapidly becomes unmanageable. On the contrary, in a Semantic Web context, paradoxes and unanswerable questions are accepted as the price that must be paid to achieve versatility.

However, decentralization raises an important issue: different identifiers could be used for the same concept, consequently it is necessary to have a way to know that different identifiers mean or refer to the same thing. For example, consider two databases that refer to *'jack'* and *'coat'* respectively, a program that wants to compare or combine information across the two databases must know that these two terms are being used to mean the same thing. Ideally, the programs must have a way to discover such common meanings for whatever databases it encounters. A solution to this problem is provided using collections of information called *ontologies*. In philosophy, an ontology is a theory about the nature of existence, of what types of things exist; ontology as a discipline studies such theories. Artificial intelligence and web service researchers have co-opted the term for their own jargon. For them, an ontology is a formal definition of the semantics of the resources and relationships. In information management, the term ontology has a particular meaning: "An ontology is an explicit specification of a conceptualization" (Gruber, 1993).

Ontology: Meaning, Usage, and Representation

Ontologies have shown to be the right answer to the Semantic Web vision, by providing a formal conceptualization of a domain that is shared and reused across domains, tasks, and group of people. Their role is to make semantics explicit. Particularly, ontologies describe domain theories for the explicit representation of the semantics of the data. The semantic structuring achieved by ontologies differs from the superficial composition and formatting of information (as data) offered by relational and XML (eXtensible Markup Language) databases. Within a database context, virtually, all the semantic content must be captured in the application logic. On the country, ontologies can provide an objective specification of domain information, by representing a consensual agreement on the concepts and relationships characterizing the way knowledge in that domain is expressed. The result is a common domain of discourse available on the Web, that can be interpreted further by inference rules and application logic. Note that ontologies put no constraints on publishing (possibly contradictory) information on the web, only on its (possible) interpretations.

Ontologies may vary not only in their content but also in their structure and implementation. Building an ontology means different things to different practitioners. Indeed, an ontology could be used for describing simple lexicons or controlled vocabularies to categorically organized thesauri and taxonomies where terms are given distinguishing properties, to full-blown ontologies where these properties can define new concepts and where concepts have named relationships. Ontologies also differ with respect to the scope and purpose of their content. The most prominent distinction is between the domain ontologies describing specific fields of endeavour like a retail textile product, and upper-level ontologies describing the basic concepts and relationships invoked when information about any domain is expressed in natural language. The synergy among ontologies (exploitable by a vertical application) springs from the cross-referencing between upper-level ontologies and various domain ontologies.

Building an ontology means to distinguish between two different structural components: The Terminological component and the *Assertional* component. The terminological component is roughly analogous to what it is known as the schema for a relational database or XML document. It defines the terms and structure of the ontology's area of interest. The assertional component populates the ontology further with instances or individuals that manifest that terminological definition. To build an ontology, a number of possible languages could be used, including general logic programming language like Prolog (PROgramming in LOGic). However, in the last few decades, the Web Ontology Language (OWL) has become the de-facto standard for the knowledge representation in the Semantic Web. It is based on a logic thought to be especially computable, known as Description Logics (DLs) (Baader & Nutt, 2003). It is a fragment of First-Order Logic.

Web Service Description of a Business Scenario

In inventory management system, different materials need to be procured for garment manufacturing supply chain management purpose. Material attribute ontology design can be viewed from higher perspectives, such as semantic meanings or logical reasoning. However, in this chapter, focus is on one pragmatic perspectives: as a definition of concepts (or taxonomies) in the domain and related relations. To illustrate the functionalities of domain ontologies, a simple business scenario has been used to demonstrate the activities.

An Example of Business Case for Matchmaking Algorithm

A simple 'retail sales scenario' is used to describe the implemented system functionalities. It envisions an application running on a mobile computer that allows its user to purchase Jackets from an online business. This example considers how a request is matched with service advertised for jacket selling service. An algorithm tries to perform semantic matching for relevant Jersey. The algorithm is shown in ALGORITHM 1, and it takes two ontological concepts, root node ($Root_N$), and the concepts graph (G) as input and computes a semantic similarity between the concepts as output. The part of ontology hierarchy used in this example is shown in Figure 4. Each node of this hierarchy represents a concept. In the experimental comparison, semantic similarity among Jersey, Waistcoat, Sweater, Vest, Cardigan, Pullover, and Jumper are considered.

Semantic Web Services and Case Based Reasoning

Semantic web service initiatives define information systems infrastructure, which enrich the human-readable data on the Web with machine-readable annotations thereby allowing the Web to evolve into the world's biggest information repository which can be accessible from anywhere, anytime throughout the world. To achieve these objectives, one main issue would be the *markup* of web services to make them computer-interpretable. Within this markup and semantically enhanced service descriptions, powerful tools should be facilitated across the *web service lifecycle* (Papazoglou, 2012). Web services lifecycle includes automatic web service discovery to find either a web service that offer a service, or a web service that is sufficiently like be used to the current service request; and automatic web service composition and interoperation that involves the run-time service selection, composition, and interoperation of appropriate web services to complete some business activity, given a high-level abstraction of service description.

At the same time, another research community has intensively been working about similarity based *retrieval* and *adaptation* of past solutions to match new problems: two main aspects in the working semantic web service lifecycle. Case-Based Reasoning (CBR) is one of thriving applied computing community is propagating the idea of finding a solution of a problem based on experience of similar type of problems. CBR systems are a analogical reasoning systems (Liang & Konsynski, 1993). It has got diverse applications in many fields, such as classification system for credit card transactions (Reategui & Campbell, 1994) and decision support systems for business acquisitions (Pal & Palmer, 1999). The aim of CBR systems is to infer a solution for a problem in hand from solutions of a set of previously solved similar problems. Attempting to imitate the way human reason, this technique solves problems by using or adopting solutions of previously-solved old problems to solve new ones. A CBR system consists of a case base, which is the set of all previously solved cases that are known to the system. The case base can be thought of as a specific kind of knowledge that contains only *cases* and their *solutions*. There are mainly four main stages in CBR life cycle and they are:

- **Case Representation:** A case is a contextualized piece of knowledge representing an experience. Since a problem is solved by recalling an experience suitable for solving the new problem in hand, the case search and matching processes need to be both effective and reasonably time efficient. Moreover, since the experience from a problem just solved must be retained in some way, these needs also apply to the method of integrating a new case into the case collection. In this way, CBR is heavily dependent on the structure and content of its collection of cases.
- **Case Storage and Indexing:** Cases are assigned indices that express information about their content, then stored in a case library.

- **Case Retrieval:** An important step in the CBR cycle is the retrieval of previous cases that can be used to solve the target problem. Whenever a new problem needs to be solved, the case library index is searched for cases which can be a potential solution. The first phase of this search is case retrieval with the aim of finding the cases which are contextually like the new problem. The case retrieval task starts with a problem description, and ends when a suitable matching previous case has been found. Its subtasks are referred to as Identify Features, Search, and Select best possible cases from the system's repository.
- **Case Matchmaking and Use:** Matchmaking performs the comparison between the similar cases and the new request to verify if the possible solution is the one applied to prior cases. The past solutions may be reused, directly or through adaptation, in the current situation.

CBR systems typically apply retrieval and matching algorithms to a case base of past problem-solution pairs. Many successful research and industry results are paving the way of CBR in software development and deployment practice. In recent years, *ontologies and descriptive logics* (DLs) have become systems of interest for the CBR community. Many multinational organizations (e.g. IBM, British Airways, Volkswagen, NASA, and so on) are using CBR technique for their knowledge intensive business operations (Watson, 1997). Moreover, some the real-world CBR applications are taking advantage of the descriptive logics (DLs) reasoning mechanisms for the processes involved in the CBR cycle. However, several different approaches are considered, they all focus on the intuition that the formal semantics and the capabilities of DLs to maintain a terminological taxonomy are interesting properties to measure similarity and to manage a case base.

In CSIA, efforts of the semantic web services lifecycle management and CBR cycle are trying to find synergies between both. Given a certain requirement describing the user goals, automatic web service discovery typically uses a dedicated inference mechanism to answer queries conforming to the logic formalism and the terms defined in the ontology.

PROPOSED SYSTEM FRAMEWORK

This section briefly presents the overall architecture of CSIA system and illustrates the interplay of the different components. The computational framework of CSIA is shown in Figure 3. It uses a relational similarity assessment measure between implicitly stated concepts. The proposed framework accepts the service consumer request which consists of the requirements of new service (e.g. input, output, precondition, and so on). Next, the user requirement information is parsed for further processing; and final semantically ranked web services are presented to service consumer. The dynamics of CSIA are as follows:

- Initially, the service repository is populated with semantically enriched web service descriptions for specific application area within a supply chain.
- The service requester inputs the service requirements using CSIA's interface.
- The service matchmaking module takes the retrieved cases and the annotation of problem description from the semantic description generator module (within the system framework), run them through a matchmaking algorithm and forward the closest match web service to the requester.

Figure 3. Diagrammatic representation of the CSIA

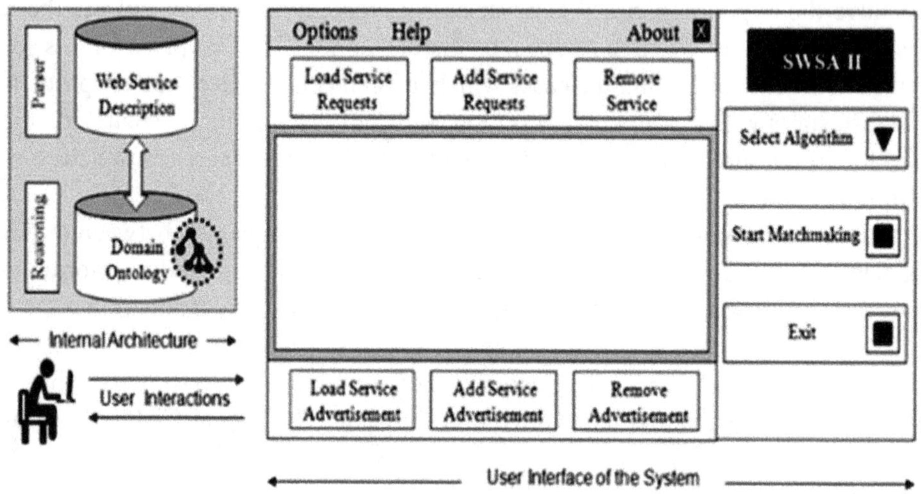

The ontologically enhanced web service descriptions are manually encoded in CSIA service repository. In the processing of ontological concept matching, when dealing with similarity between concepts, it not only considers inheritance (i.e. the relationship between supper-class and subclass) relations, but also considers the distance relationship between concepts. In CSIA, based on the comprehensive consideration of the inheritance relations and semantic distance between concepts, a concept similarity matching method based on semantic distance has been used. The CSIA uses structural case-based reasoning (S-CBR) for services and the relevant ontological concepts storage purpose; and it uses a rule-based reasoning (RBR) for service similarity assessment. The algorithm, as shown in Figure 4, is used to discover semantic web services advertised within CSIA.

In CSIA, the similarity between two concepts C_i, C_j can be expressed by a number, and it values can fall somewhere between 0 and 1. It may be viewed as a one-directional relation, and its larger values imply higher similarity between the concepts. The concept similarity is described as follows:

Concept Similarity: An ontological concept (C) similarity (∂) is considered as a *relation* and it can be defined as ∂: C x C → [0, 1]. In simple, it is a function from a pair of concepts to a real number between *zero* and *one* expressing the degree of similarity between two concepts such that:

$$\forall C_1 \in G, \ \partial\left(C_1 , C_1\right) = 1$$

$$\forall C_1, C_2 \in G, \ 0 \leq \partial\left(C_1 , C_2\right) \leq 1$$

$$\forall C_1, C_2, C_3 \in G, \ IF \ Sim_d\left(C_1, C_2\right) > Sim_d\left(C_1, C_3\right) \ THEN \ \partial(C_{1,} C_2) < \partial\left(C_1, C_3\right)$$

The above properties provide the range of semantic similarity function $\partial\left(C_i , C_j\right)$. For exactly similar concepts the similarity is $\partial\left(C_1 , C_1\right) = 1$; when two concepts have nothing in common, their similarity is $\partial\left(C_1\ ,\ C_2\right) = 0$. In this way, the output of similarity function should be in closed interval [0, 1]. Here Sim_d represents the semantic distance and (C_1, C_2, C_3) represent three concepts of graph G. In CSIA, the following semantic similarity (∂) function has been used for computation purpose:

$$\partial\left(C_1,C_2\right) \quad = \quad \frac{1}{\deg^* \ Sim_d\left(C_1,C_2\right)+1}$$

where C_1 and C_2 represent two concepts and 'deg' represents the impact degree of semantic distance on semantic similarity, and it should be between $0 < \deg \leq 1$. A weight allocation function is used, as shown below, to compute the semantic similarity between concepts:

$$w\left(C_m,\ C_n\right) \quad = \quad [\quad \max(depth(C_m)) \quad + \quad \frac{OrderNumber\left(C_n\right)}{TNodes\left(G\right)\ \ +1} \quad +1 \quad]^{-1}$$

where, $C_m\ and\ \ C_n$ represent two nodes directly connected, $\max(depth(C_m))$ represents the maximum depth of the node C_m (the depth of the root node is equal to 0 and 1 for the nodes directly connected to the root node and so on), TNodes(G) and OrderNumber(C_n) represent the total number of nodes in concept graph G and the order number of the node (C_n) between their siblings.

In Table 1, (a) is the result of synonymy similarity (Giunchiglia et al, 2004), (b) tabulates the results of Jian and Conrath similarity (Jiang & Conrath, 1997) results, (c) tabulates the results of path similarity (Varelas et al, 2005), and (d) tabulates the results of the proposed Algorithm-I used in CSIA. In this experiment a suitable value for *deg* parameter is considered.

The proposed algorithm (i.e. Algorithm-I) provides semantic similarity between concepts with a high score in comparison to Algorithm-I. In CSIA similarity measure is flexible and customizable, allowing the consideration of user preferences. This refers two aspects. Firstly, by means of advanced search interface, the user may determine the relative importance of some of similarity assessment parameter. Second, apart from presenting a single rank for each candidate service, more detailed results may also be provided (e.g. separate values for recall, the degree of match) to facilitate the user in identifying the more suitable service.

As shown in Table 1, synonymy similarity measure can only find similarity between the same concepts, and Jian and Conrath' similarity measure is better than the synonymy similarity measure. The path similarity measure and CSIA's used method are better than the above two methods. The path similarity measure can find the semantic similarity between concepts, but the similarity score is low. The CSIA's similarity method can also get the semantic similarity between concepts, and the similarity score is high.

RELATED RESEARCH WORKS

The semantic web approaches to web services give business community the ability to describe the semantics of web services and their capabilities in a formal and machine-processable manner. Most the current approaches (e.g. OWL-S, WSDL-S, and WSMO) enhancing web services using semantic tagged

Algorithm 1. Algorithm for semantic similarity computation

```
input: two concepts (C1,C2), root node (RootN), concepts graph (G)
output: semantic similarity value between two concepts
  begin
 if  C1 and C2 are same concept then Simd  = 0
  else
      if C1 and C2 are directly connected then Simd  = w (C1, C2)
       else
          if idirect path connection exist then
                Spath01 = ShortestPath (G, C1, RootN)
                Spath02 = ShortestPath (G, C2, RootN)
                Simd = w(Spath01) + w(Spath02) - 2*w(CSPath]
           end if
          ∂(C1,C2) = 1 / (deg* Simd + 1)
       end if
  end if
  return ∂
  end
```

Figure 4. The hierarchical concept relationships

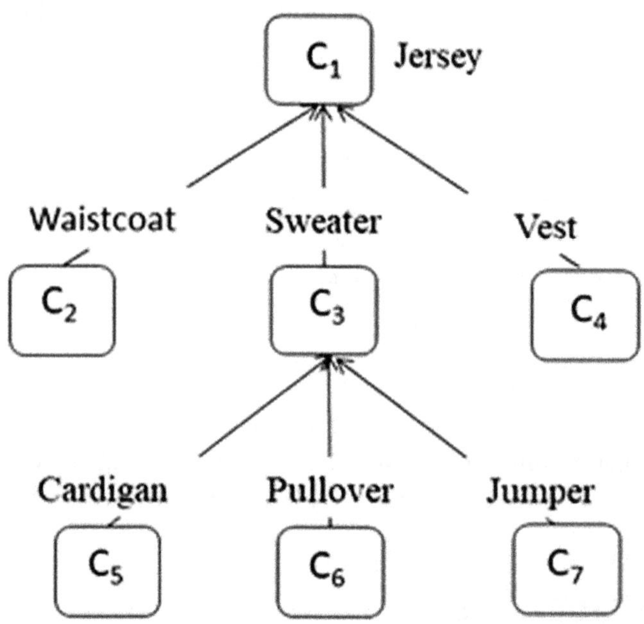

Table 1a. The results of various similarity measures: Synonymy similarity

	C_1	C_2	C_3	C_4	C_5
C_1	1.00	0.00	0.00	0.00	0.00
C_2	0.00	1.00	0.00	0.00	0.00
C_3	0.00	0.00	1.00	0.00	0.00
C_4	0.00	0.00	0.00	1.00	0.00
C_5	0.00	0.00	0.00	0.00	1.00

Table 1b. The results of various similarity measures: Jian & Conrath similarity

	C_1	C_2	C_3	C_4	C_5
C_1	1.00	0.60	0.41	0.97	0.52
C_2	0.42	1.00	0.81	0.60	0.36
C_3	0.97	0.81	1.00	0.68	0.44
C_4	0.60	0.60	0.68	1.00	0.53
C_5	0.52	0.36	0.44	0.53	1.00

Table 1c. The results of various similarity measures: Path similarity

	C_1	C_2	C_3	C_4	C_5
C_1	1.00	0.25	0.50	0.20	0.20
C_2	0.25	1.00	0.50	0.33	0.16
C_3	0.50	0.50	1.00	0.25	0.16
C_4	0.20	0.33	0.25	1.00	0.20
C_5	0.20	0.16	0.16	0.20	1.00

Table 1d. The results of various similarity measures: The proposed method

	C_1	C_2	C_3	C_4	C_5
C_1	1.0	0.48	0.65	0.51	0.38
C_2	0.48	1.0	0.65	0.51	0.38
C_3	0.65	0.65	1.0	0.71	0.48
C_4	0.51	0.51	0.71	1.0	0.59
C_5	0.38	0.38	0.48	0.59	1.0

descriptions. However, these approaches have several limitations. First, it is impractical to expect all new services to have semantic tagged descriptions. Second, descriptions of the clear majority of already existing web services are specified using WSDL and do not have associated semantics. Also, from the service requestor's perspective, the requestor may not be aware of all the knowledge that constitutes the domain. Specifically, the service requestor may not be aware of all the terms related to the service request. Because of which many services relevant to the request may not be considered in the service discovery process. Akkiraju and his research group on semantics to web services (Akkiraju et al, 2005), which is conjectured attempt to design a system which can create semantic web services is by mapping concepts in a web service description (WSDL specification) to ontological concepts. This approach is known as WSDL-S. The idea of establishing mappings between service, task, or activity description and its enrichment using domain specific ontological concepts was first introduced by Cardoso (Cardoso & Sheth, 2003). Martin has also concentrated on enhancement of service description, in work with other researchers on OWL-S (Martin et al, 2007), which is an important attempt to use lightweight web service description based on inputs, outputs and non-functional properties, to find an initial set of candidate web services for a request. Fensel has also concentrated on service enrichment and modelling, in work with Bussler on the system WSMF (Fensel & Bussler, 2002), a framework that provides the appropriate conceptual model for developing and describing web services and their composition.

Web service modelling ontology discovery framework (WSMO-DF), also contributed to sematic enrichment of web services, is based on the WSMO framework (Roman et al, 2005) for web service discovery. In the semantic web service paradigm, discovery is performed over semantic descriptions of web services. WSMO-DF and OWL-S SP (service profile) are two frameworks that generally use for service description purpose.

While these are few main style of service enrichment in semantic web service descriptions, they are by no means the only research conducted in recent past. Later work by Pengwei and his colleagues on their semantic enhancement web service research (Wang et al, 2008) has used WSMO-DF for rich web service representation purpose. In contrast to Pengwei and his research partners (Weang et al, 2008) work, CSIA follows the service profile (SP) model and uses the structural ontology information.

Li and his colleagues advocated the use of a mechanism based on the Rough sets theory to discover grid services (Li et al, 2008). The implemented software system, known as ROSSE, builds on the Rough sets theory to dynamically reduce uncertain properties when matching grid services. The evaluation results have shown that ROSSE significantly improves the precision and recall of services compared with keyword-based service matching technique and OWL-S matching. The novelty of ROSSE is in its capability to deal with uncertain properties, that is, properties that are explicitly used by one advertisement but do not appear in another service of the same category. In CSIA, only the common properties of an advertisement have been used.

Li and Horrocks have distinguished a number of things in a service matchmaking research porotype (Li & Horrocks, 2003), which uses a DL reasoner to match service advertisements and requests based on ontology enhanced service descriptions. In this project, web service descriptions are defined as complex concepts (CCs) in OWL and the matchmaking mechanism examines the subsumption relationships. FC-MATCH research project (Bianchini et al, 2006) uses the similar type of approach, performing text similarity matching using WordNet. In a research (Grimm et al., 2006), a software framework has been used to annotating web services using DLs is presents. Like CSIA, it follows the abstract web service model.

In the DAML-S/UDDI matchmaker (Sycara et al, 2003), OWL-S SP advertisements and requests refers to DAML concepts and the matching process performs inferences on the subsumption hierarchy. It uses a different definition of web service filters from CSIA and it does not consider profile taxonomies, roles or grouping filtering.

LARKS (Sycara et al, 2002) uses both syntactic and semantic matching. It uses five matchmaking filters, namely context matching, profile comparison, similarity matching, signature matching and constant matching. LARKS use its own capability description and DL language in contrast to CSIA approach.

In a hybrid, semantic web service matchmaker for OWL-S services, known as OWLS-MX (Klusch et al, 2008), researchers have used both logic-based reasoning and content-based information retrieval techniques. Experimental evaluation results show strong justification in favor of the proposition that the performance of logic-based matchmaking can be considerably improved by incorporating non-logic based information retrieval techniques into the matchmaking algorithms. However, OWLS-MX cannot handle profile taxonomies and it follows the static SP paradigm, unable to use dynamically ontology roles. iMatcher2 (Kiefer & Bernstein, 2008) follows the OWLS-MX approach, applying also learning algorithms to predict similarities. Like OWLS-MX, it uses a DL reasoner to unfold the annotation concepts, creating a vector on which the IR techniques are applied. iMatcher2 does not follow a standard matchmaking algorithm, which is defined through an iSPARQL strategy. WSMO-MX (Kaufer & Klusch, 2006) is a hybrid approach based on rich WSM service descriptions.

There are plenty of other approaches that are based on inputs/outputs, for example (Cardoso, 2006) (Pathak et al, 2005) (Skoutas et al, 2007). These approaches retrieve directly the input/output annotations and any taxonomical knowledge from special properties, such as service categorization. Moreover, they do not consider roles, using static annotation concepts, and do not apply further filtering on results (grouping filtering). METEROR-S (Verma et al, 2005) follows the WSDL-S approach, where WSDL constructs point to ontology concepts.

Thakker, Osman and Al-Dabass have reported their research in which the web service execution experiences are modeled as cases that represent the web service properties in a specific area described using OWL semantic description (Osman et al, 2006a) (Osman et al, 2006b). In this research, the service repository administrator performs the storing of service description using ontology enhanced semantics for case representation. This representation is used to semantically annotate the users' queries looking for adequate services as well as the web service execution experiences in the given area. The proposed system uses frame structures to model its cases. These structures are not generalized and depend heavily on the application domain so that the constituent elements of these structures more precisely 'slots' differ from one application to another.

Lamjmi and co-researchers have proposed the WeSCo_CBR approach based mainly on ontologies and case-based reasoning meant for the web service composition (Lajmi et al, 2006a) (Lajmi et al, 2006b). They have created an ontology that describes various features of a web service using OWL representation formalism in order to bring a semi-automatic guidance for the user. To facilitate the processing, they proceed by transforming the user's query into an ontological formulation combining a set of ontology concepts. For each received new query, the reuse process consists on retrieving similar prior stored cases and eventually evaluating and storing the new case. In WeSCo_CBR, a case comprises the following three elements: a problem, a solution and an evaluation. The discovery of web service meeting the client's needs is accomplished by using similarity measures designed in according with the formalization of the problem. The most relevant case is usually determined according to its similarity with the new problem case.

To improve the web service discovery, Wan and Cao (Wang & Cao, 2007) have introduced an additional case-based reasoning component called CBR/OWL-S Matching Engine (Wang & Cao, 2007). To find out the desired web service, this matching engine uses ontologies for semantic similarity measure.

Cassar and research colleagues (Cassar et al, 2014) presented a method, that using probabilistic machine learning techniques, is capable of extracting hidden features from semantic web services description documents such as OWL-S and WSMO. They used these features to build a model that represents different types of service descriptions in a vector which enables heterogeneous service descriptions to be represented, compared or discovered.

Many web services exist without explicit associated semantic descriptions that lead to not being considered as relevant to specific user's request during web service discovery. Paliwal et al. in (Paliwal et al, 2012) tried to categorize web service descriptions semantically and enhance semantics of user's request.

Semantic web services, as the most revolutionary technology, promises to enable the machine to machine interaction on the web and to automate the discovery and selection of the most suitable web services in building service-based applications. But most of the web service descriptions are using conventional web service annotations, e.g. WSDL that lacks semantic annotations. Farrage et al. in (Farrage et al., 2013) proposed a mapping algorithm that facilities redefinition of conventional web service annotations using semantic annotations such as OWL-S.

The CSIA work has been motivated by object-oriented structural matching techniques that are used in the domain of Structural Case-Based Reasoning (SCBR) (Bergman & Schaaf, 2003), with the use of a DL reasoner to handle semantic web service descriptions and to apply an extended matchmaking algorithm. In addition, the method of allocating the weight value to concept node has been used for similarity assessment purpose.

FUTURE RESEARCH DIRECTIONS

A major limitation of the Web Service technology is that finding and composing services still requires manual effort. This becomes a serious burden with the increasing number of Web Services. To address this problem, Semantic Web researchers proposed to augment Web services with a semantic description of their functionality in order to facilitate their discovery and integration. This technology, combining Web Services and Semantic Web techniques, is referred to as Semantic Web Services.

Semantic Web Service descriptions rely on two kinds of ontologies. First, a generic Web Service ontology specifies the main aspects used to describe Web Services regardless of the domain in which they operate. For example, such an ontology provides the vocabulary to describe Web Services inputs, outputs and complex internal data flows. Second, a domain ontology provides concept from the domain of the Web Service to populate the generic description template built with the concepts of the generic ontology. These concepts denote entities in the domain of the Web Service (e.g. raw materials, transportation) as well as functionalities that can be performed by services in the given domain (e.g. OrderRawMaterial, ArrangeTransport). Researchers collectively refer to generic and domain ontologies employed for Web Service descriptions as Web Service ontologies.

While domain ontologies play an important role when building Semantic Web Service descriptions, little research has concentrated on ways to automate their acquisition. No guidelines and no tools exist to support the acquisition of these ontologies. In future this research aims to concentrates supply chain management ontology acquisition of Web Service domain ontologies. It will start by analysing the problem of ontology learning in the context of Web Services. It will try to identify a set of characteristics that constrain the development of an automated solution and design a framework for ontology learning by considering these characteristics.

CONCLUSION

The availability of sophisticated Web Service discovery mechanisms is an essential perquisite for increasing the levels of efficiency and automation of distributed business processes in global supply chain management. The Semantic Web is well recognized as an effective infrastructure to enhance visibility of knowledge on the Web. The core of the Semantic Web is ontology, which is used to explicitly represent real-world business concept conceptualizations. Ontology engineering in the Semantic Web is primarily supported by languages such as RDF, RDFS and OWL. This chapter discusses the requirements of ontologies in the context of the Web Service, and it presents an ontological concept similarity assessment algorithm. In addition, this chapter describes the main features of an ontology-based Web Service framework, known as CSIA, for integrating distributed business information systems in a global apparel supply chain. The CSIA framework uses a hybrid knowledge-based system, which consists of Structural Case-Based Reasoning (S-CBR), Rule-Based Reasoning (RBR), and an ontological concept similarity assessment algorithm. This architecture uses ontology enhanced web service descriptions; object-oriented S-CBR knowledge representation, description logic (DL) for service formalization, and an algorithm to measure ontological concept similarity based on semantic distance. This algorithm considers not only the inheritance relation between concepts, but also the level of concepts in ontology hierarchy; and an experimental evaluation of the proposed algorithm is presented. In this architecture, ontological concepts play an important role in the development of the semantic web as a means for defining share terms in web resources in supply chain business process automation purpose. Today, business partners within global supply chains are generating huge amount raw-data; and they are collecting some of this data for business intelligence purpose. This data is commonly referred to as 'big data' – because of its volume, the velocity with which it arrives in global supply chain environment, and the variety of forms it takes. Big data is ushering a new era of corporate business intelligence promise; but one of main bottlenecks is that how to capture this raw-data and analyze it to generate meaningful information. In the future, this research will extend the current concept-based ontological reasoning framework and investigate additional mapping techniques to express service descriptions much more enhanced form by which richer service discovery will be possible for globalized supply chain world.

REFERENCES

Akkiraju, R., Farrell, J., Miller, J., Nagarajan, M., Schmidt, M., Sheth, A., & Verma, K. (2005). *Web Service Semantics - WSDL-S, A joint UGA-IBM Technical Note, version 1.0.* Retrieved from http://lsdis.cs.uga.edu/projects/METEOR-S/WSDL-S

Antoniou, G., & Harmelen, F. V. (2008). *A Semantic Web Primer*. Cambridge, MA: The MIT Press.

Baader, F., & Nutt, W. (2003). Basic Description Logics. In The description logic handbook (pp. 43-95). Cambridge University Press.

Bergmann, R., & Schaaf, M. (2003). Structural Case-Based Reasoning and Ontology-Based Knowledge Management: A Perfect Match? *Journal of Universal Computer Science, UCS, 9*(7), 608–626.

Berners-Lee, T., Hendler, J., & Lassila, O. (2001). The Semantic Web. *Scientific American, 284*(May), 34–43. doi:10.1038cientificamerican0501-34 PMID:11323639

Bianchini, D., Antonellis, V. A., Melchiori, M., & Salvi, D. (2006). Semantic Enriched Service Discovery. *International Conference in Data Engineering Workshop*, 38.

Cai, M. Zhang, W. Y., Zhang, K. & Li, S. T. (2010). SWMRD: A Semantic Web-based manufacturing resource discovery system for cross-enterprise collaboration. *International Journal of Production Research, 48*(120), 3445-3460.

Cardoso, J. (2006). Discovering Semantic Web Services with and without a Common Ontology Commitment. *IEEE Service Computing Workshop*, 183-190.

Cardoso, J., & Sheth, A. (2003). Semantic e-Workflow Composition. *Journal of Intelligent Information Systems, 21*(3), 191–225. doi:10.1023/A:1025542915514

Cassar, G., Barnaghi, P., & Moessner, K. (2014). Probabilistic matchmaking methods for automated service discovery. *IEEE Transactions on Services Computing, 7*(4), 654–666. doi:10.1109/TSC.2013.28

Copacino, W., & Anderson, D. (2003). Connecting with the Bottom Line: A Global Study of Supply Chain Leadership and its Contribution to the High Performance Business. *Accenture*, 1.

Domingue, J., Cabral, L., Galizia, S., Tanasescu, V., Gugliotta, A., Norton, B., & Pedrinaci, C. (2008). IRS-III: A Broker-based Approach to Semantic Web Service. *Journal of Web Semantics, 6*(2), 109–132. doi:10.1016/j.websem.2008.01.001

Farrag, T. A., Saleh, A. I., & Ali, H. A. (2013). Toward SWSS discovery: Mapping from WSDL to OWL-S based on ontology search and standardization engine. *IEEE Transactions on Knowledge and Data Engineering, 25*(3), 1135–1147. doi:10.1109/TKDE.2012.25

Fensel, D., & Bussler, C. (2002). The Web Service Modeling Framework WSMF. *Electronic Commerce Research and Applications, 1*(2), 113–137. doi:10.1016/S1567-4223(02)00015-7

Fugate, B., Sahin, F., & Mentzer, J. T. (2006). Supply Chain Management Coordination Mechanisms. *Journal of Business Logistics, 27*(2), 129–161. doi:10.1002/j.2158-1592.2006.tb00220.x

Ganeshan, R., & Harrison, T. P. (1995). *An introduction to supply chain management, in Supply Chain Management, Version 1*. Available from http://silmaril.smeal.psu.edu/misc/supply_chain_ intro.html

Giunchiglia, F., Shvaiko, P., & Yatskevich, M. (2004). S-Match: an algorithm and an implementation of semantic matching. *Proceedings of 1st European Semantic Web Symposium (ESWS)*, 3053, 61-75. 10.1007/978-3-540-25956-5_5

Grimm, S., Monk, B., & Preist, C. (2006). Matching Semantic Service Descriptions with Local Closed-World Reasoning. *European Semantic Web Conference*, 575-589. 10.1007/11762256_42

Gruber, T. R. (1993). *A Translation Approach to Portable Ontology Specifications*. Stanford University, Computer Science Department, Knowledge Systems Laboratory, Technical Report KSL 92-71.

Guarino, N., & Giaretta, P. (1995). *Ontologies and Knowledge Base: Towards a Terminological Classification Toward Very Large Knowledge Base: Knowledge Building and Knowledge Sharing*. Amsterdam: IOS Press.

Jiang, J. J., & Conrath, D. W. (1997). Semantic Similarity Based on Corpus Statistics and Lexical Taxonomy. *Proceedings of International Conference Research on Computational Lingustics.*

Kalakota, R., & Whiston, A. (1997). *Electronic commerce: a manager's guide*. Reading, MA: Addison Wesley.

Kaufer, F., & Klusch, M. (2006). WSMO-MX: A Logic Programming Based Hybrid Service Matchmaker. *European Conference on Web Services*, 161-170. 10.1109/ECOWS.2006.39

Kiefer, C., & Bernstein, A. (2008). The Creation and Evaluation of iSPARQL Strategies for Matchmaking. *European Semantic Web Conference*, 463-477. 10.1007/978-3-540-68234-9_35

Klusch, M., Fries, B., & Sycara, K. (2008). OWLS-MX: A Hybrid Semantic Web Service Matchmaker for OWL-S Services. *Journal of Web Semantics.*

Lajmi, S., Ghedira, C., & Ghedira, K. (2006a). How to apply CBR method in web service composition. In *Second International Conference on Signal-Image Technology and Internet Based Systems (SITI'2006)*. Springer Verlag.

Lajmi, S., Ghedira, C., Ghedira, K., & Benslimane, D. (2006b). Web_CBR: How to compose web service via case based reasoning. *IEEE International Symposium on Service-Oriented Applications, Integration and Collaboration held with the IEEE International Conference on eBusiness Engineering (ICEBE 2006).*

Laliwala, Z., Khosla, R., Majumdar, P., & Chaudhary, S. (2006). Semantic and Rule Based Event-Driven Dynamic Web Service Composition for Automation of Business Processes. *Proceedings of the IEEE Service Computing Workshop (SCW06).*

Lambert, D. M., & Cooper, M. C. (2000). Issues in Supply Chain Management. *Industrial Marketing Management*, *29*(1), 65–83. doi:10.1016/S0019-8501(99)00113-3

Li, J., Sikora, R., Shaw, M. J., & Woo Tan, G. A. (2006). Strategic analysis of inter-organizational information sharing. *Decision Support Systems*, *42*(1), 251–266. doi:10.1016/j.dss.2004.12.003

Li, L., & Horrocks, I. (2003). A Software Framework for Matchmaking Based on Semantic Web Technology. *International Conference in World Wide Web*, 331-339. 10.1145/775152.775199

Li, M., Yu, B., Rana, O. F., & Wang, Z. (2008). Grid Service Discovery with Rough Sets. *IEEE Transactions on Knowledge and Data Engineering*, *20*(6), 851–862. doi:10.1109/TKDE.2007.190744

Li, S. H. (2002). *An Integrated Model for Supply Chain Management Practice, Performance and Competitive Advantage* (PhD Dissertation). University of Toledo, Toledo, OH.

Liang, T., & Konsynski, B. R. (1993). Modeling by analogy: Use of analogical reasoning in model management systems. *Decision Support Systems*, *9*(1), 113–125. doi:10.1016/0167-9236(93)90026-Y

Martin, D., Burstein, M., Mcdermott, D., Mcilraith, S., Paolucci, M., Sycara, K., ... Srinivasan, N. (2007). Bringing Semantics to Web Services with OWL-S. *World Wide Web (Bussum), 10*(3), 243–277. doi:10.100711280-007-0033-x

Martin, D., Paolucci, M., McIlraith, S., Burstein, M., McDermott, D., McGunness, D., ... Sycara, K. (2004). Bringing Semantics to Web Services: The OWL-S Approach. *Proceeding of First International Workshop Semantic Web Services and Web Process Composition.*

Matskin, M., Maigre, R., & Tyugu, E. (2007). Computational logical semantics for business process language. *Proceedings of second international conference on Internet and Web applications and services (ICIW 2007)*, 526-531.

McIlraith, S., & Martin, D. (2003). Bringing Semantics to Web Services. *IEEE Intelligent Systems, 18*(1), 90–93. doi:10.1109/MIS.2003.1179199

OASIS. (2004). *Introduction to UDDI: Important Features and Functional Concepts.* Organization for the Advancement of Structured Information Standards.

OMG. (2009). *Business Process Model and Notation.* Retrieved from http://www.omg.org/spec/BPMN/1.2/(2009)

Osman, T., Thakker, D., & Al-Dabass, D. (2006a). Semantic-Driven Matching of Web services using Case-Based Reasoning. *Fourth IEEE International Conference on Web Services (ICWS 2006)*, 29-36.

Osman, T., Thakker, D., & Al-Dabass, D. (2006b). S-CBR: Semantic Case Based Reasoner for Web services discovery and matchmaking. *20th European Conference on Modeling and Simulation (ECMS2006)*, 723-729.

Pal, K. (2018). A Big Data Framework for Decision Making in Supply Chain Management. In Emerging Applications in Supply Chains for Sustainable Business Development. IGI Global.

Pal, K., & Karakostas, B. (2014). A Multi Agent-based Service Framework for supply Chain Management, The Fifth International Conference on Ambient System, Networks and Technologies (ANT-2014), Hasselt, Belgium. *Procedia Computer Science, 32*, 53–60. doi:10.1016/j.procs.2014.05.397

Pal, K., & Palmer, O. (2000). A decision-support systems for business acquisition. *Decision Support Systems, 27*(4), 411–429. doi:10.1016/S0167-9236(99)00083-4

Paliwal, A. V., Shafiq, B., Vaidya, J., Xiong, H., & Adam, N. (2012). Semantics-based automated service discovery. *IEEE Transactions on Services Computing, 5*(2), 260–275. doi:10.1109/TSC.2011.19

Papazoglou, M. (2012). *Web Services and SOA: Principles and Technology.* Pearson.

Pathak, J., Koul, N., Caragea, D., & Honavar, V. G. (2005). A Framework for Semantic Web Services Discovery. *ACM International Workshop on Web Information and Data Management*, 45-50. 10.1145/1097047.1097057

Patil, A., Oundhaka, S., Sheth, A., & Verma, K. (2004). METEOR-S Web service Annotation Framework. *The Proceedings of the Thirteenth International World Wide Web Conference*, 553-562.

Reategui, E. B., & Campbell, J. A. (1995). A Classification System for Credit Card Transactions. In J. P. Haton, M. Keane, & M. Mango (Eds.), *Advances in Case-Based Reasoning: Second European Workshop (EWCBR-94)*. Springer. 10.1007/3-540-60364-6_43

Roman, D., Keller, U., Lausen, H., de Bruijn, J., Lara, R., Stollberg, M., ... Fensel, D. (2005). Web service modeling ontology. *Applied Ontology*, *1*(1), 77–106.

Roman, D., Kopecky, J., Vitvar, T., Domingue, J., & Fensel, D. (2015). WSMO-Lite and hRESTS: Lightweight semantic annotations for Web services and RESTful APIs. *Journal of Web Semantics*, *31*, 39–58. doi:10.1016/j.websem.2014.11.006

Shingo, S. (1988). *Non-Stock Production*. Cambridge, UK: Productivity Press.

Skoutas, D., Simitsis, A., & Sellis, T. (2007). A Ranking Mechanism for Semantic Web Service Discovery. *IEEE Congress on Services*, 41-48.

Studer, R., Benjamins, V. R., & Fensel, D. (1998). Knowledge engineering: Principles and methods. *Data & Knowledge Engineering*, *25*(1-2), 161–197. doi:10.1016/S0169-023X(97)00056-6

Sycara, K., Widoff, S., Klusch, M., & Lu, J. (2002). LARKS: Dynamic Matching Among Heterogeneous Software Agents in Cyberspace. *Autonomous Agents and Multi-Agent Systems*, *5*(2), 173–203. doi:10.1023/A:1014897210525

Sycara, K. P., Paolucci, M., Ankolekar, A., & Srinivasan, N. (2003). Automated Discovery, Interaction and Computation of Semanticmweb Services. *Journal of Web Semantics*, *1*(1), 27–46. doi:10.1016/j.websem.2003.07.002

Varelas, G., Voutsakis, E., Raftopoulou, P., Petrakis, E. G. M., & Milios, E. (2005). Semantic Similarity methods in WordNet and their application to information retrieval on the Web. *Proceedings of the 7th annual ACM international workshop on web information and data management*. 10.1145/1097047.1097051

Verma, K., Sivashanmugam, K., Sheth, A., Patil, A., Oundhakar, S., & Miller, J. (2005). METEOR-S WSDI: A Scalable P2P Infrastructure of Registries for Semantic Publication and Discovery of Web Services. *Information Technology Management*, *6*(1), 17–39. doi:10.100710799-004-7773-4

Vrijhoef, R., & Koskela, L. (1999). Role of supply chain management in construction. *Proceedings of the Seventh Annual Conference of the International Group for Lean Construction*, 133-146.

Wang, L., & Cao, J. (2007). Web Services Semantic Searching enhanced by Case Based Reasoning. *18th International Workshop on Database and Expert Systems Applications*.

Wang, P., Jin, Z., Liu, L., & Cai, G. (2008). Building Toward Capability Specifications of Web Services Based on an Environment Ontology. *IEEE Transactions on Knowledge and Data Engineering*, *20*(4), 547–561. doi:10.1109/TKDE.2007.190719

Watson, I. (1997). *Applying Case-Based Reasoning: Techniques for Enterprise Systems*. Morgan Kaufman.

Wooldridge, M., & Jennings, N. (1995). Intelligent Agents: Theory and Practice. *The Knowledge Engineering Review*, *10*(2), 115–152. doi:10.1017/S0269888900008122

WSDL-S. (2005). Retrieved from http://www.w3.org/Submission/WSDL-S/

Zhang, W. Y., Cai, M., Qiu, J., & Yin, J. W. (2009). Managing distributed manufacturing knowledge through multi-perspective modelling for Semantic Web applications. *International Journal of Production Research*, *47*(23), 6525–6542. doi:10.1080/00207540802311114

KEY TERMS AND DEFINITIONS

Concept: A concept is a cognitive unit of meaning. Concepts are introduced to describe services in this chapter.

Ontology: Information sharing among supply chain business partners using information systems is an important enabler for supply chain management. There are diverse types of data to be shared across supply chain, namely – order, inventory, shipment, and customer service. Consequently, information about these issues needs to be shared in order to achieve efficiency and effectiveness in supply chain management. In this way, information-sharing activities require that human and / or machine agents agree on common and explicit business-related concepts (the shared conceptualization among hardware / software agents, and service providers) are known as explicit ontologies; and this help to exchange data and derived knowledge out of the data to achieve collaborative goals of business operations.

Supply Chain Management: A supply chain consists of a network of key business processes and facilities, involving end users and suppliers that provide products, services and information.

Web Service Composition: The web service composition problem aims at identifying a set of web services (and work-flow therein) such that the composition of those web services can satisfy users' goals as much as possible. Therefore, a natural objective is to identify the best composition of web services that optimizes the customer's needs in a best possible way (e.g. cost, execution time which is the process time of a service, reliability, and so on).

Web Service: A web service is a set of related activities that produce a function to consumer of the service. Web service provided via Internet or Intranet. In a simplistic sense, a web service is described by pre-conditions (inputs) and post-conditions (outputs).

XML: Extensible Markup Language (XML) is a simple, very flexible text format derived from SGML (Standard Generalized Markup Language). While XML was originally designed to meet the challenges of large-scale electronic publishing, it plays an increasingly significant role in the exchange of a wide variety of data on the web.

Chapter 9
Study on Effect of Barriers in Green Supply Chain Management Using Modified SAW Technique:
A Case Study

Jagannath Reddy
National Institute of Technology Silchar, India

Biplab Das
National Institute of Technology Silchar, India

Jagadish
National Institute of Technology Raipur, India

ABSTRACT

Nowadays along with the rapid development of industrialization across the globe, the environmental and ecological impacts of products have become a serious issue. Taking into account purely the economic impacts of industrial decisions, and excluding their ecological impacts, make the human beings and animals more at risk to many threats such as global warming, ozone layer depletion, toxic environments, and natural resources depletion. To minimize the environmental effect, implementation of green supply chain management (GSCM) is much more essential for industries in the environmental and social point of view. The purpose of this chapter is to analyze barriers to an implementation of green supply chain management in a stone crushing plant of Southern India by using modified simple additive weighting (SAW) to rank approaches. Further, this study will help the small-scale industries to understand the factors affecting implementation of GSCM in their organizations.

DOI: 10.4018/978-1-5225-8223-6.ch009

INTRODUCTION

The developing issues in the global market for green concerns and a shortage of natural resources have constrained official to observe supply chain systems from an environmental point of view. Stone crushing industry supply chain with high environmental risks, because of the mining process involves like drilling, blasting, braking, transportation and crushing process produces pollutions like air pollution, water pollution, soil pollution, noise pollution Mali et.al (2016). At present, environmental issues become real critical barriers for many supply chain corporations concerning the sustainability of their businesses. (Gilbert, 2000) the green supply chain management is developing field inspired by the requirement for environmental awareness. In this context, several studies have been proposed by both academia and industry trying to develop a new technique for analysis of barriers related to the implementation of green supply chain management (GSCM).

In the literature, researchers have been recommended various experimental criteria for green supply chain (GSC) performance measurement and projected both qualitative and quantitative frameworks. Mathiyazhagan et.al (2013b) has analyzed pressures of GSCM among 36 barriers. Wang and Chan (2013) proposed the fuzzy hierarchical approach to identify an improvements area when implementing when supply chain initiatives green. However, these are mainly operational in nature and specific to the focal company. Prior studies have applied the multi-criteria decision making (MCDM) methods used to address a few issues of existing methods. The application of MCDM has grown extensively in the past few decades, and so has the number of techniques to evaluate the alternatives and select the best of them. The scientific literature mentions that over more than 70 MCDM techniques are available (Sun & Li, 2010), to help the decision maker take an appropriate decision in different phases of life-cycle, spreading its wings through the fields of sciences, business, production engineering. The collection of data from a literature survey or published articles and mainly depends on experts opinion, this kind of data information involves many interrelation process parameters are uncertainty, inexact and ambiguity in nature this kind of data or information will affect the ranking the parameters. The most of the MCDM methods having rank reversal problem Sameer and Muzasikkr (2018). The ranking of parameters is essential to improve the green supply chain management. The addition/removal of GSCM factors leads to rank reversal; the improper ranking of parameters affects the performance of the whole green supply chain management. In this chapter, introduces a modified simple additive weighting (SAW) of MCDM technique to rank the parameters to improve the GSCM performance.

The aim of this chapter is to study and analysis of barriers in GSCM using MCDM methods. In this context, the present chapter proposed an MCDM based approach, i.e. modified simple additive weighting (SAW) method. In this chapter, first, a brief introduction of Green Supply Chain Management and its barriers to implementation in the industry has been addressed. The second section, a brief introduction to multi-criteria decision making (MCDM) techniques and issues of rank reversal illustrated. Next section, proposed a modified SAW method which uses linear deviation and helps in minimizing rank reversal chances compare to other existing methods. Thereafter, critical analysis of the implementation of GSCM carried out real-time data of the southern part of Indian (Bangalore) industry. And finally, the chapter concludes with important guidelines concerning the application of modified MCDM methods for effective utilization of process parameters for any green supply chain management and future direction of work.

BACKGROUND

A report on mining and metals has pointed out that India has immense natural resources. The mining sector is an important sector in the Indian financial system as the nation is endowed with rich mineral resources. The growth of mining industries in India, resulting from suitable policy and investment climates and favorable market demand, has unfortunately intensified environmental degradation Barve and Muduli (2011). In India, a small- scale mine is generally defined as one where maximum production capacity is < 50,000 tonnes/annum Ghose (2003a). The Indian Government, which has devoted significant attention to small-scale industries at the regulatory level, has paid little attention to its small-scale mining industry Ghose (2003c). The most of the activities involved in the quarry industry do results to health hazards are the inhalation of fumes smokes and dust in the quarry site cause a lot of diseases to workers, third party, people who live close to quarry sites.

The applications of aggregate resources they are the major raw materials used in the construction of roads, rail lines, bridges, hospitals, schools, airports, factories and homes Langer et al. (2004). Infrastructural development ensures the growth of the country's economy and revenue for the government through royalties, employment, and taxation especially of the rural population Divya et al. (2012). Thus the importance of the quarry industry provides opportunities for both skilled and unskilled workers thereby supporting many families like urban and rural. On the other hand, quarrying raises various environmental pollutions including land pollution, air pollution, noise and ground vibrations due to the blasting process and transportation of vehicles Langer et al. (2004).

The increasing environmental pollutions, social pressures to towards green supply chain management in their business. The GSCM defines environmental concern in their activities like design, production, transportation, and recycling of products Zsidisin and Siferd(2001), Diabat and Govind(2011). The industrial environmental act and practices that support the adoption of GSCM are life cycle analysis, green manufacturing, environmental management systems and eco-design Seuring (2004), Sarkis (2012). The number of articles featuring special issues on GSCM practices and several authors stated that GSCM as sustainable supply chain Jayaraman et al. (2007), Litnton et al. (2007), Seuring (2008). Literature survey reveals that mining industries are implementing environmental management systems which are also parallel practices to GSCM (Zhu et al.2010) so that they may obey with governmental policies (Ghose 2003a) obtain social license to operate and attracts financial groups and increase their eco-efficiency Hilson and Nayee (2002), Nikolaou and Evangelinos (2010), Azapagic (2004), Berkel(2007).

Various authors have contributed to research on GSCM practices in mining industries. Hilson (2000) identified various barriers to eco-friendly production in mining industries. Ghose (2003b) investigates the schematics of environmental management plans adopted by small-scale mines in India. Chaulya (2004) study focuses on environmental health and safety management practices in Indian mining industries. The rank reversal was first identified by Belton and Gear (1983) which leads to a tremendous significant argument about the validity and accuracy of analytic hierarchy process (AHP) and also demonstrated that rank reversal happens in analytic hierarchy process when alternatives added. Wang and Luo (2009) shown that the rank reversal may occur in SAW and technique for order of preference by similarity to ideal solution (TOPSIS) tool. Macharis et al. (2004) illustrated that rank reversal can happen in Preference Ranking Organization Method for Enrichment of Evaluations (PROMETHEE) family. The research of Zankis et al. (1998) and Wang and Traintaphyllou (2008) investigated that rank reversal problem in ELECTRE method. Huszak and Imre (2010) rank reversal problem may occur in grey relational analysis (GRA). Kong et al. (2016) explained that rank reversal problem in analytic network process (ANP).

Each multi criteria decision making (MCDM) method has certain strengths and weakness and has been developed to analysis with different types of issues. But the major weaknesses of these methods, like analytic hierarchy process (AHP), technique for order of preference by similarity to ideal solution (TOPSIS), simple additive weighting (SAW), analytic network process (ANP) etc., are facing rank reversal. Rank reversal phenomenon where the alternatives order of preference is altered when new alternatives are added/remove from the decision problem.

METHODOLOGY

Modified Simple Additive Weight

The most commonly used multi-criteria decision method is SAW. This method is simplest; the overall score of criteria and sub-criteria is evaluated by the weighted sum of all attributes values. The main drawback of SAW method is the rank reversal, which occur when the alternatives add or remove in the decision matrix. This proposed method approaches to minimizing the rank reversal in SAW method by preference index value. The proposed method consists of three steps: pairwise comparison of sub-criteria and criteria matrix, normalization of the pair-wise comparison matrix, Weighted normalized matrix, overall preference index value and final ranking based on preference index value.

Step 1: Pair-wise comparison of matrix

A method of constructing pairwise comparison matrix in decision making and rate the criteria on relative scale of importance by using Saatys (1980), 1-9 scale of pairwise comparisons are '1' is equal importance, '3' is moderate importance, '5' is strong importance, '7' is very strong importance, '9' is extreme importance, '2','4','6','8', are intermediate values. After that, calculating the consistency check, compute the consistency index (CI) for each matrix order n using Eq. (1). Based on the CI and Random consistency ratio (CR) is calculating using Eq. (2). The CI and RI define as follows Saatys (2008).

$$CI = \frac{(\lambda_{\max} - n)}{n - 1} \quad (1)$$

$$CR = \frac{CI}{RI} \quad (2)$$

where, CI is Consistency Index, $\lambda_{\max}$ is Max, n is the order of the matrix, RI is Random Index

Step 2: Normalization of the pairwise comparison matrix

In this step, normalization is to remove dimensional unit variations among the criteria by divided criteria value of the matrix by its column total by the following Eq. (3). Normalize criteria weighted matrix divide the sum of normalized matrix column of the matrix by the number of criteria used (n) to

Table 1. Random index for matrixes order 1 to 8

N	1	2	3	4	5	6	7	8
R.I.	0	0	0.58	.90	1.12	1.24	1.32	1.41

generate the criteria weighted matrix by using Eq.(4), based on the criteria weight assign importance of criteria, higher and lower the criteria weight values are first and last importance respectively.

$$K_{ij} = \frac{C_{ij}}{\sum_{i=1}^{n} Z_{ij}} \tag{3}$$

where C_{ij} is criteria, K_{ij} is Normalization matrix, $\sum_{i=1}^{n} Z_{ij}$ is the sum of column total of pair-wise matrix

$$U_{ij} = \frac{\sum_{i=1}^{n} K_{ij}}{n} \tag{4}$$

where U_{ij} is normalization matrix, $\sum_{i=1}^{n} K_{ij}$ is the sum of rows, n is number of criteria's

Step 3: Ranking based on the overall preference index value

The overall preference index value of each criterion or sub-criteria is determined using Eq. (5). This value shows the overall closeness of criteria with the best possible criteria for a given decision problem. The overall preference index value assists in establishing a consistent ranking hierarchy capable of minimizing the rank reversal phenomenon. The criteria or sub-criteria with lowest overall preference value will have the highest preference with the best possible criteria. The criteria with least value of P_{ij} being ranked first, followed by alternatives with increasing P_{ij}.

$$P_{ij} = \frac{\sum_{i=1}^{n} K_{\max} - K_i}{\sum_{i=1}^{n} K_i - K_{\min}} \tag{5}$$

where $\sum_{i=1}^{n} K_{\max} - K_i$ are benefits criteria, $\sum_{i=1}^{n} K_i - K_{\min}$ is nonbeneficial, P_{ij} is overall preference index value.

Stone Crushing Plant

The stone crushing industry is an essential mechanical area in the nation occupied with creating a squashed stone of different sizes relying on the fundamentals which have the different development of exercises, for example, Development of streets, Highways, Bridges, Buildings, Canals and etc. The supply chain management of stone quarry consists of raw materials from mines to crushing plant through transportation like trucks, dumpers, in the crushing process with help of primary jaw crushers, reduces the size of the rocks, with the help of belt conveyors materials transfers to secondary jaw crusher and after that screeners to separates the different size of rocks. the final product is separated based on the size, as a requirement of the customer.

There are extensive varieties in the type of stone crusher setup the people over relying upon topographical areas, kind of interest for crushed items, closeness to urban regions, type of crude materials, accessibility of plant and hardware locally and so on. Essentially the stone crusher industry division could be isolated in three classifications small, medium and large.

CASE STUDY: STUDY ON EFFECT OF BARRIERS IN GREEN SUPPLY CHAIN MANAGEMENT USING MODIFIED SAW TECHNIQUE

Background

Sunil Hi-Tech India Infra Pvt. Ltd., Karnataka this unit is located at the Manchenahalli, Gowribidanur Tq, Chikkaballapur dist in Karnataka state, India. The material is rock. The unit normally operates in 3 shifts/day. The products are used in the construction of roads, bridges, & buildings, etc. The basic detail of the unit is given below. Type of Raw material: Rock stone from open cast mines. Size and Type of primary crushers: (36″X24″), Jaw, 3 No. Size and Type of secondary crushers: (30″X15″), Jaw, 3 No. Size and Type of Tertiary crusher: (30″X6″), Jaw, 1 No, crushing capacity per day: 400 Tonnes per day

Figure 1. Stone crushing plant Supply chain

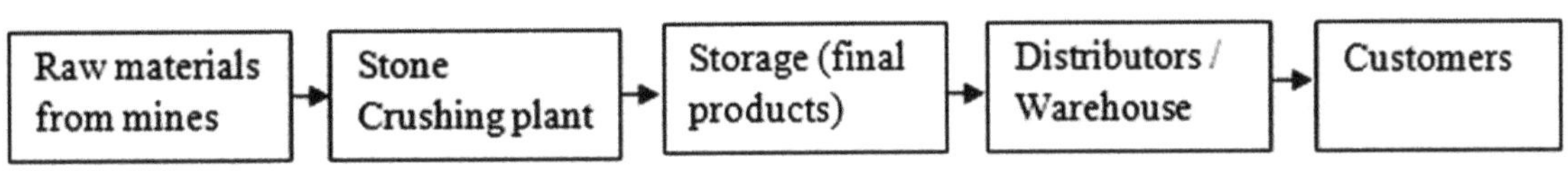

Table 2. Sizes and Capacity of stone crusher

Sl. No	Size	Capacity	Important machines
1	Small	3TPH to 25TPH	Single jaw, vibratory screens, belt conveyors.
2	Medium	25TPH to 100TPH	One/two primary & secondary crushers, vibratory screens, belt conveyors.
3	Large	100TPH to 200TPH	2-3 number of crushers i.e., primary, secondary, & tertiary crusher. 2-3 vibrating screen, automatic unloading & loading conveyor process.

(16.66TPH). Type and Screens: Mesh, 4 Nos. A number of conveyors: 6. A number of cone crusher: 2. Electricity consumed: 200KW per day. Final products are 6mm, 12mm, 20mm and powder, local names of these products dust, aggregate, kappchi, and dust.

The process description of this plant, the raw material as a rock from mines by the process of blasting, breaking and transportation to primary hopper. The raw material is feed to the primary feeder, after primary crushing the material size is reduced by the crushing process, these materials conveyed to secondary crusher by conveyor belts. Further, it transfers to the screener; the different sizes of stones can be separated by screening. The oversize material from screener sends it back to tertiary crusher by conveyor belts. Again oversize material is transferred back to secondary crusher; it goes again screening and separation of different sizes of material. The hopper stored dust is directly unloaded to tractors or trucks. The different sizes of products are discharge from screen to stockpiles by various belt conveyors. From stockpiles, the product is loaded to trucks or tractors with the help of loaders. It can be delivered to the end point of use or customers. The process flow diagram of the stone crushing plant shown in Figure 2.

Barriers to Implementation of GSCM

The adoption and development of green supply chain management activities are not significantly reduced because implementing these activities or process faces a lot of challenges or barriers. To list out these barriers, or collection of data through primary and secondary sources. The primary sources are literature survey and secondary sources are industry experts, production manager and skilled workers having more than 8 year's experience from the industry. From this process this paper considered five main criteria's or barriers in the perspectives organization commitment, Environmental concern, capacity constrained, social awareness and government regulations and 20 sub-criteria are identified as depicted in Table 3.

Application of Proposed Methodology

Analysis of these barriers by proposed methodology has been applied to study and rank the importance of various GSCM adoption barriers in the above-mentioned stone crushing industry in a Southern part of India. The collection of data regarding the above factors from the industry with the help of a ques-

Figure 2. Process flow diagram of a stone crushing plant

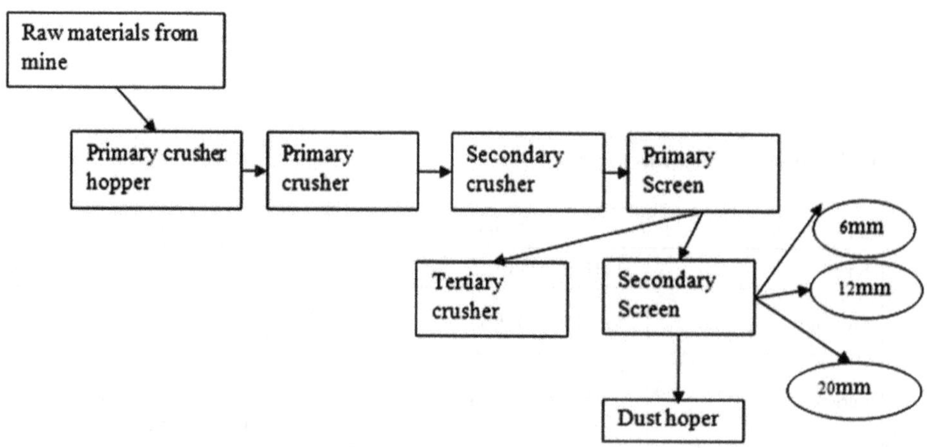

Table 3. Barriers to GSCM adoption

Factors and sub-factors	Item
1. Lack of organization Commitment	
1.Lack of employee commitment	LOC1
2.Lack of employee awareness about occupational health hazards	LOC2
3. Lack of awareness regarding cost savings due to the implementation of GSCM	LOC3
4. Lack of top management commitment to the adoption of GSCM.	LOC4
5. Lack of management awareness about new technologies	LOC5
2. Lack of Environmental Concern	
1. Reduction in air emissions, liquid & solid wastes	LEC1
2. The extent of recycling & reuse	LEC2
3. The decrease in the use of harmful/hazardous materials/ components	LEC3
3. Lack of financial constraints	
1.Increase in cost of purchasing environmentally friendly materials/products	LFC1
2. Increasing training cost	LFC2
3.Increase in investment	LFC3
4.Increase in operational cost	LFC4
4. Lack of Society Awareness	
1. Lack of community pressure	LSA1
2.The resistance offered by employees to the adoption of modern technology	LSA2
3. The absence of green consumer awareness	LSA3
4. Lack of employee motivation	LSA4
5. Lack of Government rules	
1. Lack of enforcement	LGR1
2. Lack of financial support	LGR2
3.Lack of strict supervision	LGR3
4. Changing regulations due to changing political climate	LGR4

tionnaire formulated on the basis of weighting. By using SAW method to rank the barriers or priority weighting by pairwise comparison matrix, normalization of the matrix and the final ranking by overall preference index value.

Step 1: Pair-wise comparison of criteria's and sub-criteria.

First, a pairwise comparison of sub-criteria of lack of organizational commitment, Lack of environmental concern, lack of financial investment, lack of government rules and lack of social awareness is done and their results are shown Tables 4-8 (Saatys (1980). In this step, pairwise comparison matrix is generated using 1-9 Saaty scale in which digit, 1 denotes “equally important”, 3 for “slightly more important”, 5 for “strongly more important”, 7 for “demonstrably more important” and 9 for “absolutely more important” and 2, 4, 6, 8 for “slightly differing judgment”

Table 4. Lack of organization commitment

	LOC1	LOC2	LOC3	LOC4	LOC5
LOC1	2	1	0.5	1	0.5
LOC2	1	0.5	0.33	0.5	0.33
LOC3	3	2	1	2	1
LOC4	2	1	0.5	1	2
LOC5	3	2	1	0.5	1
	11	6.5	3.33	5	4.83

Table 5. Lack of environmental concern

	LEC1	LEC2	LEC3
LEC1	1	3	1
LEC2	0.33	1	0.5
LEC3	1	2	1
	2.33	6	2.5

Table 6. Lack of financial constraints

	LFC1	LFC2	LFC3	LFC4
LFC1	1	2	0.25	0.33
LFC2	0.5	1	0.25	0.5
LFC3	4	4	3	3
LFC4	3	2	1	1
	8.5	9	4.5	4.83

Table 7. Lack society awareness

	LSA1	LSA2	LSA3	LSA4
LSA1	1	2	0.25	0.33
LSA2	0.5	1	0.25	0.5
LSA3	4	4	1	3
LSA4	3	2	0.33	1
	8.5	9	1.83	4.83

Then, a consistency check is done for all the pair-wise comparison matrices via determining the CI and CR values using Eqs. (1-2). The consistency check is done to measure the accuracy of the relative values of the pairwise comparison matrix. The result of consistency check for all the pair wise matrices is shown in Table 9.

Table 8. Lack of government rules

	LGR1	LGR2	LGR3	LGR4
LGR1	1	0.33	2	0.5
LGR2	3	1	3	2
LGR3	0.5	0.33	1	0.5
LGR4	2	0.5	2	1
	6.5	2.16	8	4

Table 9. Consistency check for all pair-wise comparison matrices

Criteria	λ_{max}	CI	CR
Lack of organizational commitment	5.223	0.055	0.049
Lack of environmental concern	3.018	0.009	0.015
Lack of financial constraints	4.159	0.053	0.059
Lack of government rules	4.071	0.023	0.026
Lack of social awareness	4.159	0.053	0.059

Step 2: Normalization of pairwise comparison

In this step, normalization is to remove dimensional unit variations among the criteria by divided criteria value of the matrix by its column total by the following Eq. (3) and sub-criteria weighting by using Eq. (4). Thereafter, sub-criteria weight value higher the value is most important(first priority) and lower the weight is least important(least priority) of lack of organizational commitment, lack of environmental concern, lack of financial constraints, lack of social awareness, and lack of government rules are shown in the following Tables10-14.

The importance of the organizational commitment of sub-criteria shown in Table 10. In this lack of awareness regarding cost saving due to the implementation of GSCM is most important sub-criteria that organization focuses on cost saving by the adoption of GSCM in their organization. Second most important is awareness of new technologies to reduce cost, time and environmental issues, followed by third and fourth importance is top management commitment towards implementation and continuous improvement of GSCM in their organization and finally works about occupational health hazards.

Table 10. Normalization and importance of lack of organizational commitment sub-criteria

	LOC1	LOC2	LOC3	LOC4	LOC5	Criteria weight	Importance
LOC1	0.181	0.153	0.150	0.200	0.103	0.157	4
LOC2	0.090	0.076	0.099	0.100	0.068	0.087	5
LOC3	0.272	0.307	0.300	0.400	0.207	0.297	1
LOC4	0.181	0.153	0.150	0.200	0.414	0.219	3
LOC5	0.272	0.307	0.301	0.100	0.207	0.236	2

Table 11. Lack of environmental concern

	LEC1	LEC2	LEC3	Criteria weight	Importance
LEC1	0.429	0.500	0.400	0.443	1
LEC2	0.141	0.166	0.200	0.169	2
LEC3	0.429	0.333	0.400	0.387	1

The environmental concern is the most important criteria in this organization and the importance of each sub criteria's shown in Table11. The most influential sub-criteria are a reduction in air pollution, water pollution, soil pollution, and solid wastes are most important to reduce emissions by their process like mining, crushing etc. And equal important of criteria, the use of hazardous materials used in their process special in blasting process and even sound pollution dangerous for wildlife and human being nearby the mining area. Least important of criteria recycling and reuse materials in the stone crushing process.

The lack of financial constraints importance of criteria weights are shown in Table 12. Increasing in investment to the implementation of GSCM is most important sub-criteria and second importance is an increase in operational cost for manpower, new equipment, and technologies implementation towards environmental friendly operations. Third importance is purchasing of environmentally friendly materials from suppliers for the production of the final product and final importance is training cost for an employee to know about new technologies and operations involved in the implementation of GSCM and continuous improvement.

In the lack of Social awareness importance, sub-criteria show in Table 13. The most important sub-criteria is lack of green consumers due to lack of awareness of green products and recycling of the product after used and by using green products can reduce environmental effects. Second, important is motivation to industry people to implement green supply chain management in their organizations and

Table 12. Lack of financial constraints

	LFC1	LFC2	LFC3	LFC4	Criteria weight	Importance
LFC1	0.117	0.222	0.056	0.069	0.116	3
LFC2	0.058	0.111	0.056	0.104	0.083	4
LFC3	0.470	0.445	0.667	0.623	0.550	1
LFC4	0.353	0.223	0.223	0.208	0.251	2

Table 13. Lack of society awareness

	LSA1	LSA2	LSA3	LSA4	Criteria weight	Importance
LSA1	0.118	0.223	0.056	0.069	0.116	3
LSA2	0.059	0.112	0.056	0.104	0.083	4
LSA3	0.470	0.445	0.667	0.622	0.551	1
LSA4	0.359	0.223	0.223	0.208	0.252	2

at the same time community pressure to reduce environmental pollution by using harmful products in their production process and conducting awareness programs for eco-friendly production process and recycling of used products. Finally, the resistance offered by employees to the adoption of modern technology to reduce environmental issues.

In the lack of government rules, the most importance sub-factor is lack of financial support from the government to implement green supply chain management in their organization which helps them balancing the financial constraints and subsidies to those who implement green supply chain management. Second important factor is changing the rules and regulations are a huge impact on the industries due to a change in government (changing political climate) and third important is lack of enforcement from the government to maintain proper environmental concerns, lack of frequent auditing for environmental management systems in industries and finally lack of monitoring system from government supervision.

The criteria's level pair-wise Comparison of lack of organizational commitment, lack of environmental concern, lack of financial constraints, lack of social awareness and lack of governmental rules are shown in Table 15. Thereafter normalization of pair-wise comparison of criteria's by using Eq.(3) and criteria weights by using Eq.(4) to assign importance of the criteria for their organization to improve (focus) are shown in Table 15.

The importance of pair-wise comparison of criteria's show in Table 16, higher the criteria weight is first importance, The lack of environmental concern is a most important barrier to the adoption of GSCM, and second higher value, the lack of organizational commitment towards green supply chain management implementation. The third most important criteria lack of financial constraints to implementing Green supply chain management and fourth and fifth criteria's are lack of government rules and lack of social awareness about green supply chain management implementation which will reduce environmental impacts.

Table 14. Lack of government rules

	LGR1	LGR2	LGR3	LGR4	Criteria weight	Importance
LGR1	0.154	0.153	0.250	0.125	0.171	3
LGR2	0.462	0.463	0.375	0.500	0.449	1
LGR3	0.077	0.153	0.125	0.125	0.119	4
LGR4	0.308	0.232	0.250	0.250	0.259	2

Table 15. Pair-wise comparison of criteria

	LEC	LOC	LGR	LFC	LSA
LEC	1	2	3	2	4
LOC	0.5	1	2	1	3
LGR	0.33	0.5	1	0.5	3
LFC	0.5	1	2	1	2
LSA	0.25	0.33	0.33	0.5	1
	2.58	4.83	8.33	5	13

Table 16. Normalization of criteria

	LEC	LOC	LGR	LFC	LSA	Criteria weight	Importance
LEC	0.389	0.415	0.360	0.400	0.308	0.374	1
LOC	0.198	0.208	0.240	0.200	0.231	0.215	2
LGR	0.128	0.104	0.120	0.100	0.231	0.137	4
LFC	0.194	0.208	0.241	0.200	0.154	0.199	3
LSA	0.097	0.069	0.039	0.100	0.077	0.077	5

Step 3: Ranking based on overall preference index value

The overall preference index value determines each criterion is determined using equation (5) by adding the weighted preference values corresponding to each criterion. The criteria with least overall preference value will have the first rank and criteria with the highest value are the last rank. The Table 17 shows that lack of environmental concern is the lowest value and lack of social awareness is highest of overall preference index value.

It is observed from Table 17, that there is no change on the relative ranking of the criteria when solved using proposed method. The modified SAW method minimizing the rank reversal and a reliable MCDM method for decision making problems. The modified SAW method having only three simple steps and is independent of the number of alternatives in a decision problem. The proposed method revealed that is robust against rank reversal phenomenon, and the cases where it occurs, alteration is minimized compared to other methods.

RESULTS AND RECOMMENDATIONS

The present chapter provides identification and ranking of barriers of GSCM implementation in Stone crushing plant. A comprehensive literature review was conducted to the identification of factors affecting the adoption of GSCM and group discussion and brainstorming carried out to make pairwise comparison of these identified factors. SAW methodology has been used for ranking (Table 18) of these identified factors.

Table 17. Preference index value ranking

	LEC	LOC	LGR	LFC	LSA	Preference Index Value	Rank
LEC	0.000	0.000	0.000	0.000	0.000	0.000	1
LOC	0.194	0.207	0.120	0.200	0.077	0.798	2
LGR	0.259	0.310	0.240	0.300	0.077	1.188	4
LFC	0.194	0.208	0.120	0.200	0.247	0.968	3
LSA	0.290	0.346	0.321	0.300	0.231	1.488	5

Figure 3. Ranking of barriers to adopting GSCM

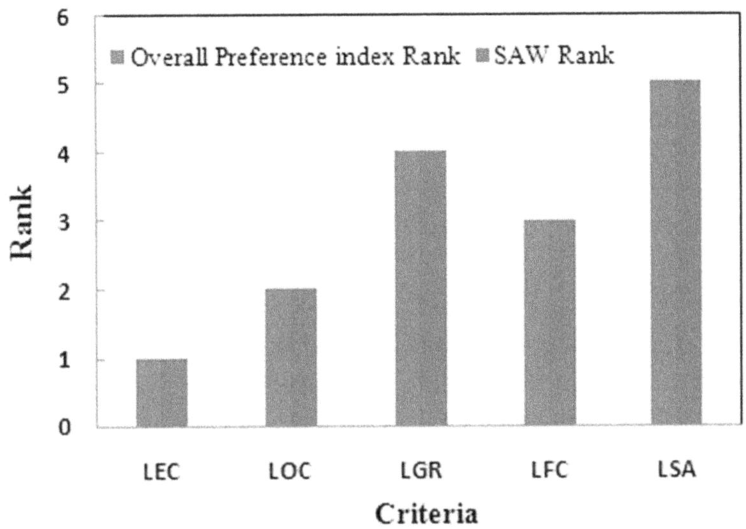

Table 18. Ranking of criteria by SAW and Preference index value

S.No	Criteria	Criteria weight	SAW Rank	Preference index value	Rank
1	**LEC**	0.374	1	0.000	1
2	**LOC**	0.215	2	0.798	2
3	**LGR**	0.137	4	1.188	4
4	**LFC**	0.199	3	0.968	3
5	**LSA**	0.077	5	1.488	5

Environmental concern has been found highest ranking or first priority and lack of social awareness is least ranking. In the environmental concerns sub-criteria's are minimization of air emissions; water pollution, soil pollution, and solid wastes through mining activities and also decrease in the use of hazardous materials/components used for blasting purpose have been found most important factors/berries. The extent of recycling and reuse activities least important. In the lack of organizational commitment are top management commitment towards the adoption of GSCM is ranked first due to capital investment and proper awareness about the green supply chain implementation benefits, lack of knowledge middle and lower management towards GSCM. The third most rank or important criteria is lack of financial constraints towards implementation of GSCM due to investment towards advanced equipment, technology, software and manpower and they worried about returns. The fourth importance is lack of government rules and regulations to monitor the environmental management system in industries and subside for implementation of GSCM in their organization. Finally, the lack of social awareness towards green supply chain management implementation in the industry for the production of eco-friendly products. Society actively involved in awareness and encouraging implementing GSCM.

FUTURE RESEARCH DIRECTIONS

This chapter considered real-time data of a particular industry, future research direction is considering different types of industries for analysis barriers of GSCM and analysis of interrelation between criteria's of Green supply chain management. Further performance measurements of the environmental system in the organization.

CONCLUSION

This case study may play an important role to understand various barriers perspectives and measures. Ranking of these barriers will help to understand the importance of environmental, financial, social, operational barriers to implement green supply chain in their organization. The proposed SAW methodology for ranking of criteria's and sub-criteria based on priority weight to improve towards green supply chain management. The pairwise comparisons in SAW have been made experts opinions. The opinion of experts may depend on the type of industry, size of the industry, a region of industry, etc. Further studies may be applied to different multi-criteria decision-making models and results can be compared.

ACKNOWLEDGMENT

The authors acknowledge to Sunil Hitech India Infra Pvt. Ltd., Nagpur, Maharashtra, India, for providing the necessary resources and other facilities during the research work and also, thanks to Dr. M. Sreenivasa Reddy, Principal and Professor in Department of Mechanical Engineering, R.L. Jalappa Institute Technology-Doddaballapur, Banglaore, for his valuable guidance during research work.

REFERENCES

Azapagic, A. (2004). Developing a framework for sustainable development indicators for the mining and minerals industry. *Journal of Cleaner Production, 12*(6), 639–662. doi:10.1016/S0959-6526(03)00075-1

Barve, A., & Muduli, K. (2011). Challenges to environmental management practices in Indian mining industries. *Proceedings of 2011 International Conference on Innovation, Management and Service*, 297-302.

Belton, V., & Gear, T. (1985). The legitimacy of rank reversal da comment. *Omega, 13*(3), 143–144. doi:10.1016/0305-0483(85)90052-0

Berkel, R. V. (2007). Eco-efficiency in the Australian minerals processing sector. *Journal of Cleaner Production, 15*(8-9), 772–781. doi:10.1016/j.jclepro.2006.06.017

Chaulya, S. K. (2004). Improving EHS management in India's small-scale mining sector. *Environmental Quality Management, 13*(3), 65–80. doi:10.1002/tqem.20005

Diabat, A., & Govindan, K. (2011). An analysis of the drivers affecting the implementation of green supply chain management. *Resources, Conservation and Recycling*, *55*(6), 659–667. doi:10.1016/j.resconrec.2010.12.002

Divya, C.M., Divya, S., Ratheesh, K. & Volga, R. (2012). *Environmental issues in stone crushers.* Academic Press.

Ghose, M. K. (2003a). Promoting cleaner production in the Indian small-scale mining industry. *Journal of Cleaner Production*, *11*(2), 167–174. doi:10.1016/S0959-6526(02)00036-7

Ghose, M. K. (2003b). Indian small-scale mining with special emphasis on environmental management. *Journal of Cleaner Production*, *11*(2), 159–165. doi:10.1016/S0959-6526(02)00035-5

Ghose, M. K. (2003c). A perspective on small-scale mining in India. In G. M. Hilson (Ed.), *The Socio-economic Impacts of Artisanal and Small-scale Mining in Developing Countries* (pp. 449–457). Balkema Publishers. doi:10.1201/9780203971284.ch26

Gilbert, S. (2000). *Greening supply chain: Enhancing competitiveness through the top forum on enhancing competitiveness green productivity.* Report of through green productivity held in The Republic of China.

Hilson, G. (2000). Barriers to implementing cleaner technologies and cleaner production (CP) practices in the mining industry a case study of the Americas. *Minerals Engineering*, *13*(7), 699–717. doi:10.1016/S0892-6875(00)00055-8

Hilson, G., & Nayee, V. (2002). Environmental management system implementation in the mining industry: A key to achieving cleaner production. *International Journal of Mineral Processing*, *64*(1), 19–41. doi:10.1016/S0301-7516(01)00071-0

Huszak, A. & Imre, S. (2010). *Eliminating Rank Reversal Phenomenon in GRA e based Network Selection Method.* doi:10.1109/ICC.2010.5502475

Jayaraman, V., Klassen, R., & Linton, J. D. (2007). Supply chain management in a sustainable environment. *Journal of Operations Management*, *25*(6), 1071–1074. doi:10.1016/j.jom.2007.01.016

Kong, F., Wei, W. & Gong, J.H. (2016). Rank reversal and rank preservation in ANP method. *Journal of Discrete Mathematics Science Crypt to gr., 19*, 821-836.

Langer, W. H., Drew, L. J., & Sachs, J. S. (2004). Aggregates and the environment. American Geological Institute.

Linton, J. D., Klassen, R., & Jayaraman, V. (2007). Sustainable supply chains: An introduction. *Journal of Operations Management*, *25*(6), 1075–1082. doi:10.1016/j.jom.2007.01.012

Macharis, C., Springael, J., De Brucker, K., & Verbeke, A. (2004). PROMETHEE and AHP: the design of operational synergies in multi-criteria analysis: strengthening PROMETHEE with ideas of AHP. *European Journal of Operational Research*, *153*, 307–317. doi:10.1016/S0377-2217(03)00153-X

Mali, A. V., Morey, N. N., & Khtri, A. P. (2016). Improvement in the efficiency of the Stone Crusher. *International Journal of Science. Engineering and Technology Research.*, *5*(6), 2265–2271.

Mathiyazhagan, K., Govindan, K., Noorul, H. A., & Geng, Y. (2013b). An ISM approach for the barrier analysis in implementing green supply chain management. *Journal of Cleaner Production*, *47*, 283–297. doi:10.1016/j.jclepro.2012.10.042

Mufazzal, S., & Muzakkir, S. M. (2018). A new multi-criterion decision making (MCDM) method based on proximity indexed value for minimizing rank reversals. *Computers & Industrial Engineering*, *119*, 427–438. doi:10.1016/j.cie.2018.03.045

Nikolaou, I. E., & Evangelinos, K. I. (2010). A SWOT analysis of environmental management practices in Greek mining and mineral industry. *Resources Policy*, *35*(3), 226–234. doi:10.1016/j.resourpol.2010.02.002

Piplani, R., Pujawan, N., & Ray, S. (2008). Sustainable supply chain management. *International Journal of Production Economics*, *111*(2), 193–194. doi:10.1016/j.ijpe.2007.05.001

Saaty, T.-L. (1980). *The analytic hierarchy process*. McGraw-Hill.

Saaty, T. L. (2008). Decision making with analytic hierarchy process. *International Journal of Services Sciences*, *1*(1), 83–98. doi:10.1504/IJSSCI.2008.017590

Sarkis, J. (2012). A boundaries and flows perspective of green supply chain management. *Supply Chain Management*, *17*(2), 202–216. doi:10.1108/13598541211212924

Seuring, S. (2004). Industrial ecology, life cycles, supply chains: Differences and interrelations. *Business Strategy and the Environment*, *13*(5), 306–319. doi:10.1002/bse.418

Seuring, S., Sarkis, J., Muller, M., & Rao, P. (2008). Sustainability and supply chain management e an introduction to the special issue. *Journal of Cleaner Production*, *16*(15), 1545–1551. doi:10.1016/j.jclepro.2008.02.002

Sun, X., & Li, Y. (2010). An intelligent multi-criteria decision support system for systems design. 10th AIAA Aviation Technology, Integration, and Operations (ATIO) Conference, 1–11. doi:10.2514/6.2010-9222

Wang, X., & Chan, H. K. (2013). A hierarchical fuzzy TOPSIS approach to assess improvement areas when implementing green supply chain initiatives. *International Journal of Production Research*, *51*(10), 37–41. doi:10.1080/00207543.2012.754553

Wang, X., & Triantaphyllou, E. (2008). Ranking irregularities when evaluating alternatives by using some ELECTRE methods. *Omega*, *36*(1), 45–63. doi:10.1016/j.omega.2005.12.003

Wang, Y. M., & Luo, Y. (2009). On rank reversal in decision analysis. *Mathematical and Computer Modelling*, *49*(5-6), 1221–1229. doi:10.1016/j.mcm.2008.06.019

Zanakis, S. H., Solomon, A., Wishart, N., & Dublish, S. (1998). Multi-attribute decision making: A simulation comparison of select methods. *European Journal of Operational Research*, *107*(3), 507–529. doi:10.1016/S0377-2217(97)00147-1

Zhu, Q., Geng, Y., Fujita, T., & Hashimoto, S. (2010). Green supply chain management in leading manufacturer's case studies in Japanese large companies. *Management Research Review*, *4*, 380–392. doi:10.1108/01409171011030471

Zsidisin, G. A., & Siferd, S. P. (2001). Environmental purchasing: A framework for theory development. *European Journal of Purchasing and Supply Management*, *7*(1), 61–73. doi:10.1016/S0969-7012(00)00007-1

Chapter 10
Improvement Proposal in Inventories at an Automotive Supply Company Settled in Mexico

Diana Sánchez-Partida
UPAEP University, Mexico

Adrian Guevara-Corona
UPAEP University, Mexico

José-Luis Martínez-Flores
UPAEP University, Mexico

Patricia Cano-Olivos
UPAEP University, Mexico

ABSTRACT

Controlling inventories and achieving effective management of them can significantly improve the profits to any company. This chapter presents a case of study of an automotive supply company that has operational problems due to over inventories. An ABC study and the economic order quantity model (EOQ) for 942 raw materials are presented promptly in order to increase the rotation efficiently, the yield of their inventories, as well as to minimize the logistics costs and maximize medium and long-term operations.

INTRODUCTION

Kiekert is a worldwide company that produces and markets latchets for automobiles. On their official site, they state that "... one of three side door latches around the world are based on a product design developed by Kiekert." Thus, Kiekert de Mexico, S.A. de C.V. is one of its eight plants. It has a great variety of suppliers and clients around Mexico and the world. The company is located in the Autopista Puebla-Orizaba Km. 14.5 Chachapa Industrial Park, Amozoc, Puebla, Mexico (Kiekert, 2017).

DOI: 10.4018/978-1-5225-8223-6.ch010

Currently, the company has a portfolio of 131 suppliers. Having significant control of its supply chain, Kiekert has divided their suppliers into three principal regions 1) NAFTA it refers the suppliers located in Mexico, United States of America, and Canada. 2) Europe, these suppliers are mainly in Germany, the Czech Republic, France, Italy, and Spain. Moreover, 3) Asia region, China, Korea, and Japan. Kiekert's spirit of innovation, awareness of quality and constant international growth are the results of this attitude. As a technology leader with more than 2,000 patents, we have become one of the leading development partners for the global automotive industry (Kiekert, 2017).

Today, Kiekert closure systems can be found in more than 50 car brands around the world. Some of the primary customers of the company are Audi, BMW, Chrysler, Ford, General Motors, Maserati, and Lincoln. The plant has a production area divided per client and by latch models such as General Motors, Ford Chrysler, Fiat, and Volkswagen –each one has a specific area. Moreover, have offices and a general warehouse where the Raw Material and Finished Product are stored.

Nowadays, the company acquires the raw materials based on the purchasing politics and experience but not based on the Economic Order Quantity that can reduce the logistics cost and find a tradeoff between the ordering and holding costs. Therefore, the main problems found were the overstocking and the lack of storage space in the warehouse, thus, in consequence, the company presents operational problems due to overstocking. Commonly, another problem that the companies used to have is the understocking of raw material to the production line. In the case of Kiekert, it was not the problem because it is preferable to buy raw material in excess into advance instead falls into this problem.

The current global crisis has affected firm's behavior on markets, forcing them to search for rational ways of planning their business. The optimal order quantity is one of the leading logistics tasks that take place in modern theory and practice. Variety of consumers with its demand and irregularity of consumption during the year exponentially increase number and times of calculation of economic order quantity. Basing on the proposed approach and collected data, the recommendation and decision support can be made for choosing order size in various external influence. For example, use own transport or hire it, use own storage or rent, what size of storage provide best efficiency of logistics system in this conditions, what vehicle's capacity should use for obtaining better results of functioning, which are aimed to sustainable developed of any logistics system (Halkin, 2017).

The effectiveness and efficiency of the inventory management can minimize the costs. Where the forecasting methods are necessary to determinate inventory and the economic order quantity model (EOQ) can reduce the inventory cost. The company can save more money or allocated for other purposes and also increase their profits (Sanny & Felicia, 2014).

The management of the supply chain and the role and responsibilities of various persons involved varies from industry to industry due to which supply chain management has become a vital issue for manufacturing organizations, professionals, and researchers (Singh Sunhal & Mangal, 2017).

Numerous Organizations face difficulties in the management of inventories. Poor inventory administration may bring about understocking, overstocking, and also high inventory total cost. The ABC analysis procedure for the inventory control framework is first used to distinguish the most important multiple things and afterward the EOQ of every item is produced to discover their inventory demonstrate condition independently. By doing ABC analysis, the items are categorized and give different level of control to different items. If the EOQ demonstrates, it is equitably utilized, with the guide of some judgment by the management, holding cost and ordering cost will turn out to be low. The utilization of this model will assist the organization with knowing the correct levels of raw materials to order and when to submit new requests for every raw material (Nishad, 2018).

In the same way, the ABC inventory classification process is an analysis of a range of distinct items referred to as stock keeping units (SKUs), such as, raw materials or finished products into three categories: A – outstandingly important; B – of average importance; C – relatively unimportant, as a basis for an inventory control scheme. Each category can and sometimes should be handled differently, with more attention being devoted to category "A," less to "B," and even less to C (Rezaei & Dowlatshahi, 2010).

However, the ABC inventory technique is a technique based on the idea that a small portion of goods in inventory represents the total value of money in inventory. ABC is a prevalent method in categories based on number and value that needs to be reviewed yearly (Oktaviani, Subawanto, & Hardi, 2017).

The purpose of this research is to determinate the optimal quantity for raw materials to support the inventory reduction and help free storage spaces. Also, it pretends to demonstrate an effective supply chain management when the company uses the appropriate quantity to request and the adequate categorization, as well as, the increase in the efficiency of its operating costs. An audit was done to the warehouse at Kiekert Company located in Puebla on October 5th, 2017, to detect strengths and weaknesses in this area. The results of this audit are presented in the section Issues, Controversies, Problems.

BACKGROUND

Inventories management has become one of the significant challenges facing managers about planning and control, especially in manufacturing companies. Although technically the inventory is an asset in the balance sheet of companies, it is considered that the holding cost it is an expense that can be significant, due to its different components, and therefore seeks to minimize it. The holding cost of inventories measures the expenses that an organization must make to maintain the items for a period. The inventory is considered as a buffer between two systems, one of supply (production or supply) and another of demand (customers or distributors). In this way, it can then be understood that the "size" of this shock absorber will depend on the behavior of these systems. The longer the supply chain, the higher the size of the inventory along with the length of the supply chain is, due to the bullwhip effect of the demand, since there are patterns that recurrently present greater uncertainty, insofar as they move away from the market (Aguilar, 2012).

A large number of studies have been focused on the inventories in the process, and the inventories of final product constitute an aspect of great importance for the organization and are starting point for the strategic decision making of the company. Taken this consideration, the inventory management for the efficient commercialization of goods and services becomes a tool to record the amounts that the company possesses, which play a fundamental role in the supply stage and the development of the demand, giving. As a result, steady states in the control of materials and products (Causado, 2015).

Recent studies have demonstrated that the control of the inventories achieve effective management of them, and can significantly improve the producer profits. The minimization of inventory costs, go through different methodologies such as operations research, mathematical formulations, and accounting theories (Cepeda Valero & Jiménez Sánchez, 2016).

Nevertheless, most of the inventory control models have been developed for deterministic or probabilistic demand patterns, with a constant average over time, generating high levels of inventory in times of low demand and low inventory levels in times of high demand, undesirable behavior from every point of view (Vélez & Castro, 2002).

In this sense; the costs generated by inventories represent one of the most significant elements that are reflected in the income statement, so when the accounts related to inventories - as a basic form - the following components are present: initial inventories, purchases, purchases returns, purchasing expenses, sales, sales returns, transit merchandise, consignment merchandise, and final inventories. The purpose of inventories is to make administrative operations more flexible. In this way, product inventories become an absolute necessity given that they allow opportunities for development and expansion in processes. Similarly, it should be noted that another of the essential functions in inventory management would be linked to the elimination of irregularities in supply, purchase or production in lots in excess, this would allow efficient and comprehensive management of the company (Cejas & Garrido, 2017).

Taken in consideration this thought; the decision-making process related to the management of inventories is fundamental, due to characteristics that generate problems; it is due to the nature of the system and the external variables that influence its behavior. Thus, it is possible that companies through the implementation of different mathematical tools obtain dynamism and flexibility to adapt to changing environmental conditions, anticipating the needs and expectations of customers, thus achieving a high level of competitiveness (Toro & Bastidas, 2011).

If the logistics management must reach the balance between supply and demand through policies and decisions that must be supported in the analysis of data and factors where objectivity and prediction are mixed, that is why to avoid shortages of products there are more important aspects than the delivery time, such as sales forecasts and orders that are sent to suppliers, which must be developed based on optimal inventory control policies and reliable and updated information that ensures proper management (Aguirre & Franco, 2005).

In the base case, the management needs to support this effort in the company an analytical model is proposed to determine the optimum production batch based on the recognized EOQ model. The value of this model is that it includes various types of costs, many of which had not been previously considered, which allows solutions closer to the real optimum (Valencia, Lambán, & Royo, 2014).

A good management practice involves the development of a commercial strategy that generally offers better offers of product services, financial resources, inventory of products, and a survey of customer demand in the market. The EOQ is widely used as a decision-making tool to control inventory in a supply chain. Most EOQ models assume that retailers pay the purchasing cost at the time of delivery (Birbil, Bülbül, & Frenk, 2009). Very often, the supplier offers the retailer a period of delay to pay the interest-free purchasing cost (Sankar, 2012).

The EOQ has been a well-known model that calculates the most favorable economic order quantity. Engineers study the EOQ model in engineering economics and industrial engineering courses. On the other hand, business discipline studies the EOQ in both operational and financial courses. In both, the EOQ model has practical and exact applications in defining concepts of cost tradeoffs. The companies are not focused entirely on the economic order quantity. In today's leading technology, many companies are not taking advantage of the fundamental inventory models. There are various software packages in aiding companies with inventory control, but if the data entered are inaccurate, it may lead to poor results (Singh Sunhal & Mangal, 2017).

In this manner, the EOQ calculate the order quantity that minimizes the balance between the holding cost of the inventory and the ordering cost. For calculating a basic EOQ, certain assumptions are necessary:

1. The demand is uniform, constant, and continues over time;
2. The delivery time is constant;

3. There is no limit on the size of the order due to the capacity of the warehouse is enough;
4. The cost of placing an order is independent of the size of the order;
5. The holding cost of a unit stored does not depend on the quantity in existence (Agarwal, 2014).

Instead to order based on the purchasing philosophy; is likely to motivate retailers to ask for more quantities because a late payment indirectly reduces the purchasing cost. Several researchers have studied commercial credit policy in the context of the deterministic EOQ model (Sankar, 2012). The primal assumption of the classical EOQ model is that the order quantity should be perfect. However, due to so many unavoidable factors, this assumption may not be valid always for most of the production environment (Jaggi & Mittal, 2011).

In the same way, the underlying assumption of these models is that 100% of items ordered or produced are of perfect quality. However, decision-makers have come to realize that this assumption is not valid in most of the production systems. Consequently, many companies need to develop a new inventory model to control the ordering and production. A large number of inventory models have been proposed by various researchers in order to show the impracticability of the assumption that all items are perfect (Öztürk, Eroglu, & Lee, 2015).

There are maximization of the benefits analyzing three EOQ models based on the inventory through geometric programming technique (GP) and find an optimal order of quantity and price for each of these models when considering the production (size of lot), and the commercialization of decisions (prices). It is important to know that EOQ can be applied to the situations with several circumstances. In this case, first, the demands are specific and constant over time. Second, the lack of items is not allowed. Third, the lead time is constant. Fourth, it is the number of orders received at one time. Fifth, there is fixed ordering cost in any orders. Sixth, the holding costs are charged for each stored item(Oktaviani, Subawanto, & Hardi, 2017).

Finally, the study in optimal inventory was applied at Brown Medical India Pvt. Ltd. The company offers Surgery, Intensive Care, Plexux Anesthesia, and Acute and Chronic Dialysis Equipment, including Syringe, Infusion Pumps, and FM Systems; Nerve Stimulators; and Dialysis Machines. This research goes throughout the process of examining the company's current inventory model and suggests an inventory model as the EOQ model in order to decide how much to order and when to order. The results indicate that the holding cost in 2016 has been reduced to 5.83% and the ordering cost has been reduced to 0.91% (Singh Sunhal & Mangal, 2017).

Another tool that can help almost any research is the ABC costing system, which has the purpose of optimizing their internal processes and the use of their resources. Thus, this ABC classification model quickly provides useful information for strategic planning and decision making. The process of implementation of the ABC system in companies requires as a necessary condition an adequate methodology to avoid failures when implementing it and thus, increase the rate of adoption of this system within companies (Cherres, 2010).

The ABC costing system emerged in the mid-1980s to satisfy the need for reliable information regarding the cost of resources assignable to products, services, customers, and distribution channels. The industry was becoming more complex every day, and the market demanded more dynamism and innovation in products and services, which led companies to think about how to grow and diversify, and costing systems had to evolve at the same pace. Thus, the ABC costing system is based on the fundamental premise that activities consume resources and products and objects of costs consume activities.

Under this scheme, the allocation of costs is made through generators or cost drivers (Marín, Ramírez, & Muñoz, 2012).

If a company decides to use fixed service levels per class, then an apparent key question is what those service levels should be. However, the literature does not provide clear guidelines. It is not even clear which class should get the highest/lowest service level. On the one hand, authors have argued that items classified as A are the most critical for a firm. Therefore, have the highest service levels in order to avoid frequent backlogs On the other hand, it has been claimed that dealing with stock outs is not worth the effort for C items and they should, therefore, get the highest service level (Teunter, Babai, & Syntetos, 2010). Even though, the experience says that all categories are essential because the categories A and B are products with high value, and the category C are products with low value but frequently with a high inventory level that no permit liberate money and storage space and, thus, the company must put attention in the levels of all the products.

METHODS USED IN THE RESEARCH

Firstly, the research will use an ABC costing system. This system is the most traditional cyclic counting method. It consists in dividing the inventory into ABC classification based on the 80-20 rule or the Pareto Law, in which articles are classified in two ways. First, by their value in money or by their frequency of use.

In many cases, a combination of the two is used. It allows distinguishing three categories of products, and each of them must be defined based on the part of the turnover that it represents. For purposes of inventory management, this allows determining that the products not be treated in the same way. By the classification, articles "A" are more frequent than articles "B," and articles "B" are more frequent than articles "C."

The development of the cyclic count method by ABC Analysis is detailed below:

1. The Pareto analysis must be done. On units of existence, using the desired method; in this case, the valuation for money will be used.
2. Decide how often each category will be counted; there is no rule to define the frequency of counting of each item, this is done taking into account the time it takes the inventory manager to count in one day.
3. Multiply the number of units of existence in each category by the desired frequency to establish the total number of counts (Olivos & Penagos, 2013).

ABC analysis is one of the most widely used tool for materials management. This analysis is considered a universal principle. It is therefore widely used in many situations of businesses and in this research the next categorization range will be used:

- Class A represents 20% of the materials in inventory and 80% of the inventory value.
- Class B represents 30% of the materials in inventory and 10% of the inventory value.
- Class C represents 50% of the materials in inventory and only 10% of inventory value.

As per ABC grouping, it recommends that the more investigation ought to be connected to materials with high inventory esteem. Class A ought to be most broadly handled, and Class C has analyzed pretty much nothing. The advantage of ABC grouping is that controlling small quantities of items adding up to 10-20% will result in the control of 75-80% of the money related estimation of the inventory held. If items in the inventory are not classified, managing and dealing with materials would be exceptionally costly since square with consideration is given to all items. Rigorous control methods ought to be utilized with "A" items and the controller ought to have incredible authority. Inventory held in safety stock should be low or none remunerated with more regular request arrangements. Utilization control and item development should be surveyed regularly, weekly or every day. Class "B" can be controlled by middle management. Low safety stock policy is applied to this class with quarterly or monthly orders. Past consumption can be used a basis for calculating order quantity. There should be two or four reliable suppliers to ensure that lead time is reduced. Power can be delegated to the user department to determine stock level. Class "C" items do not need to be highly controlled. Since the items have the lowest value compared to the class "A" and "B," orders can be placed at a greater volume to take advantage of a quantity discount. Rough estimates are sufficient to manage class "C" materials (Nishad, 2018).

Finally, the traditional ABC analysis is based on a single criterion such as annual dollar usage. However, in practice, the vast and various demands from customers lead to an increasing variety of inventory items which may not be homogeneous and hence the main differences among them may also lie in other criteria. It has been generally recognized that, for example, average unit cost, lead time, substitutability, commonality, scarcity, demand distribution, and so forth. These are also important criteria in deciding the importance of an item. When multiple criteria should be taken into account, sophisticated decision-making tools are required for inventory classification (Chen, 2012).

Secondly, the EOQ model will be used. This model is considered as the simplest and most fundamental of all inventory models, because it describes the necessary compromise between ordering cost and the holding cost, and is the basis for the implementation of much more systems complex. Figure 1 shows the tradeoff between these costs and how finding the balance point is possible to find the optimal lot size.

In this research, the following assumptions were considered:

1. The demand for the product (*D*), is in units and is known, constant, and independent;
2. Lead Time (supplier's supply time) is known and constant;
3. There are no discounts for order volume;
4. The total costs include the ordering cost, which is the cost when place an order $S*D/Q$. The purchasing cost of the article, which is the unit cost of purchase by demand $C*D$, in monetary value. Moreover, the holding cost of inventory H, in monetary value. It is equal to the cost of inventory management as a percentage unit cost of purchase of the product $i * C$, in monetary value.

Where $Q*$ is the optimal order size, which represents the equation (1) of the EOQ model below:

$$Q* = \frac{\sqrt{2SD}}{H} \tag{1}$$

Figure 1. Visual interaction between the ordering and holding costs and the lost size
Source: (Winton, 2005)

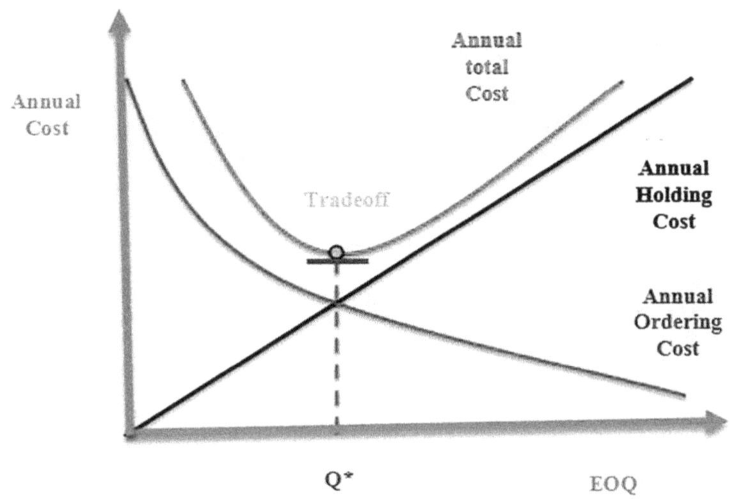

D = Total annual demand.
D/Q = The annual number of orders.
S = Fixed costs per order or ordering cost.
H = Annual holding cost per unit.

Also, it is very critical take into account all costs related to inventories; here the companies must know the ordering, purchasing, and holding cost (Causado, 2015). Similarly Reorder point would determine the sufficient stock at hand in order to satisfy the customer demand till the next order arrives (Nain Sukhia, Ashraf Khan, & Bano, 2014).

Another important thing about this model, is the assumption that the inventory is instantly replenished when it reaches zero level. Nevertheless, the company has many suppliers that handle their own delivery time or lead time and it have to make a complex decision when planning the release of the order. For that reason, in this study the reorder point was calculated with the equation 2.

$$R = LT * D \tag{2}$$

where R is the reorder point by each raw material, and LT is the lead time agreed with the suppliers. In the Figure 2 the behavior with all the important components of the model are represented. This saw graphic show how the inventory decreased uniformly and it is represented with a straight line because the demand is constant with low variations through the time (deterministic). Also, the maximum level that the inventory would be reach, the minimum inventory level of zero, the lead time promised by the supplier, the reorder point in where the order must be requested in order it arrives on time, and the cycle stock that the company must hold for attending the customers.

Figure 2. Saw graphic representing the EOQ behavior
Source: (Winton, 2005)

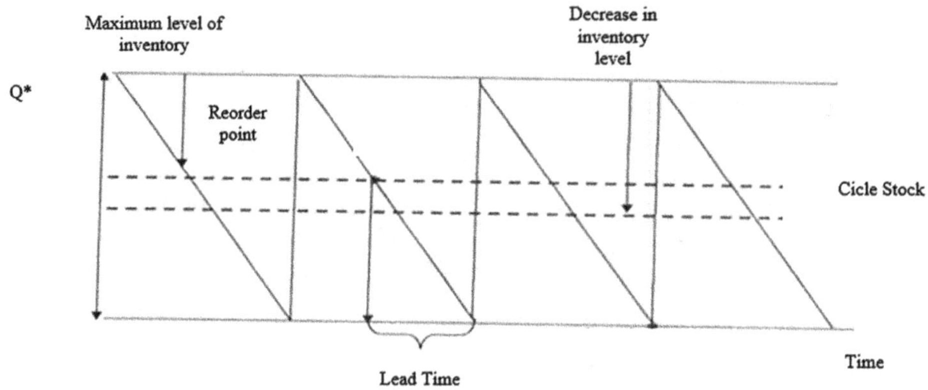

MAIN FOCUS OF THE ARTICLE

Issues, Controversies, Problems

In this section, it is presented a general overview of the current status found out in the warehouse area. The data provided by the company was the annual and the monthly forecast data 2018 for each raw materials, days of stock, delivery pending orders, unit product costs, material descriptions, ABC classification, and the purchasing politics for each raw material. This information was used to perform the data analysis and giving the improvement proposal.

As the first step, an audit in the in the warehouse area was applied, and its result is presented in the following Table 1, like a Strengths, Weaknesses, Opportunities, and the Threats (SWOT) analysis.

So with this analysis is possible to observe that the main problem of the company is that the overstocking produce that the plant has operational problems and the missing of space in the warehouse for the raw material that is arriving.

Then, the information on materials (standard materials or raw materials) was analyzed. The raw material for the latches manufacturing is a total of 942 items. The company handles a classification ABC, but there are not exist how this classification work, is that, nobody knows which are the criteria because it was made in 2007. Also, it was found that various raw materials did not have an ABC classification. The current status of the ABC classification is presented in Table 2. As it can be seen, there were 68 raw materials without classification.

Also, this research was found that there was a large number of raw materials that had high inventories. The number of stock days reached or exceed the company´s goal of 27.5 days in almost all the year, and thus, these raw materials could not be ordered again in 2018. The average stock days were 39.7 days. In Table 3, is shown a summary of the current status of the raw materials.

In addition, the lead time of the suppliers were considered for the analysis. The company has many suppliers around the world that handle their own delivery time and it have to make a complex decision when planning the release of the order. The lead time is so important because it lead the amount of inventory that the company must to hold at least in the warehouse when the new order is requested in order to satisfy the customer. The next Table 4 show the average lead time per regions permitted by the company, and which will be considered for the calculations.

Table 1. SWOT Analysis

Strengths	Weaknesses
Order: It has the order proposed in the audit format.	Inventory levels: They have inventory problems since they have an average of 39.7 days of inventory in hand when the goal is 27.5 days
Personnel: Knows and coordinates their work areas and processes, well identified and stratified to perform their tasks.	Contingency plans: Supply problems due to outdated suppliers.
Areas: Well identified, clean and with adequate signage.	
Materials: Properly identified and sheltered on the shelves.	
Security: It is adequate based on their needs.	
System: Correct use of the MRP for the arrangement and disposition of its articles.	
Opportunities	**Threats**
Materials Dispatch The Production: Orders are not planned, the department sends them indistinctly at any time of the day and any number of orders.	Entry operation: The receipt area is found with items not placed. In many times the warehouse does not have enough space for the number of materials that arrive. Financial area: operational problems because the capital are invested in unnecessary raw material.
Internal: The lamps give little light since the warehouse is growing upwards. There are no alarms for the area; they only have one speaker.	

Table 2. Raw materials classification made by the company

ABC classification	
Raw materials without ABC classification	68
With ABC Classification	874
Total raw materials	942

Table 3. Raw materials: the day of coverage

Raw materials – day coverage	
Days of coverage	**Number of raw materials**
From 1 to 10	149
From 10 to 20	212
From 20 to 30	223
More than 30 days	359
Total	942

Table 4. Lead times permitted per regions

Suppliers Lead Time	
Region	**Lead time**
Asia	6 weeks
Europe	5 weeks
Nafta	1 week

More Issues, Controversies, Problems

Over the audit were detected another minor issue. For these problems, we gave the company the observations in order to take the proper actions with good practices.

- Earthquake alarm system. The company did not have this kind of alarm system in their warehouse building. They installed an earthquake alert to avoid any risk with their personal. The action was coordinated with "Security people."
- Over the inspection, it was detected a damaged pallet. The company will take actions to move pallets immediately to the proper area. It was created the area called: "Detained Area" with the objective to move there temporally the damaged pallets.
- Lack of light between the racks. The company will install appropriate lights to illuminate the areas between the racks.
- There was a few spaces on the warehouse floor. In the next section, we will explain the reason for the lack of floor´s warehouse.
- The ABC analysis was done for the last time in 2007. New analysis will be provided to the company to get the best overview of the raw materials management.

RESEARCH METHODOLOGY

For the elaboration of the proposal that was made for Kiekert de Mexico, S.A. of C.V. the following methodology was followed promptly:

1. An audit was applied to the warehouse area of the company for determining the problems of the company. During this visit, it was identified that the opportunity for improvement was the inventory levels, the details of that audit were explained in the SWOT analysis.
2. The new classification of raw materials is made in base on their purchasing cost. The results were used to identify the current context and were changed in the information system. Analyzing was possible to observe that the raw materials needed a new classification because there were some products classified as A now are classified as B, see Table 5. The results show that the company needed to focus a significant control over 113 raw materials.
3. It was asked to the company a forecast of its articles in order to analyze the pattern of the data and define the models to be applied. The following Table 6 presents like the example, sixteen of the 942 raw materials without classification and with overstocking. In the same way, the forecast of eight months and the amount of purchase policy for the material is presented.

Table 5. Summary of material classification

Classification	Initial frequency	Final frequency	Total
A	0.0317066372	0.6754896674	113
B	0.6773537961	0.9500882975	333
C	0.9503759073	1.0000000000	496
Total of raw materials			942

If we analyze, this information is possible to detect promptly that these materials have issues with overstocking because these are bought to the suppliers based on the purchasing policy. This policy does not have the same lot size with the economic order quantity. In the following Table 7, is presented the total demand for the same materials.

For determinating, the EOQ is necessary to have the ordering and holding costs explained. The integration of the ordering and holding costs that handled the company are explained in Table 8 and 9. These costs were established and provided by the company, and they are calculated for every 1000 pieces.

4. After that, the average demand, standard deviation, and coefficient of variability (CV) were calculated. The calculations are shown in Table 10.
5. Based on this information, it was determined that the EOQ model would be applied, since the articles under study had a coefficient of variability less than or equal to 20%, for determining the

Table 6. Example characteristics are taken into account each raw material

Material	Classification	Days of coverage	may-18	jun-18	jul-18	ago-18	sep-18	oct-18	nov-18	dic-18	Purchasing policy
114425015804	A	25.9	251,992	355,584	290,804	318,737	317,856	309,810	280,560	290,804	1,500
4700211402	NO Classification	26.7	30,180	26,460	26,640	25,620	24,000	31,800	22,123	24,896	1,000
4582413803	B	23.2	55,962	45,501	43,065	53,146	52,640	69,832	54,407	43,406	16,800
4198602600	A	169.3	7,896	7,827	7,827	7,995	7,638	5,722	7,995	7,638	13,500
4652018001	B	24.5	126,860	176,288	143,257	159,851	154,620	159,539	135,501	135,501	15,000
4764117100	B	121	1,864	1,296	1,864	1,728	1,296	1,728	1,864	1,864	198
4558411900	B	45.3	4,720	5,354	4,850	5,651	4,295	5,744	4,982	4,949	1,000
4501602600	B	122.2	2,080	2,840	1,483	2,821	2,664	2,736	2,736	2,736	4,032
4702115101	A	43.1	2,796	1,952	1,884	1,984	1,944	2,376	2,376	2,376	1,500
4552414500	B	296.3	2,314	2,474	2,278	2,750	2,059	2,789	2,307	2,649	250
4113115104	B	119.2	1,536	1,768	1,768	1,536	1,152	1,920	1,920	1,920	2,400
4520115103	B	26.5	4,339	3,940	3,598	3,153	3,096	4,312	3,791	3,406	1,600
4720411010	B	181.3	212	247	270	325	256	384	284	284	1,250
4594413700	NO Classification	122.5	2,109	1,969	2,080	2,413	1,616	2,261	2,183	1,837	12,000
4520515001	A	122.7	3,456	3,456	2,728	3,645	4,025	3,500	3,483	3,360	10,000
4547015800	B	108.2	960	960	760	680	640	640	640	640	4,000

Table 7. Total demand for the periods

Material	Total Demand
114425015804	2,416,147
4700211402	211,719
4582413803	417,959
4198602600	60,538
4652018001	1,191,417
4764117100	13,504
4558411900	40,545
4501602600	20,096
4702115101	17,688
4552414500	19,620
4113115104	13,520
4520115103	29,635
4720411010	2,262
4594413700	16,468
4520515001	27,653
4547015800	5,920

Table 8. Integration of ordering cost

Ordering Cost	3.4% per each 1000 pieces
Freights, Inbound	1.1
Duty	1.3
Packaging	1.0

Table 9. Integration of holding cost

Holding Cost	1.6% per each 1000 pieces
Administrative	0.6
Light, rent, handling	0.6
Damages, Obsolescence	0.4

CV, it was taken into account the forecast provided by the company, holding and ordering costs. The EOQ formula was applied, obtaining the following results. See Table 11; the EOQ results for the raw material taken like the example, number of orders and pieces per year.

6. Then the annual holding, ordering, purchasing, and the annual total costs are presented in Table 12; these costs are Mexican pesos. For further explanations and comparative analysis see the section "Results and Discussion."

Table 10. Average demand, standard deviation, and coefficient of variability (CV)

Raw material	Total demand	Demand average	Standard deviation	CV
4121124405	1,044,762	130,595	17,250	13%
4118117108	2,882,134	360,267	64,957	18%
114125017905	2,424,468	303,059	30,587	10%
4100311006	336,608	42,076	7,111	17%
4117117108	1,811,799	226,475	41,366	18%
8999322100	1,241,110	155,139	26,682	17%
4161117103	2,373,144	296,643	55,134	19%
4174016010	604,754	75,594	12,016	16%
4748413903	868,404	108,551	20,892	19%
4747413903	823,284	102,911	19,756	19%
4101122103	487,955	60,994	10,858	18%
4405116801	7,191,525	898,941	148,708	17%
4125015803	1,486,369	185,796	30,160	16%
4171016010	329,925	41,241	4,802	12%
4100414703	693,779	86,722	15,097	17%
4102127303	467,471	58,434	6,204	11%
4510120801	5,488,157	686,020	70,571	10%
4101127303	507,838	63,480	6,636	10%
4105122507	532,951	66,619	9,328	14%
4174122602	473,414	59,177	9,113	15%

RESULTS AND DISCUSSION

According to the results obtained by applying the EOQ model to the materials of the company Kiekert de Mexico, S.A de C.V. it was found that for some raw materials, the number of requested orders increases and for others decreases, for example, of 2 to 185 orders in the case of local suppliers or of 200 to 116 orders in the case of some suppliers of Asia. When the frequency increases it means that the raw materials have to be requested more times in the year that the current context. The situation of increase is given due to the holding cost is more expensive than the ordering cost. It means that the company should make more orders since it is more beneficial to have the materials in transit than to have them in the warehouse. In this way, the overstocking is reduced, there is a warehouse with enough space for the storage of raw materials, the turnover of the inventory is incremented, and the capital cost is released.

Conversely, when the frequency decreases mean that the ordering cost are more expensive than the holding cost. Even though, this decrease does not mean something negative. Annually, the company decreased the order frequency 15%. The order frequency estimated is 84,471times per year, versus 99,203 times per year to the methodology applied by the company based on the purchasing policy. It means that the critical difference between the methodology proposed and current are the frequency and thus, the lot size to request.

Table 11. Economic order quantity, number of annual orders, and yearly demand to satisfy

Raw material	Economic order quantity	Number of annual orders	Yearly demand to satisfy
4121124405	8866	118	1,046,230
4118117108	24010	121	2,905,182
114125017905	20272	120	2,432,619
4100311006	3017	112	337,939
4117117108	15182	120	1,821,814
8999322100	10554	118	1,245,324
4161117103	19842	120	2,381,091
4174016010	5224	116	605,988
4748413903	7404	118	873,654
4747413903	7036	118	830,239
4101122103	4274	115	491,559
4405116801	59580	121	7,209,174
4125015803	12548	119	1,493,211
4171016010	2965	112	332,119
4100414703	5966	117	698,036
4102127303	4092	115	470,625
4510120801	45599	121	5,517,501
4101127303	4424	115	508,751
4105122507	4633	116	537,484
4174122602	4147	115	476,918

The company estimated the annual total costs of MXN 24,046,479,880.04, and the annual total estimated with the EOQ model was MXN 23,999,361,553.52 obtaining annual recurring savings of MXN 47,118,326.52. In Table 13 evidently shows the results.

CONCLUSION

Through the development of the present research, a model of the EOQ was applied for the company Kiekert de Mexico, S.A. of C.V. This will help the company to manage their inventories correctly as well as having a higher turnover of them. It is necessary for the company to focus its efforts on the correct application of the methodology since this will allow it to know the exact amounts to request. The most important thing is that this study will act as a facilitator of daily work within corporate operations. To avoid shortages and improve performance with customers.

It can be assumed that the correct application of the EOQ model helps to determine an effective purchasing policy. Thus managing to leverage the competitiveness of the company and manage correctly the problems faced by the company. It is necessary that the purchasing policy change because establishing the optimal targets (lot size) the company would have the explained benefits. Therefore, this study

Table 12. Annual holding, ordering, and annual total costs

Raw material	Annual holding cost	Annual ordering cost	Annual total cost
4121124405	$ 590,416,660.44	$ 33,242,395.68	$ 623,659,056.12
4118117108	$ 1,469,222,948.90	$ 29,986,419.56	$ 1,499,209,368.46
114125017905	$ 1,042,231,093.78	$ 25,287,077.99	$ 1,067,518,171.77
4100311006	$ 183,363,158.98	$ 32,043,410.07	$ 215,406,569.05
4117117108	$ 579,180,876.30	$ 18,804,217.97	$ 597,985,094.27
8999322100	$ 445,097,250.30	$ 21,095,786.19	$ 466,193,036.49
4161117103	$ 749,711,449.34	$ 18,583,226.93	$ 768,294,676.27
4174016010	$ 181,423,547.04	$ 17,646,800.77	$ 199,070,347.82
4748413903	$ 262,230,902.60	$ 17,762,869.83	$ 279,993,772.43
4747413903	$ 234,759,978.00	$ 16,773,568.38	$ 251,533,546.39
4101122103	$ 163,491,248.67	$ 19,709,055.70	$ 183,200,304.37
4405116801	$ 1,493,031,046.82	$ 12,212,341.01	$ 1,505,243,387.84
4125015803	$ 322,417,705.88	$ 12,759,784.01	$ 335,177,489.89
4171016010	$ 74,072,879.58	$ 13,206,723.38	$ 87,279,602.95
4100414703	$ 166,843,636.50	$ 14,146,192.89	$ 180,989,829.39
4102127303	$ 107,651,226.63	$ 13,546,134.62	$ 121,197,361.24
4510120801	$ 865,440,625.74	$ 9,276,023.28	$ 874,716,649.02
4101127303	$ 126,421,133.03	$ 14,643,522.61	$ 141,064,655.65
4105122507	$ 111,290,157.43	$ 12,283,455.42	$ 123,573,612.84
4174122602	$ 73,295,225.32	$ 9,107,216.61	$ 82,402,441.92
4405118402	$ 839,217,119.81	$ 7,659,532.96	$ 846,876,652.77

Table 13. Comparative of the current context against the proposed context

Compared components	Current context	Proposed context
Annual frequency	84,471	99,203
Annual holding cost	$ 22,985,336,364.18	$ 22,947,262,578.86
Annual ordering cost	$ 1,061,143,515.86	$ 1,052,098,974.66
Annual total cost	$ 24,046,479,880.04	$ 23,999,361,553.52

will help the company to focus its activities in order to have the necessary in its warehousing and in a successful way to control the management of their operations.

Also, the parameters into the system must be changed because the lot size will tend to be modified in accordance to the proposal, as well as, the reorder point, which is closely related to the time of supply (lead time).

Finally, the ABC study helped determine the raw materials. The company has started with the application of the EOQ in their operations and decided to start with the raw materials classified as “A” because they represent the major level of the inventory, the major purchasing costs, and required the performance of a strategic decision. The managers want to replicate the model to the raw materials type "B" and "C" after to see and evaluate the results of the materials "A."

ACKNOWLEDGMENT

This research was developed in the Master's program "Logistics and Supply Chain Management" awarded with the standard of excellence and supported with resources granted by the National Science and Technology Council (Consejo Nacional de Ciencia y Tecnología) – CONACyT. Scholarship holder number: 774796

REFERENCES

Agarwal, S. (2014). Economic order quantity model, a review. *VSRD International Journal of Mechanical, Civil, Automobile and Production Engineering*, 233.

Aguilar, P. A. (2012). Un modelo de clasificación de inventarios para incrementar el nivel de servicio al cliente y la rentabilidad de la empresa. *Pensamiento & Gestión*, 152-153.

Aguirre, S., & Franco, C. (2005). Diseño de un modelo de inventarios para la operación logística de una compañía farmacéutica. *Ingeniería y Universidad*, 30.

Birbil, S. I., Bülbül, K., & Frenk, J. (2009). *On The Economic Order Quantity Model With Transportation Costs*. Istanbul: H.M. Mulder.

Causado, E. (2015). Modelo de inventarios para control económico de pedidos en empresa comercializadora de alimentos. *Revista Ingenierías Universidad de Medellín*, 165.

Causado, E. (2015). Modelo de inventarios para control económico de pedidos en empresa comercializadora de alimentos. *Revista Ingenierías Universidad de Medellín*, 165.

Cejas, M., & Garrido, I. Y. (2017). La gestión del inventario como factor estrátegico en la administración de las empresas. *Revista Científica Electrónica de Ciencias Gerenciales*, 112-113.

Cepeda Valero, Ó. M., & Jiménez Sánchez, L. F. (2016). Modelo de control óptimo para el sistema Producción-Inventarios. *Ingeniería Industrial. Actualidad y Nuevas Tendencias*, 35-36.

Cherres, S. L. (2010). Un caso de aplicación del sistema ABC en una empresa peruana: Frenosa. *Contabilidad y Negocios*, 30.

Halkin, A. (2017). Estimation of Economic Order Quantity with Variable Parameters (Ukraine Case Study). *Science Research*, 1-4.

Jaggi, C. K., & Mittal, M. (2011). Economic Order Quantity Model for Deteriorating Items with Imperfect Quality. Delhi: Revista Investigación Operacional.

Marín, C., Ramírez, G. S., & Muñoz, J. A. (2012). Sistema de costeo ABC para empresas del sector eléctrico que actúen como operadores de red. *Sciences et Techniques (Paris)*, 78–79.

Nain Sukhia, K., Ashraf Khan, A., & Bano, M. (2014). Introducing Economic Order Quantity Model for inventory control in web based point of sale applications and comparative analysis of techniques for demand forecasting in inventory management. *International Journal of Computers and Applications, 107*(19), 1–8. doi:10.5120/18856-7385

Nishad, I. (2018). Inventory Management "ABC Analysis and Economic Order Quantity Approach": A case study. *Jasc: Journal of Applied Science and Computations*, 132-137.

Oktaviani, A., Subawanto, H., & Hardi, H. (2017). The implementation of the ABC Classification and (Q,R) with Economic Order Quantity (EOQ) model on the travel agency. *ResearchGate*, 45.

Olivos, S., & Penagos, J. W. (2013). Modelo de Gestión de Inventarios: Conteo Cíclico por Análisis ABC. *INGENIARE, Universidad Libre-Barranquilla*, 108-109.

Öztürk, H., Eroglu, A., & Lee, G. M. (2015). An economic order quantity model for lots containing defective items with rework option. *International Journal of Industrial Engineering*.

Sankar, S. (2012). An Economic Order Quantity Model for Nonconforming Quality Products. *Service Science*, 331–332.

Sanny, L., & Felicia, M. (2014). Strategy of optimization inventory: Case of Study in private manufacturing in construction field company in Indonesia. *Journal of Applied Sciences (Faisalabad), 14*(24), 3538–3546. doi:10.3923/jas.2014.3538.3546

Singh Sunhal, A., & Mangal, D. (2017). Analysis of inventory management in a supply chain by using economic order quantity (EOQ) model. *International Journal of Engineering sciences & Research Technology*, 303-308.

Teunter, R. H., Babai, M. Z., & Syntetos, A. A. (2010). *ABC Classification: Service Levels and Inventory Costs*. Production and Operations Management Society.

Toro, L. A., & Bastidas, V. E. (2011). Metodología para el control y la gestión de inventarios en una empresa minorista de electrodomésticos. *Sciences et Techniques (Paris)*, 85.

Valencia, J., Lambán, M. P., & Royo, J. (2014). Modelo analítico para determinar lotes óptimos de producción considerando diversos factores productivos y logísticos. Red de Revistas Científicas de América Latina, el Caribe, España y Portugal, 62.

Vélez, M., & Castro, C. (2002). Modelo de Revisión Periódica para el Control del Inventario en Artículos con Demanda Estacional una Aproximación desde la Simulación. Red de Revistas Científicas de América Latina, el Caribe, España y Portugal, 23-24.

Chapter 11
A MILP and Genetic Algorithm Approach for a Furniture Manufacturing Flow Shop Scheduling Problem

Marcell S. Kalman
Universidad Panamericana, Mexico

Omar G. Rojas
https://orcid.org/0000-0002-0681-3833
Universidad Panamericana, Mexico

Elias Olivares-Benitez
Universidad Panamericana, Mexico

Samuel Moisés Nucamendi-Guillén
Universidad Panamericana, Mexico

ABSTRACT

A MILP and genetic algorithm optimization model for the sequencing of jobs in a medium-sized factory, dedicated to the manufacturing of home furniture, where different categories and types of articles are produced and whose routes and manufacturing processing times vary widely, are proposed. Different scenarios are considered for the objective function based on minimizing makespan and tardiness. The results of the optimization for an instance of 24 jobs on five machines, chosen as a representative instance of the order sizes that are handled by the company, show important reductions in the productive system's usage times, oscillating between 10% and 20% with respect to a random initial sequence in the production plan. Improvements were similar in both techniques, the main difference being the solution time of each one.

DOI: 10.4018/978-1-5225-8223-6.ch011

INTRODUCTION

The task-sequencing problem within a factory, also known as a Job-Shop Scheduling Problem (JSSP), is very common and of great importance for companies in the manufacturing sector. Due to globalization and increasing market competitiveness, there is an urgent need for companies to reduce their manufacturing costs. This implies that productive system's technological resources, machinery and processes should be used in the most efficient possible way, attempting to maximize *throughput* and minimize *makespan* for a given production program.

In general, the benefits of considering the optimal sequence in which work must be carried out on the production floor, include revealing bottlenecks and imbalances within the system, giving certainty when acquiring delivery commitment dates with customers, reducing costs due to the use of machinery and infrastructure, reducing inventories of work in process, synchronizing the arrivals of raw materials with the production program, improving the use of floor space, avoiding loss of contracts with customers for non-compliance in delivery dates of merchandise, amongst others.

The above becomes more complicated when it comes to a make-to-order manufacturing system with a batch scheme where there is a high mix of products with medium or low volume demand, as is the case of study described in this chapter; therefore, for companies of this type to achieve their cost objectives, it becomes essential to calculate the production or task sequence within the system that allows to take the most advantage of their technological resources.

Currently, many medium-sized companies, and even some large ones, give little or no importance to job sequencing; they simply release production orders to the floor and, in most cases, these are executed in ascending chronological order, without any kind of analysis. Depending on the number of jobs to be carried out, the probability that the first selected job sequence is the best option decreases drastically.

The main objective of this chapter is to propose a mathematical model and make use of quantitative techniques that would allow the company's operations management team to consider sequencing the jobs in their production system, as a fundamental factor to achieve efficiency and guarantee customer satisfaction. Similarly, the pros and cons of using Mixed Integer Linear Programming (MILP) vs Genetic Algorithms (GA) to find solutions to different instances of the problem under study are presented. Additionally, it is intended to prove that, even though GAs by their stochastic nature do not necessarily converge to exact solutions, it might be convenient to rely on them in contexts such as the one presented in this chapter. This is because they can produce solutions that are close enough and with much more acceptable run-times for instances such as the ones that the company in question handles in its planning horizons.

BACKGROUND

One of the techniques used in this study is MILP, which is one of the most important techniques in operations research and mathematical optimization, where mathematical models are used to describe various types of problems where optimal decisions are to be made. Some of these decision include how to allocate resources of a given system in the most efficient manner (Kondili, Pantelides, & Sargent, 1993; M. L. Pinedo, 2008) To be able to use such a technique, the following conditions are required (Hillier & Lieberman, 2014; Rao, 2009).

- The resources of the system are limited or have constraints.
- There is an objective or goal to achieve (objective function).
- There is a linear relationship between the objectives and the constraints.

The other technique used in the present chapter is based on GAs, which are stochastic optimization techniques introduced by Holland (1992). These algorithms work through semi-random search methods by means of mechanisms based on simplifications of the theory of evolution by natural selection observed in nature (Whitley, 1994). In contrast to other stochastic techniques, GAs work from an initial population of randomly generated solutions, unlike other stochastic methods that do so from a single initial solution, as is the case of the simulated annealing method. Once the initial population of solutions is established, the GA assigns a value to each individual within the population based on the parameters and specific objectives of each problem. A survival-of-the-fittest mechanism is executed to choose the individuals that will produce and be part of the next generation of solutions; subsequently, a stage of reproduction is reached where crossover and mutation operators are applied to individuals. The above gives rise to a new generation with more fit individuals than the previous one. To obtain the maximum benefit from GAs, the specific properties of each problem must be analyzed, as well as the correct representation of its variables and objective function (Cheng, Gen, & Tsujimura, 1996; Cortez, 2014). It is also important to build an appropriate method for the generation of the initial population and the choice of crossover and mutation parameters. The procedure for constructing a GA is illustrated in Fig. 1.

Job-shop scheduling problems are combinatorial optimization problems that are usually NP-hard -at least as hard as an NP-problem (nondeterministic polynomial time) problem- and it's still debated whether it may be NP-complete -NP problem, i.e., verifiable in nondeterministic polynomial time and NP-hard- as well, see (Fortnow, 2009; Sotskov & Shakhlevich, 1995). Given their great importance in industrial settings, several researchers have formulated different ways of treating them based on different scenarios and assumptions (Asadzadeh, 2015; Blackstone, Phillips, & Hogg, 1982; Hillier & Lieberman, 2014). Some models deal with situations of high volume and low product mix where there are several machines of the same type and the jobs are processed in parallel (Mason, Fowler, & Matthew Carlyle, 2002; Mönch, Schabacker, Pabst, & Fowler, 2007; Moreno, Rojas, Olivares-Benitez, Nucamendi-Guillén, & Garcia de Alba Valenzuela, 2018); other approaches aim to minimize the total weighted delay where due dates are stipulated for each job and delays are penalized (Abdul-Razaq, Potts, & Van Wassenhove, 1990; Asano & Ohta, 2002; Bierwirth & Kuhpfahl, 2017; Essafi, Mati, & Dauzère-Pérès, 2008; M. Pinedo & Singer, 1999).

In its most general form, this problem can be described as the set of n jobs or batches of the same item to be produced $\left\{J_i\right\}_{i=1}^{n}$, to be processed in a set of m machines or processes $\left\{M_j\right\}_{j=1}^{m}$. Each job follows a logical sequence within the production system, which must be carried out in an uninterrupted manner and abide its precedence relations. Usually, within the context of manufacturing, this sequence is known as the *manufacturing route.* The processing of a job J_i in a machine M_j, is defined as *operation* O_{ij} which takes a processing time P_{ij} previously measured by means of a time-motion study, for each item type. From the above, the calculation of a *job schedule* is derived, which is nothing but the set of *completeness times,* $\left\{C_{ij}\right\}$, for all jobs listed in the production order that satisfy the system's constraints. The main objective of this problem is to minimize the *makespan* of the entire order or total time $L = C_{nm}$. Fig. 2 shows an example of a schedule representation for a 4 job-4 machine system.

Figure 1. Flow diagram of a GA

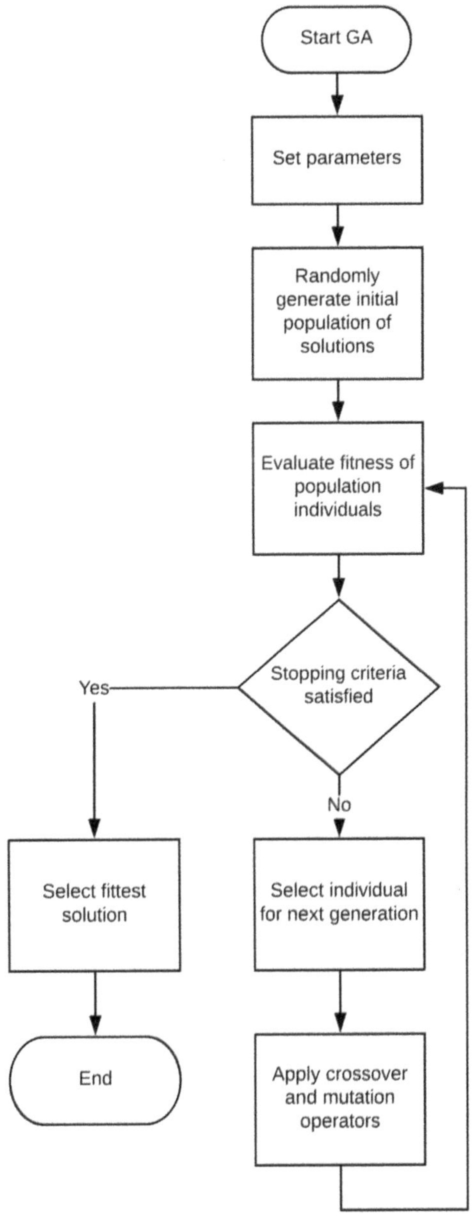

Fig. 3 illustrates the differences between the job and the technological sequences. The latter must always be respected for this particular production system where as the former may vary as to optimize the total makespan of the system. This problem belongs to the category of combinatorial optimization problems whose main characteristic is the exponentially increasing difficulty of the calculation as the instance size grows (Blum & Roli, 2003; Papadimitriou & Steiglitz, 1998).

The problem of sequencing tasks in the production line has different versions and has been modeled and solved using diverse techniques, both exact and metaheuristic. The first method in which we will focus on will be MILP due to the accuracy of its solutions. Such method might consider both discrete

Figure 2. Schedule representation for a 4 job-4 machine system

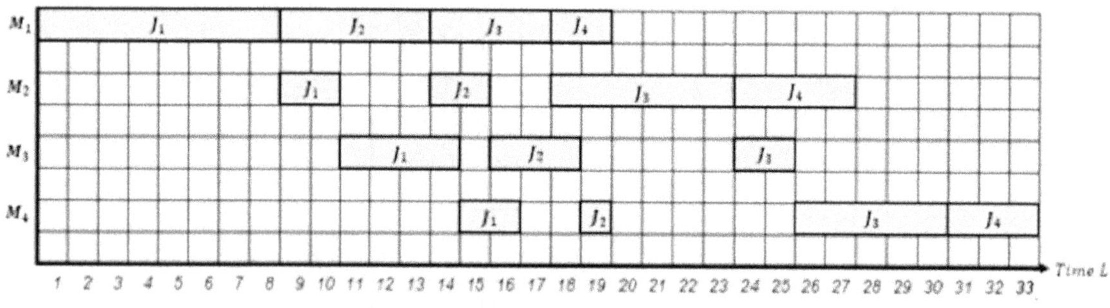

Figure 3. Technological sequence for a 4 job-4 machine system

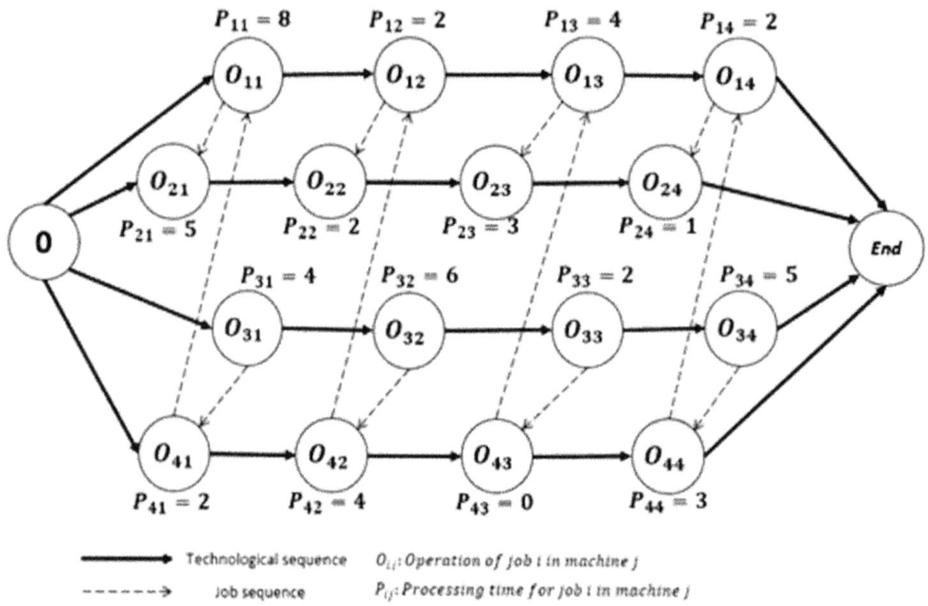

and continuous variables (Floudas & Lin, 2005). Normally, these types of problems base their solution on the branch and bound algorithm (Brucker, Jurisch, & Sievers, 1994; Lawler & Wood, 1966; Vila & Pereira, 2014) or the cutting plane algorithm (Györgyi & Kis, 2018; Kelley James E, 1960; Mokotoff & Chrétienne, 2002; Nemhauser & Savelsbergh, 1992). Another important aspect is that MILP problems are regarded as NP-hard therefore, they are intractable for large instances. The computational complexity function for the problem in question is $O(n!)$ (Eiselt & Sandblom, 2007).

Another variant of the problem is the Flexible Job-Shop scheduling Problem (FJSP) which is an extension of the JSSP and in which each operation may be processed by any of the machines in the system (Pezzella, Morganti, & Ciaschetti, 2008; Zhang, Gao, & Shi, 2011). Compared to the JSSP, with the FJSP it is necessary to make more machine assignments and to determine an appropriate processing sequence. This creates two sub-problems within the main problem: to find an optimal machine assignment schedule and to find the optimal sequence for the list of jobs to be carried out. The aforementioned implies an even bigger solution space than that of the JSSP. The FJSP is a difficult problem by nature and one tackled successfully with a hierarchical optimization approach by means of a competitive genetic algorithm (Ishikawa, Kubota, & Horio, 2015). This approach consists in dividing the complex system, i.e. the production system in question, hierarchically into smaller simpler subsystems, and each level of hierarchy is then optimized.

A different approach (Gao et al., 2015; Li, Pan, & Tasgetiren, 2014) for the FJSP, was to apply an artificial bee colony algorithm to schedule jobs with the variant of new job insertions. This approach was used to solve the case where the production schedule changes due to the arrival of new jobs, thus creating the need for rescheduling. Xiong, Fan, Jiang, & Li (2017) proposed a simulation-based study with special focus on dispatching rules: four new dispatching rules to consider upon scheduling and applying them to technical precedence constraints between operations in the sequence. Yazdani, Zandieh, Tavakkoli-Moghaddam, & Jolai (2015) presented a variation of the problem defined as the Dual-Resource Constrained FJSP to minimize makespan; this particular approach deals with systems where there exist both machines and workers which are treated as constraints in the problem and the solution has to do with assigning operations to compatible machines, selecting the best worker among a group of skilled workers for performing the operation on the selected machine and only then sequencing the operations on the machines. The simulated annealing metaheuristic was utilized to tackle this scenario which according to the authors has the benefit of a computational low cost with acceptable results. Also, a Taguchi experimental design is utilized for parameter and operation and tuning.

CASE OF STUDY

The company referred to in this chapter is a medium-sized factory focused on the design, production and distribution of home furnishing products under a wholesale make-to-order scheme. As a result of an increase in the demand due to the acquisition of new customers, there is a need to resort to quantitative techniques and implement models that lead to an improvement in the use of resources, and the service level for key customers while lowering manufacturing costs. The wide range of products offered by the company, as well as the diversity in size of the manufactured lots, generate important challenges when developing an adequate production schedule. This is because the nature of the product types vary widely in terms of material composition, number of sub-assemblies and manufacturing routes per item type.

One of the main challenges within the company is to maintain an adequate balance of the production system capabilities, a fact that is mainly derived from an incorrect or non-existing job sequencing within the manufacturing floor. The main production processes of the company's manufacturing system is illustrated in Fig. 4.

Figure 4. Flow diagram of current manufacturing system

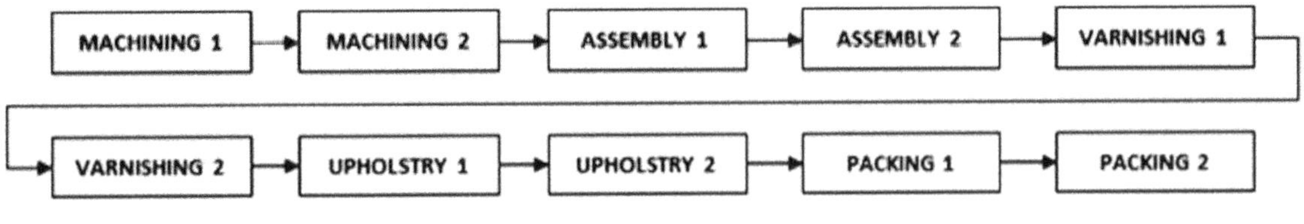

To solve the problem described, a computational tool that assists the planner in generating production programs based on the best possible sequence was developed. For this, a comparison between two possible paths was explored, one through a MILP model, which was executed with the help of LINGO and another path through a GA which was applied with the use of the Analytic Solver platform.

Empirical Data

Table 1 shows the unitary processing times P_{ij} measured in the time-motion study for a subset of items produced by the company. The main purpose of this information is to serve as an example to show the structure of the data for which the algorithms where applied. This collection of items corresponds to the 24×5 instance also tackled in this chapter. The demand values for each item pertain to a particular multicustomer order scenario of the company at the time the study was performed. According to the management team of the company this example is representative of a typical instance throughout the business year.

Unitary processing times for an item i on process j, defined as the parameter P_{ij} are shown. However it is important to point out that the total time for the hole batch (i.e. job) given the demand for the item can be calculated by multiplying the unitary processing time for the item in each process times the demand of said item, $P_{ij}D_i$. Having looked at the data structure of the problem, it is pertinent to move on to the next section where the mathematical framework for the MILP is presented.

Mathematical Model

The description of the MILP model is presented in this section. Table 2 presents the variables, parameters and sets considered.

Having explained the functionality and benefits of production sequencing by means of MILP and taking into account the declaration of variables and parameters, the mathematical model is presented.

$$Z_{min} = C_m^n \qquad m \in M,\ n \in J, K \tag{1}$$

Subject to:

$$\sum_{i=1}^{n} Y_i^k = 1 \qquad \forall k,\ i \in J \tag{2}$$

$$\sum_{k=1}^{n} Y_i^k = 1 \qquad \forall i,\ k \in J, K \tag{3}$$

Table 1. Processing times for a 24×5 *instance*

			$j=1$	$j=2$	$j=3$	$j=4$	$j=5$
			MACHINING	**ASSEMBLY**	**VARNISHING**	**UPHOLSTRY**	**PACKING**
DEMAND D_i **(units)**		**MODEL / ITEM TYPE**	P_{i1} **(Hours)**	P_{i2} **(Hours)**	P_{i3} **(Hours)**	P_{i4} **(Hours)**	P_{i5} **(Hours)**
D_1	111	Arty Mirror	0.12	0.16	0.13	0.00	0.10
D_2	89	Arty Table	0.12	0.08	0.19	0.00	0.13
D_3	159	Arty Sideboard	0.20	0.19	0.32	0.00	0.13
D_4	146	Arty Dining chair	0.00	0.00	0.11	0.09	0.05
D_5	137	Chest Armchair	0.15	0.11	0.13	0.15	0.05
D_6	157	Elena Headboard	0.07	0.08	0.05	0.15	0.05
D_7	149	Elena Dresser	0.15	0.19	0.19	0.00	0.10
D_8	185	Eliot Table	0.12	0.08	0.24	0.00	0.13
D_9	124	Eliot Sideboard	0.15	0.19	0.24	0.00	0.10
D_{10}	166	Foster Nightstand	0.15	0.13	0.11	0.00	0.11
D_{11}	123	Foster Bed	0.30	0.29	0.21	0.33	0.13
D_{12}	126	Foster Dresser	0.15	0.16	0.19	0.00	0.10
D_{13}	70	Foster Mirror	0.15	0.16	0.13	0.00	0.10
D_{14}	119	Foster Sofa	0.20	0.19	0.16	0.27	0.10
D_{15}	61	Foster Loveseat	0.20	0.16	0.13	0.24	0.08
D_{16}	175	Foster Sectional	0.27	0.29	0.24	0.45	0.16
D_{17}	85	Kross Nightstand	0.15	0.16	0.13	0.00	0.10
D_{18}	180	Kross Bed	0.25	0.27	0.19	0.33	0.13

continued on following page

Table 1. Continued

			$j=1$	$j=2$	$j=3$	$j=4$	$j=5$
			MACHINING	**ASSEMBLY**	**VARNISHING**	**UPHOLSTRY**	**PACKING**
DEMAND D_i **(units)**		**MODEL / ITEM TYPE**	P_{i1} **(Hours)**	P_{i2} **(Hours)**	P_{i3} **(Hours)**	P_{i4} **(Hours)**	P_{i5} **(Hours)**
D_{19}	153	Kross Sideboard	0.20	0.19	0.27	0.00	0.10
D_{20}	78	Kross Dining chair	0.02	0.03	0.11	0.09	0.05
D_{21}	143	Madison Nightstand	0.12	0.13	0.11	0.00	0.10
D_{22}	118	Madison Bed	0.25	0.27	0.21	0.33	0.16
D_{23}	163	Madison Dresser	0.12	0.16	0.16	0.00	0.08
D_{24}	160	Madison Mirror	0.05	0.08	0.05	0.00	0.10

Table 2. Statement of variables and parameters for the MILP model

Response variable	
C_j^k	
Decision variable	
Y_i^k	$= \begin{cases} 1, \text{ if job } i \text{ is assigned to position } k \text{ in the sequence} \\ 0, \text{ Otherwise} \end{cases}$
Parameters	
P_{ij}	*Unitary processing time for item I on machine j (hrs)*
D_i	*Demand in i (units)*
m	*Number of machines or processing units of the system*
n	*Number of jobs to produce for a given order*
k	*Number of possible positions within a sequence for a given order*
Sets	
M	*The set of all machines and processing units*
J	*The set of all jobs*
K	*The set of all positions*

$$C_1^1 \geq \sum_{i=1}^{n} Y_i^1 D_i P_{i1} \qquad \forall j = 1,\ \forall k = 1,\ \forall i = 1,\dots,n,\ i \in J \tag{4}$$

$$C_j^k \geq C_{j-1}^k + \sum_{i=1}^{n} Y_i^k D_i P_{ij} \qquad \forall j = 2,\dots,m,\ \forall k = 1,\dots,n,\ i \in J \tag{5}$$

$$C_j^k \geq C_j^{k-1} + \sum_{i=1}^{n} Y_i^k D_i P_{ij} \qquad \forall j = 1,\dots,m,\ \forall k = 2,\dots,n,\ i \in J \tag{6}$$

$$C_j^k \geq 0 \tag{7}$$

$$Y_i^k \in \{0,1\} \tag{8}$$

The objective function of equation (1) minimizes the completion time for whatever job happens to fall into the last position n on the last machine m, i.e., the makespan of the entire production program. The constraints in equations (2) and (3) ensure that no more than one job i gets assigned to a position k within the sequence and that no job i gets to occupy more than a single position k. For the constraint in equation (4), the evaluation for the job completion time for machine $j = 1$ at position $k = 1$ is made. This constraint is proposed since this position in the sequence has no predecessors in neither the stream of jobs nor the technological sequence. The constraints in equations (5) and (6) calculate the completion time C_j^k of the jobs for the remaining positions and machines $\neq 1$ based on a validation between two possible scenarios; either a job waits for the availability of the next machine where it is to be processed which may still be in the middle of processing the job in $k-1$, or a machine waits for the job assigned in position k to be processed by machine $J-1$ to which C_j^k will take the greater value between the two. Equation (7) establishes the non-negativity for the response variable C_j^k and finally, equation (8) states that the decision variable Y_i^k is binary.

Next, the model was programed and executed in LINGO 17 linear engine in order to evaluate its performance for several instances provided by the company. After execution, the program yielded results in the form of a job sequence that ensures the minimum makespan possible for any particular order evaluated which, as mentioned earlier, will be a guaranteed global minimum within the solution space.

Genetic Algorithm

In this section a GA approach is used to find an approximation for the minimum makespan. It is important to remember that given the stochastic nature of GAs, the solutions obtained with this method are not guaranteed to achieve a global optimum, since the solution space is not explored in its entirety; nonetheless, as it will be shown further on, this method features some important benefits for this particular problem as the instances grow in size as is the case for this furniture company. To this end, the definition of parameters and initial conditions is shown in Table 3.

Based on the above, the following equations are proposed to execute the algorithm (Werner, 2013).

$$FT(i) = \begin{cases} F' - F_i(S_i), \; if \, F_i(S_i) < F' \\ 0, \; Otherwise \end{cases} \tag{9}$$

Table 3. Parameters and initial conditions for the GA

Parameter	Value
Number of generations	10,000
Mutation rate	0.075
Population size	100
Max Runtime without improvement	30 seconds
Seed	Random
Crossover operator	Order Crossover type 1 (OX1)
Crossover cut-off point	Random number $\in \{1,\ldots,n\}$
Mutation operator	Inversion
Cut-off points for inversion	Random1, Random2 $\in \{1,\ldots,n\}$
Parent selection mechanism	Selection by roulette
Reference value for fitness evaluation	1000
Replacement strategy for individuals	Generational fitness evaluation
F'	Reference value to evaluate fitness
S_i	Permutation sequence for individual i
$F_i(S_i)$	Objective function´s value for a sequence S_i
TP	Population size
$FT(i)$	Fitness value for individual $i \in TP$
$P(i)$	Probability of choosing the $i-th$ individual

$$P(i) = \frac{FT(i)}{\sum_{k=1}^{TP} FT(k)} \qquad (10)$$

Because the application of GAs deal with maximizing fitness functions, even though the problem in question is a minimization one, the fitness function must be considered in terms of a difference between the value of the objective function, evaluated for a given sequence S_i from the population and a reference value F', selected with bias towards the improvement of fitness. The aforementioned is expressed in equation (9). Equation (10) calculates the probability of selecting an individual within the population based on its aptitude to subsequently apply the crossover operators. We defined the first sequence randomly to start the GA to obtain a reference for the choice of F'. Next, using the parameters, data and fitness function described above, the GA is executed utilizing the Analytic Solver platform for Microsoft Excel. The instances presented in this work were attempted to be solved under both approaches in order to compare accuracy and runtimes of the algorithms.

Manufacturing Costs

At this stage of the problem and once the optimal production sequence was obtained, the calculation of the corresponding manufacturing costs for each scenario, both the initial random sequence of the order and the optimized one via the two methods studied in this work, is performed. The foregoing is for comparison purposes between these scenarios and to gauge monetary savings percentage accomplished. Hourly cost data for the factory processes are presented in Table 4.

The utilization costs shown above were provided by the company's management team and are the result of averaging several measurements performed throughout the various business cycles that take place during the year. Variables, parameters and the cost function are stated in Table 5.

Table 4. Costs per hour per processing unit

Machine/Processing unit	Utilization costs per hour
MACHINING 1	$521.00
MACHINING 2	$548.00
ASSEMBLY 1	$667.00
ASSEMBLY 2	$651.00
VARNISHING 1	$771.00
VARNISHING 2	$784.00
UPHOLSTRY 1	$875.00
UPHOLSTRY 2	$850.00
PACKING 1	$375.00
PACKING 2	$363.00

Table 5. Variables and parameters for manufacturing cost function

Response Variables	
$MC\left(C_j^{*k}\right)$	*Manufacturing cost as a function of usage time*
$C_j^{*k=n}$	*Optimized completion time for machine j on position k = n*
Parameters	
UC_j	*Utilization cost per hour per machine / process j*
m	*Number of machines or processing units*
n	*Number of jobs to carry out*
k	*Number of possible positions to occupy within a sequence*

Equation (11) gives the manufacturing cost as a function of usage time for this factory. This function is fed with the optimal completion time calculated by the means described above, for each machine in the last position of the sequence, i.e., the total usage time for each processing unit. The calculation yields the cost for the entire system per production order.

$$MC\left(C_j^{*k}\right) = \sum_{j=1}^{m} UC_j C_j^{*k} \qquad \forall k = n \tag{11}$$

SOLUTIONS AND RECOMMENDATIONS

This section shows the results obtained from the optimizations performed on the different instances using MILP and GA. The instances were artificially created using a sample of 50 items extracted from the product catalogue offered by the company. The standard processing time data and costs were provided by the company's engineering team. A selection of 13 instances was considered for the comparison of results between both MILP and GA methods. The selection of the instances was designed so different number of machines and demand for articles was taken into consideration. Such instances ranged between 32 variables with 34 constraints to 3,000 variables with 1,042 constraints. The structures of variables and constraints for each instance are shown in Table 6.

The MILP model was solved using LINGO 17 linear optimization software and the GA by means of Analytic Solver add-in for Excel. Both were executed on a DELL XPS 13 computer with Intel® Core ™ processor i7-7560U CPU @ 2.40 GHz 16GB RAM, Windows 10 operating system 64-bit platform. In the case of the 24×5 instance, the MILP had a runtime of about 14 hours while the GA for the same instance did it in less than 60 seconds approximately.

Table 6. Model structure for the evaluated instances for the model

Instance	Variables	Constraints
4×4	32	34
5×5	50	52
10×5	150	107
10×10	200	202
15×5	300	162
20×5	500	217
20×8	560	334
24×5	696	261
24×10	816	496
30×6	1,080	386
40×5	1,800	437
50×8	2,900	844
50×10	3,000	1,042

Considerable improvements are shown for the utilization time of the production system compared to the results prior to the optimization process. To illustrate this, the instance 24×5 in Table 7 is taken as an example, showing a 14.5% improvement for the makespan with respect to the initial random sequence of jobs and applying the MILP; however, the runtime of the linear algorithm is considerably high at almost 14 hours.

Table 8 shows the results of makespan for the GA, which also yielded significant improvements over the pre-optimized production schedules; however, unlike with the linear model, there is a decrease in the precision as the instance sizes grow. On the other hand, runtimes are much faster than with the MILP suggesting that the GA is a more efficient technique for this problem.

Figs. 5 and 6 illustrate graphically the contrasting results between the two techniques regarding their two main performance indicators, the runtime and the accuracy.

A makespan comparison for both methods with respect to the current production scheduling paradigm of the company is shown in Fig. 7. A significant improvement can be appreciated upon applying optimization tools to the production schedule. Both methods show acceptable accuracies according to the company´s standards.

As for the manufacturing cost savings, the optimization tools proposed help achieve sizable contributions specially for medium to big instances which occur for the most part of the year according to management. This trend is appreciated in Fig. 8.

Table 7. MILP results for optimal makespan

Instance	Pre-optimized Makespan	Runtime (sec)	Accuracy	Optimized Makespan (hrs)	Makespan improvement
4×4	81.98	0.56	100%	72.73	11.28%
5×5	294.13	0.39	100%	278.83	5.20%
10×5	361.37	5.05	100%	323.71	10.42%
10×10	672.14	4.78	100%	655.11	2.53%
15×5	411.51	113.83	100%	384.39	6.59%
20×5	521.48	264.78	100%	440.88	15.46%
20×8	710.69	9,130.69	100%	631.51	11.14%
24×5	**670.15**	**50,372.92**	**100%**	**572.97**	**14.50%**
24×10	894.23	117,913.99	100%	758.19	15.21%
30×6	812.15	360,109.01	100%	738.45	9.07%
40×5	861.02	Intractable	-	-	-
50×8	1350.59	Intractable	-	-	-
50×10	1388.43	Intractable	-	-	-

Fig. 9 shows that a manufacturing cost improvement was achieved with both methods for all instances analyzed in this work, showing the benefits of applying optimization tools.

Given the noticeable benefits of using GA over MILP due to the efficiency in its performance, a confidence interval of the cost saving results was computed for all 13 instances looking to back up the assumption of a 7 to 12% expected value in savings, regardless of the instance size. The analysis results show that the data is normally distributed with a p-value of 0.251 and a confidence interval for the mean of $7.987\ \% < \mu < 12.737\ \%$ at a 95% significance level. Furthermore, a Monte Carlo experiment of 110 trials was conducted for two of the most representative instances according to the company's record, to be able to establish a confidence interval for the mean of the accuracy for the GA.

Each trial in the experiment consisted of calculating a random initial sequence for the instance, then, applying the GA engine to that instance and then measure its performance and accuracy against the known global optimum found by the MILP, for all 110 iterations. This was achieved by means of programing Microsoft Excel through VBA. The results of the simulation are illustrated in Fig. 10. Also, with the data collected from the Monte Carlo experiment, a two sample t-test was performed in order to test the hypothesis that the expected value of the improvement percentage on the makespan, when applying the GA, remains the same regardless of the instance. The results of the test backup the null hypothesis that the means are equal with a p-value of 0.455

Table 8. GA results for optimal makespan

Instance	Pre-optimized Makespan	Runtime (sec)	Accuracy	Optimized Makespan (hrs)	Makespan improvement
4×4	81.98	42.82	100%	72.73	11.28%
5×5	294.13	33.67	100%	278.83	5.20%
10×5	361.37	4.67	100%	323.71	10.42%
10×10	672.14	2.91	100%	655.11	2.53%
15×5	411.51	16.19	100%	384.39	6.59%
20×5	521.48	16.42	100%	442.4	15.16%
20×8	710.69	32.30	100%	634.46	10.73%
24×5	**670.15**	**47.66**	**99%**	**578.04**	**13.74%**
24×10	894.23	41.00	99%	762.58	14.72%
30×6	812.15	47.19	97%	757.61	6.72%
40×5	861.02	74.625	-	746.46	13.31%
50×8	1350.59	102.047	-	1180.77	12.57%
50×10	1388.43	94.985	-	1225.58	11.73%

Figure 5. Runtime comparison between MILP and GA

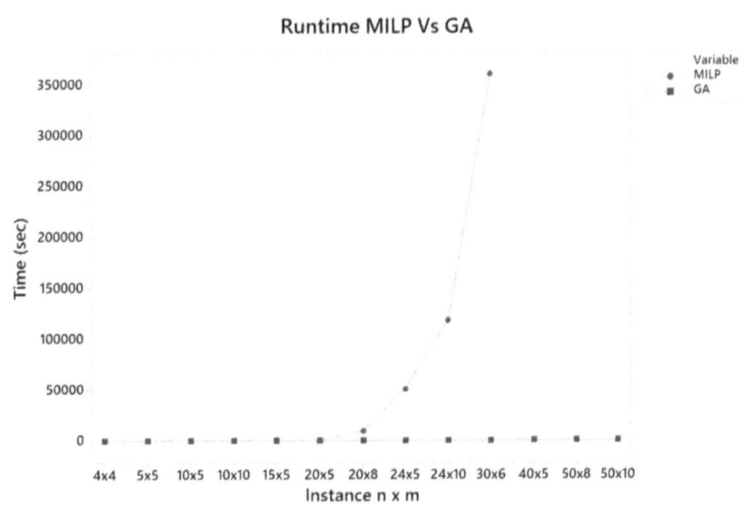

Figure 6. Accuracy comparison between MILP and GA

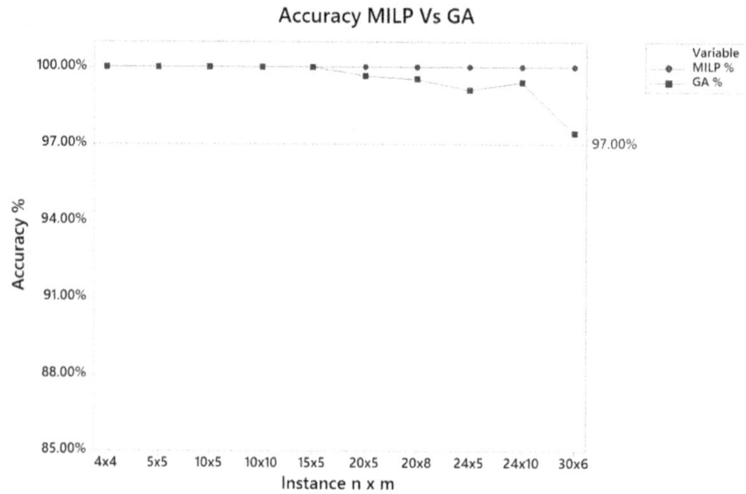

Figure 7. Makespan comparison between current scheme, MILP and GA

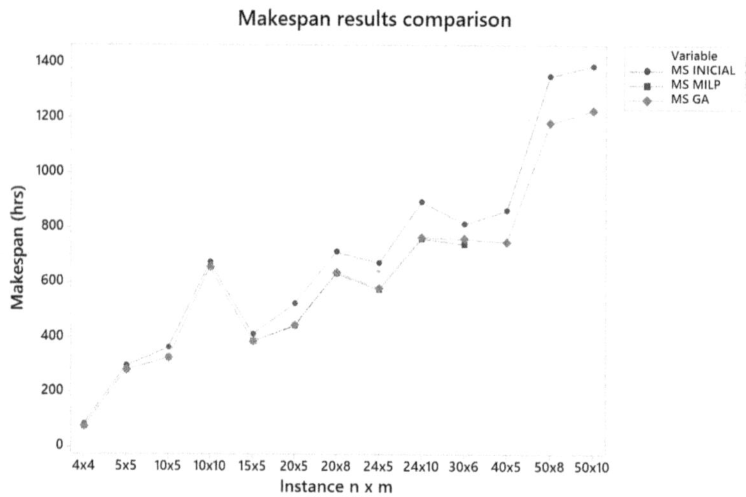

CONCLUSION

According to the results shown in this work, the hypotheses proposed can be verified with respect to time and cost savings. It becomes clear that merely by applying optimization tools to the production schedules, significant improvements on both regards take place. As for which optimization tool is more suitable, the hypotheses are also verified that for this specific company, the differences in the results between the MILP and the GA in the saving of costs and usage time are marginal, despite the inherent disadvantages of the evolutionary methods. It can also be concluded that, due to the exponential trend in runtimes for the MILP for instances greater than 30 jobs, it becomes evident that the genetic algorithm is the most appropriate production programming tool for this particular type of manufacturing system.

Figure 8. Manufacturing cost comparison between current scheme, MILP and GA

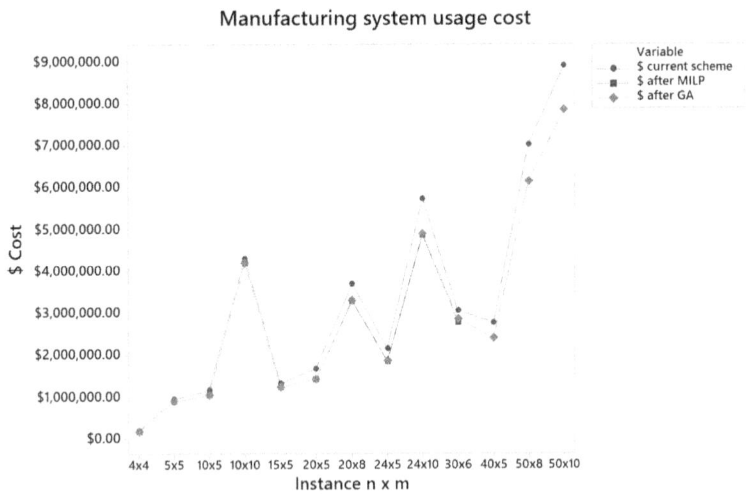

Figure 9. Manufacturing cost savings comparison between MILP and GA

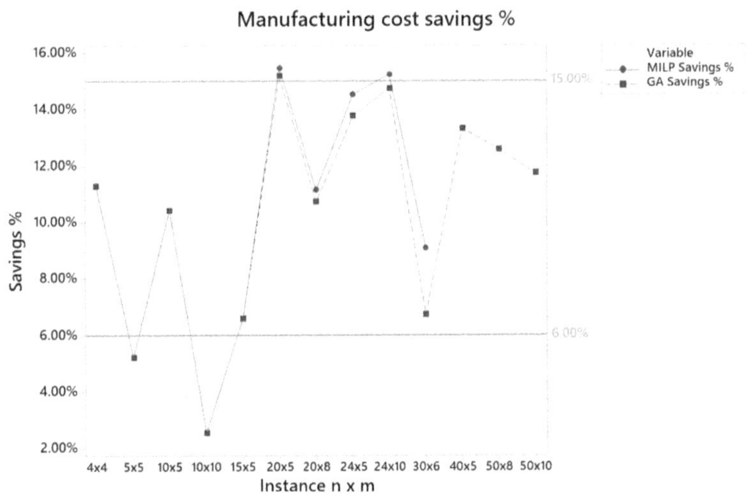

In general, the economic savings shown for instances greater than 20 items are considered very important, since they are in the order of hundreds of thousands and are the most frequently occurring instance sizes among the production horizon of the company. This work shows that annual savings ranging between 7% and 12% can be projected, which is undoubtedly substantial.

In addition to the obvious and palpable benefits discussed in this work, the application of the sequencing criterion to the production programs means that the company's manufacturing system is available for more hours to meet additional work orders, creating an increase in throughput and, consequently, potential revenue growth for the company.

Figure 10. Expected value for improvement in makespan over 110 trials

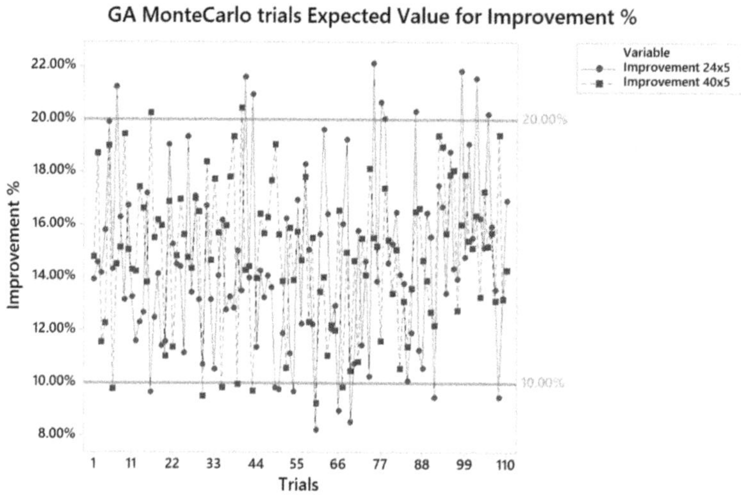

FUTURE RESEARCH DIRECTIONS

It is intended to deepen in the sequencing models proposed in this work by using different dispatching rules and applying different prioritization criteria such as tardiness or customer class. Another path would be to implement constraints on complete batches per customer order; i.e., a production order can contain orders from several customers and it is preferable that the sequences contemplate termination by order to improve the output rate of factory orders.

The consideration of the set-up times for each process is not explicit in this work, but it is implicit in the unit processing times for each item type. It is also intended to break down these times and add the relevant constraints to the model, to be able to account for optimal batch size production. Adaptations to the model can also be made, that allow for scenarios where there can be parallel work systems, mainly in bottlenecks that have been identified. In the same way, it is desirable to implement and perform a factorial experimental design that allows to find with statistical significance, the optimal combination or fine-tunning of the GA hyper-parameters, see Lujan-Moreno, Howard, Rojas, & Montgomery (2018).

ACKNOWLEDGMENT

This research received no specific grant from any funding agency in the public, commercial, or not-for-profit sectors.

REFERENCES

Abdul-Razaq, T. S., Potts, C. N., & Van Wassenhove, L. N. (1990). A survey of algorithms for the single machine total weighted tardiness scheduling problem. *Discrete Applied Mathematics*, *26*(2), 235–253. doi:10.1016/0166-218X(90)90103-J

Asadzadeh, L. (2015). A local search genetic algorithm for the job shop scheduling problem with intelligent agents. *Computers & Industrial Engineering*, *85*, 376–383. doi:10.1016/j.cie.2015.04.006

Asano, M., & Ohta, H. (2002). A heuristic for job shop scheduling to minimize total weighted tardiness. *Computers & Industrial Engineering*, *42*(2–4), 137–147. doi:10.1016/S0360-8352(02)00019-0

Bierwirth, C., & Kuhpfahl, J. (2017). Extended GRASP for the job shop scheduling problem with total weighted tardiness objective. *European Journal of Operational Research*, *261*(3), 835–848. doi:10.1016/j.ejor.2017.03.030

Blackstone, J. H., Phillips, D. T., & Hogg, G. L. (1982). A state-of-the-art survey of dispatching rules for manufacturing job shop operations. *International Journal of Production Research*, *20*(1), 27–45. doi:10.1080/00207548208947745

Blum, C., & Roli, A. (2003). Metaheuristics in combinatorial optimization: Overview and conceptual comparison. *ACM Computing Surveys*, *35*(3), 268–308. doi:10.1145/937503.937505

Brucker, P., Jurisch, B., & Sievers, B. (1994). A branch and bound algorithm for the job-shop scheduling problem. *Discrete Applied Mathematics*, *49*(1–3), 107–127. doi:10.1016/0166-218X(94)90204-6

Cheng, R., Gen, M., & Tsujimura, Y. (1996). A tutorial survey of job-shop scheduling problems using genetic algorithms—I. representation. *Computers & Industrial Engineering*, *30*(4), 983–997. doi:10.1016/0360-8352(96)00047-2

Cortez, P. (2014). *Modern Optimization with R*. London: Springer. doi:10.1007/978-3-319-08263-9

Eiselt, H., & Sandblom, C.-L. (2007). Computational Complexity. In *Linear Programming and its Applications* (pp. 31–44). Berlin: Springer.

Essafi, I., Mati, Y., & Dauzère-Pérès, S. (2008). A genetic local search algorithm for minimizing total weighted tardiness in the job-shop scheduling problem. *Computers & Operations Research*, *35*(8), 2599–2616. doi:10.1016/j.cor.2006.12.019

Floudas, C. A., & Lin, X. (2005). Mixed Integer Linear Programming in Process Scheduling: Modeling, Algorithms, and Applications. *Annals of Operations Research*, *139*(1), 131–162. doi:10.100710479-005-3446-x

Fortnow, L. (2009). The Status of the P Versus NP Problem. *Communications of the ACM*, *52*(9), 78–86. doi:10.1145/1562164.1562186

Gao, K. Z., Suganthan, P. N., Chua, T. J., Chong, C. S., Cai, T. X., & Pan, Q. K. (2015). A two-stage artificial bee colony algorithm scheduling flexible job-shop scheduling problem with new job insertion. *Expert Systems with Applications*, *42*(21), 7652–7663. doi:10.1016/j.eswa.2015.06.004

Györgyi, P., & Kis, T. (2018). Minimizing the maximum lateness on a single machine with raw material constraints by branch-and-cut. *Computers & Industrial Engineering*, *115*, 220–225. doi:10.1016/j.cie.2017.11.016

Hillier, F. S., & Lieberman, G. J. (2014). *Introduction to Operations Research* (10th ed.). New York: McGraw-Hill.

Holland, J. H. (1992). *Adaptation in Natural and Artificial Systems: An Introductory Analysis with Applications to Biology, Control and Artificial Intelligence*. Cambridge, MA: MIT Press.

Ishikawa, S., Kubota, R., & Horio, K. (2015). Effective hierarchical optimization by a hierarchical multi-space competitive genetic algorithm for the flexible job-shop scheduling problem. *Expert Systems with Applications*, *42*(24), 9434–9440. doi:10.1016/j.eswa.2015.08.003

Kelley James, E. (1960). The cutting-plane method for solving convex programs. *Journal of the Society for Industrial and Applied Mathematics*, *8*(4), 703–712. doi:10.1137/0108053

Kondili, E., Pantelides, C. C., & Sargent, R. W. H. (1993). A general algorithm for short-term scheduling of batch operations—I. MILP formulation. *Computers & Chemical Engineering*, *17*(2), 211–227. doi:10.1016/0098-1354(93)80015-F

Lawler, E. L., & Wood, D. E. (1966). Branch-and-bound methods: A survey. *Operations Research*, *14*(4), 699–719. doi:10.1287/opre.14.4.699

Li, J.-Q., Pan, Q.-K., & Tasgetiren, M. F. (2014). A discrete artificial bee colony algorithm for the multi-objective flexible job-shop scheduling problem with maintenance activities. *Applied Mathematical Modelling*, *38*(3), 1111–1132. doi:10.1016/j.apm.2013.07.038

Lujan-Moreno, G. A., Howard, P. R., Rojas, O. G., & Montgomery, D. C. (2018). Design of Experiments and Response Surface Methodology to Tune Machine Learning Hyperparameters, with a Random Forest Case-Study. *Expert Systems with Applications*, *109*, 195–205. doi:10.1016/j.eswa.2018.05.024

Mason, S. J., Fowler, J. W., & Matthew Carlyle, W. (2002). A modified shifting bottleneck heuristic for minimizing total weighted tardiness in complex job shops. *Journal of Scheduling*, *5*(3), 247–262. doi:10.1002/jos.102

Mokotoff, E., & Chrétienne, P. (2002). A cutting plane algorithm for the unrelated parallel machine scheduling problem. *European Journal of Operational Research*, *141*(3), 515–525. doi:10.1016/S0377-2217(01)00270-3

Mönch, L., Schabacker, R., Pabst, D., & Fowler, J. W. (2007). Genetic algorithm-based subproblem solution procedures for a modified shifting bottleneck heuristic for complex job shops. *European Journal of Operational Research*, *177*(3), 2100–2118. doi:10.1016/j.ejor.2005.12.020

Moreno, M. A., Rojas, O., Olivares-Benitez, E., Nucamendi-Guillén, S., & Garcia de Alba Valenzuela, H. R. (2018). Production Planning for a Company in the Industry of Compact Discs Mass Replications. In *New Perspectives on Applied Industrial Tools and Techniques* (pp. 497–516). Cham: Springer. doi:10.1007/978-3-319-56871-3_24

Nemhauser, G. L., & Savelsbergh, M. W. P. (1992). A cutting plane algorithm for the single machine scheduling problem with release times. In *Combinatorial optimization* (pp. 63–83). Springer. doi:10.1007/978-3-642-77489-8_4

Papadimitriou, C. H., & Steiglitz, K. (1998). *Combinatorial optimization: algorithms and complexity*. Courier Corporation.

Pezzella, F., Morganti, G., & Ciaschetti, G. (2008). A genetic algorithm for the flexible job-shop scheduling problem. *Computers & Operations Research*, *35*(10), 3202–3212. doi:10.1016/j.cor.2007.02.014

Pinedo, M., & Singer, M. (1999). A shifting bottleneck heuristic for minimizing the total weighted tardiness in a job shop. *Naval Research Logistics*, *46*(1), 1–17. doi:10.1002/(SICI)1520-6750(199902)46:1<1::AID-NAV1>3.0.CO;2-#

Pinedo, M. L. (2008). *Scheduling. Theory, Algorithms and Systems (3rd ed.)*. New York: Springer.

Rao, S. S. (2009). *Engineering Optimization. Theory and Practice (4th ed.)*. New York: Wiley. doi:10.1002/9780470549124

Sotskov, Y. N., & Shakhlevich, N. V. (1995). NP-hardness of shop-scheduling problems with three jobs. *Discrete Applied Mathematics*, *59*(3), 237–266. doi:10.1016/0166-218X(95)80004-N

Vila, M., & Pereira, J. (2014). A branch-and-bound algorithm for assembly line worker assignment and balancing problems. *Computers & Operations Research*, *44*, 105–114. doi:10.1016/j.cor.2013.10.016

Werner, F. (2013). A survey of genetic algorithms for shop scheduling problems. In Heuristics: Theory and Applications. Nova Science Publishers.

Whitley, D. (1994). A genetic algorithm tutorial. *Statistics and Computing*, *4*(2), 65–85. doi:10.1007/BF00175354

Xiong, H., Fan, H., Jiang, G., & Li, G. (2017). A simulation-based study of dispatching rules in a dynamic job shop scheduling problem with batch release and extended technical precedence constraints. *European Journal of Operational Research*, *257*(1), 13–24. doi:10.1016/j.ejor.2016.07.030

Yazdani, M., Zandieh, M., Tavakkoli-Moghaddam, R., & Jolai, F. (2015). Two meta-heuristic algorithms for the dual-resource constrained flexible job-shop scheduling problem. *Scientia Iranica. Transaction E. Industrial Engineering (American Institute of Industrial Engineers)*, *22*(3), 1242.

Zhang, G., Gao, L., & Shi, Y. (2011). An effective genetic algorithm for the flexible job-shop scheduling problem. *Expert Systems with Applications*, *38*(4), 3563–3573. doi:10.1016/j.eswa.2010.08.145

Chapter 12
GUI Toolbox for Approximation of Fractional Order Parameters With Application to Control of pH Neutralization Process

Bingi Kishore
https://orcid.org/0000-0002-1039-4707
Universiti Teknologi PETRONAS, Malaysia

Rosdiazli Ibrahim
Universiti Teknologi PETRONAS, Malaysia

Mohd Noh Karsiti
Universiti Teknologi PETRONAS, Malaysia

Sabo Miya Hassan
Abubakar Tafawa Balewa University, Nigeria

Vivekananda Rajah Harindran
PETRONAS Group Technical Solutions, Malaysia

ABSTRACT

Fractional-order systems have been applied in many engineering applications. A key issue with the application of such systems is the approximation of fractional-order parameters. The numerical tools for the approximation of fractional-order parameters gained attention recently. However, available toolboxes in the literature do not have a direct option to approximate higher order systems and need improvements with the graphical, numerical, and stability analysis. Therefore, this chapter proposes a MATLAB-based GUI for the approximation of fractional-order operators. The toolbox is made up of four widely used approximation techniques, namely, Oustaloup, refined Oustaloup, Matsuda, and curve fitting. The toolbox also allows numerical and stability analysis for evaluating the performance of approximated transfer function. To demonstrate the effectiveness of the developed GUI, a simulation study is conducted on fractional-order PID control of pH neutralization process. The results show that the toolbox can be effectively used to approximate and analyze the fractional-order systems.

DOI: 10.4018/978-1-5225-8223-6.ch012

INTRODUCTION

The modeling and controlling possibilities of fractional-order systems and controllers gained a lot of attention in the last two decades. However, a key issue with the implementation of such systems is the approximation of the fractional-order parameters (Kar & Roy, 2018; Machado, Kiryakova, & Mainardi, 2011; Matušů, 2011; Shah & Agashe, 2016). For an effective approximation of such parameters, researchers have proposed several time domain and frequency domain techniques (Deniz, Alagoz, Tan, & Atherton, 2016; Du, Wei, Liang, & Wang, 2017; Krishna, 2011; Li, Liu, Dehghan, Chen, & Xue, 2017; Monje, Chen, Vinagre, Xue, & Feliu-Batlle, 2010; Vinagre, Podlubny, Hernandez, & Feliu, 2000). However, it is very difficult to select the best method among these proposals. This is because, each of these methods has its strength and weakness which could be in terms of the order of approximation, the accuracy of frequency and time responses (Djouambi, Charef, & BesançOn, 2007; Tepljakov, 2017b). Furthermore, researchers have developed few toolboxes such as Commande Robuste d'Ordre Non Entier (CRONE) (Malti, Melchior, Lanusse, & Oustaloup, 2011, 2012; Oustaloup, Melchior, Lanusse, Cois, & Dancla, 2000), Non-Integer (Ninteger) (de Oliveira Valério, 2005; Valério & Da Costa, 2004), Fractional Order Modeling and Control (FOMCON) (Tepljakov, 2017b; Tepljakov, Petlenkov, & Belikov, 2011; Tepljakov, Petlenkov, Belikov, & Finajev, 2013) and fractional-order PID (Lachhab, Svaricek, Wobbe, & Rabba, 2013) for fractional-order modeling and control applications. A common feature of these toolboxes is that they are based on MATLAB/SIMULINK software.

The CRONE toolbox was developed by CRONE research team and is dedicated to the application of fractional-order derivatives in science and engineering. The toolbox allows for mathematical modeling, system identification and CRONE control design (Lanusse, Malti, & Melchior, 2013; Malti et al., 2011, 2012; Oustaloup et al., 2000). On the other hand, the Ninteger toolbox developed by Valerio presents the implementation of non-integer PID controllers and CRONE controllers (Valério & Da Costa, 2004). The toolbox also allows for the approximation of non-integer order derivatives, functions for model development and frequency response plots. Unlike the aforementioned toolboxes, the FOMCON toolbox incorporates advanced features such as system identification, fractional-order PID control, real-time implementation etc. The toolbox is an extension of the fractional-order transfer function (FOTF) toolbox for fractional-order system identification and PID controller design. Similarly, the authors in (Lachhab et al., 2013) developed a fractional-order PID (FOPID) toolbox for designing and tuning the controller using the steepest descent method. A summary of these toolboxes explaining the features and limitations is given in Table 1.

A key disadvantage of these toolboxes is that there is no direct option to approximate higher order systems. Moreover, the available techniques are mostly limited to Outsaloup and refined Oustaloup approximations. Another improvement needed on the toolboxes is in the graphical, numerical and stability analysis. Therefore, the contribution of this chapter is to develop a GUI for the approximation of fractional-order differentiator, integrator, PID controller and transfer function. The development of GUI can be done in any simulation software. However, in this paper MATLAB is used. Furthermore, four widely used approximation techniques as against the two techniques will be incorporated into the design of the toolbox. The toolbox will also allow for graphical, numerical and stability analysis for evaluating the performance of approximated fractional-order systems.

The remaining sections of the chapter are organized as follows: the definition of fractional-order differentiator/integrator operator, PID controller, transfer function and the overview of Oustaloup, refined Oustaloup, Matsuda and curve fitting approximation algorithms are presented in Section 2. An overview of the toolbox is provided in Section 3. The simulation study on pH neutralization process for demonstrating the functionality of toolbox is given in Section 4. Final conclusions are given in Section 5.

FRACTIONAL CALCULUS AND APPROXIMATION TECHNIQUES

In this section, a brief introduction to fractional calculus and an overview of the approximation techniques are presented.

Fractional Calculus

In this subsection, the fractional-order definitions of differintegral operator, PID controller and transfer function are presented. First, the fractional-order differintegral is a combined differentiator and integrator operator that generalizes the notations for differentiator ($\alpha > 0$) and integrator ($\alpha < 0$) (Monje et al., 2010; Valério, Trujillo, Rivero, Machado, & Baleanu, 2013). Thus, the differintegral operator for the function *f(t)* is defined as follows:

$$ {}_aD_t^{\alpha} f(t) = \begin{cases} \dfrac{d^{\alpha} f(t)}{dt^{\alpha}}, \alpha > 0 \\ f(t), \alpha = 0 \\ \int\limits_0^t f(\tau) d\tau^{\alpha}, \alpha < 0 \end{cases} \tag{1} $$

where t and a are the lower and upper bounds of the operator and $\alpha \in R$ is the fractional-order parameter. The Laplace transform of equation (1) at zero initial condition is defined as

$$ L\{{}_aD_t^{\alpha} f(t); s\} = s^{\alpha} F(s) \tag{2} $$

In the equation, the approximation of s^{α} for ($\alpha > 0$) and $1/s^{\alpha}$ for ($\alpha < 0$) will be done using the developed GUI. Then, with the use of these operators, the transfer function of FOPID controller is defined as (Chen & Atherton, 2007; Monje et al., 2010):

$$ C(s) = K_p + \frac{K_i}{s^{\lambda}} + K_d s^{\mu}, 0 < \lambda, \mu, 2 \tag{3} $$

where K_p, K_i and K_d are the proportional, integral and derivative constant gains, λ and μ are the fractional-order integration and differentiation. Furthermore, as reported in (Monje et al., 2010), (Xue, 2017) and (Chen & Atherton, 2007), the generalized fractional-order transfer function in the form of polynomial with fractional powers is given as:

Table 1. Summary of numerical toolboxes for fractional-order systems and PID controller

Name	Typical Usage	Features & Limitations	Software	Syntax	Download	Reference
CRONE	System identification and control	• Non-integer differentiation, differential equations, system identifications, inverse Laplace transform and CRONE control • Not applicable for time-delay and MIMO systems	Matlab/ Simulink	-	✓	(Malti & Victor, 2015; Yousfi, Melchior, Rekik, Derbel, & Oustaloup, 2012)
Ninteger	GUI for controller design	• Fractional-order approximation techniques, system identification function and analysis, GUI for controller design. • Not updated and many inbuilt functions are prompting errors because of overload editing	Matlab/ Simulink	nipid()	✓	(Valério & Da Costa, 2004)
FOPID	Design of FOPID controller	• FOPID design, Approximation techniques, Step, Nyquist, Bode, Nichols plots. • There is no publicly available source for download.	Matlab	-	✗	(Lachhab et al., 2013)
@fotf	Design of fractional-order systems	• Overload functions was developed for fractional transfer function, time and frequency domain, stability analysis. • Lack of complete FOPID module with design, tuning and analysis.	Matlab	fotf()	✓	(Chen, Petras, & Xue, 2009)
FSST	Fractional-order state-space design and analysis	• Simulink blocks for stochastic and non-stochastic fractional-order systems, fractional Kalman filter. • Developed only Simulink blocks and another major drawback is step size has the high large impact on results.	Simulink	-	✓	(Dzielinski & Sierociuk, 2008; Sierociuk, 2005)
FOMCON	Modeling and control of fractional-order systems	• System identification in both time and frequency domains, FOPID design, tuning and implementation, Step, Bode, Nyquist and Nichols plots • Uses only Oustaloup and refined Oustaloup approximation techniques	Matlab/ Simulink	fpid()	✓	(Tepljakov, 2017a, 2017b; Tepljakov et al., 2011)
Sysquake FOPID	Analysis and design of FOPID	• Integrative tool for FOPID design and analysis • Developed in Sysquake: A programming language based numerical computing tool	Sysquake	GUI	✗	(Dormido, Pisoni, & Visioli, 2010, 2012; Pisoni, Visioli, & Dormido, 2009)
DFOC	Discrete FOPID controller	• Discrete version of FOPID controller • Provides only a transfer function of FOPID with the given parameters	Matlab	DFOC()	✓	(Petrás, 2011; Petráš, 2011)
FIT	Fractional-order integration	• Riemann-Liouville type of differentiation and integration • Contains only fractional-order differentiation and integration	C++, Matlab	-	✗	(Marinov, Ramirez, & Santamaria, 2013)

The 'Download' column denotes whether the toolbox is publicly available for download either from MATLAB file exchange or other websites/forums.

The 'Syntax' column denotes the MATLAB command for FOPID controller.

$$G(s) = \frac{b_m s^{\alpha_m} + b_{m-1} s^{\alpha_{m-1}} + \cdots + b_1 s^{\alpha_1} + b_0 s^{\alpha_0}}{a_m s^{\eta_m} + a_{m-1} s^{\eta_{m-1}} + \cdots + a_1 s^{\eta_1} + a_0 s^{\eta_0}} \tag{4}$$

where $b_i, a_i \in R$ are the coefficients of numerator and denominator of FOTF and $\alpha_i, \eta_i \in R$ are the fractional-order powers. In the equation, α_0 and η_0 are usually set to zero. Therefore, the static gain of the system is defined as $K = \frac{b_0}{a_0}$.

Overview of the Approximation Algorithms

In this subsection, the four standard and widely used frequency domain approximation algorithms for s^α are presented. The considered algorithms are Oustaloup, refined Oustaloup, Matsuda and curve fitting. Thus, the Oustaloup approximation of s^α is defined as:

$$s^\alpha \approx \omega_h^\alpha \prod_{k=1}^{N} \frac{s + \omega_k^{'}}{s + \omega_k}, 0 < \alpha < 1 \tag{5}$$

where the zeros $\left(\omega_k^{'}\right)$ and poles $\left(\omega_k\right)$ are computed as

$$\omega_k^{'} = \omega_l \left(\frac{\omega_h}{\omega_l}\right)^{\frac{2k-1-\alpha}{2N}} ; \omega_k = \omega_l \left(\frac{\omega_h}{\omega_l}\right)^{\frac{2k-1+\alpha}{2N}} \tag{6}$$

where *α* is the order of fractional-order differentiator, *N* is the order of approximation and $\left(\omega_l, \omega_h\right)$ is the frequency range of interest. The Oustaloup approximation algorithm is the most widely used technique. However, for practical applications, it is limited by the desired frequency range and cannot fit the wide range of frequencies (Monje et al., 2010). Therefore, the refined Oustaloup approximation of s^α in the desired frequency range $\left(\omega_l, \omega_h\right)$ is defined as:

$$s^\alpha \approx \left(\frac{d\omega_h}{b}\right)^\alpha \left(\frac{ds^2 + b\omega_h s}{d(1-\alpha)s^2 + b\omega_h s + d\alpha}\right) \prod_{k=1}^{N} \frac{s + \omega_k^{'}}{s + \omega_k}, 0 < \alpha < 1 \tag{7}$$

where $\left(\omega_k^{'}\right)$ and $\left(\omega_k\right)$ are the respective zeros and poles given in equation (6). The constants *b* and *d* are selected as 10 and 9 respectively to achieve good approximation (K. Bingi, R. Ibrahim, M. N. Karsiti, & S. M. Hassan, 2017; Kishore Bingi, Rosdiazli Ibrahim, Mohd Noh Karsiti, & Sabo Miya Hassan, 2017; Chen & Atherton, 2007; Xue, 2017). However, this method results in a very high integer order transfer function. On the other hand, Matsuda approximation of s^α will be done in two stages. First, by using the continued fraction expansions, a rational model of s^α will be obtained. Then, fitting of the original function in a desired frequency points $\omega_0, \omega_1, \cdots, \omega_N$ is performed. Thus, the approximated transfer function of s^α is defined as:

$$s^\alpha \approx d_0(\omega_0) + \cfrac{s-\omega_0}{d_1(\omega_1) + \cfrac{s-\omega_1}{d_2(\omega_2) + \cfrac{s-\omega_2}{\cdots}}} \tag{8}$$

where

$$d_0(\omega) = \left|(j\omega)^\alpha\right|; d_{k+1}(\omega) = \frac{\omega - \omega_k}{d(\omega) - d_k(\omega_k)}, k = 0,1,2,\ldots,N \tag{9}$$

In this method, the sum of the total number of zeros and poles is known as the order of approximation *N*. Here, *N* should be an even number, else the approximation technique results in an improper transfer function (Deniz et al., 2016; Vinagre et al., 2000). Similarly, the procedure for the integer-order approximation of s^α based on the curve fitting of frequency response data is as follows:

1. Obtain the frequency response data for integer-order part of s^α within the desired frequency range using the function frd().
2. Obtain the exact frequency response data of s^α by powering the data obtained in previous step with α.
3. Choose the order of the approximate *N* for the integer-order model
4. Obtain the state space model of exact function response data using the function fitfrd() and then convert the state space model to transfer function using ss2tf().

Overview of the Toolbox

The GUI based toolbox allows the approximation of fractional-order based systems and controllers. The main window of the GUI is shown in Figure 1. As shown in the figure, the GUI is divided into three modules namely fractional-order function, approximation method and results. Furthermore, the organization of the developed GUI is shown in Figure 2.

Fractional-Order Function

This is the first module of the GUI. Here, the users will have the option to select among the following four types of fractional-order functions namely differentiator, integrator, PID controller and transfer function. Upon selecting the function, a brief description of it can be obtained from the "info" button. An example of such description for the fractional-order transfer function is shown in Figure 3 where the definition of the fractional-order transfer function in polynomial form with fractional powers is clearly

Figure 1. Main window of the toolbox

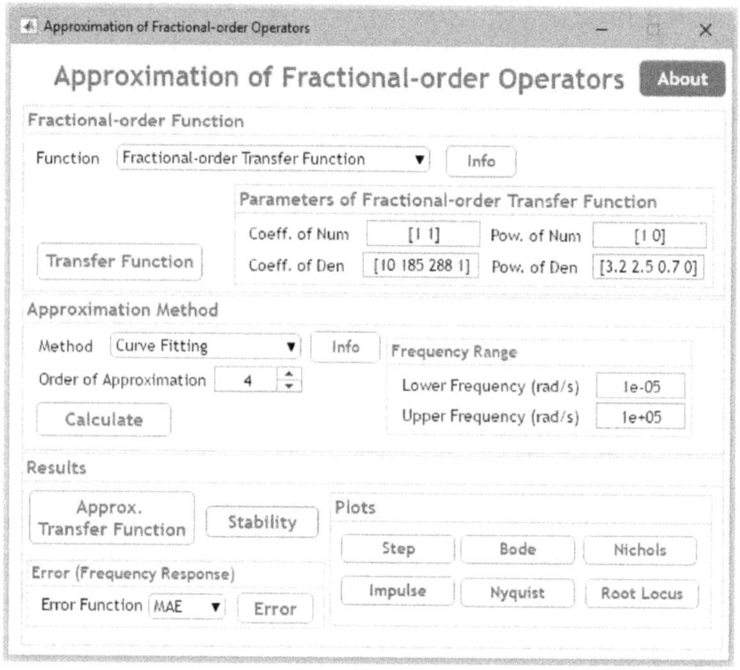

Figure 2. Organization of the MATLAB based GUI toolbox

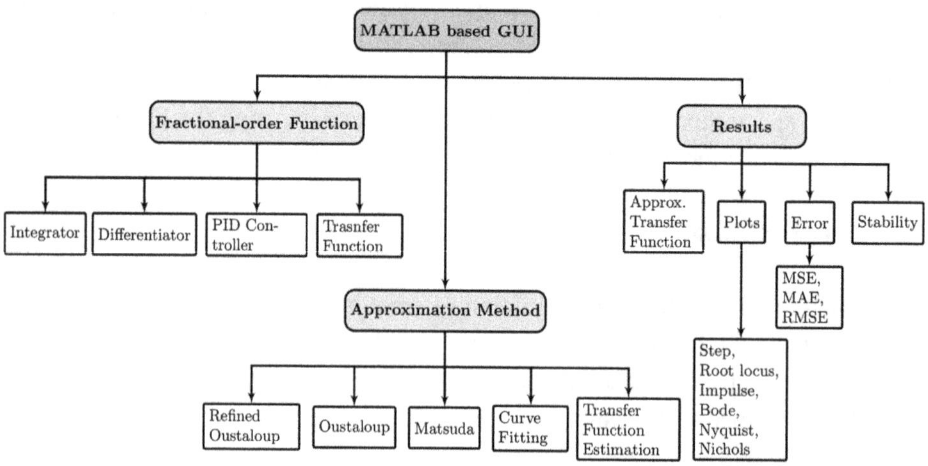

seen. Furthermore, the GUI will also allow users to enter the parameters of the selected function. This can be seen clearly in the first module of Figure 1. Consequently, the coefficients and powers of the numerator and denominator of the FOTF can be entered directly in the form of an array. Similarly, the parameter window for the fractional-order PID controller is shown in Figure 4. Once defined, the "Transfer function" button will display the selected function with the defined parameters in the command window.

Figure 3. Details of the fractional-order Transfer Function

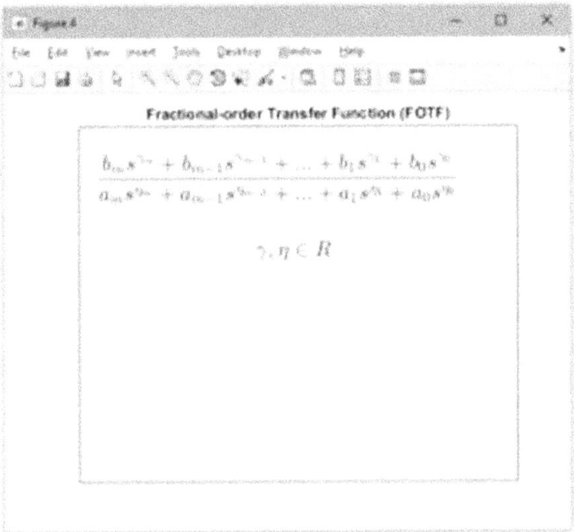

Figure 4. Parameters window for the fractional-order PID controller

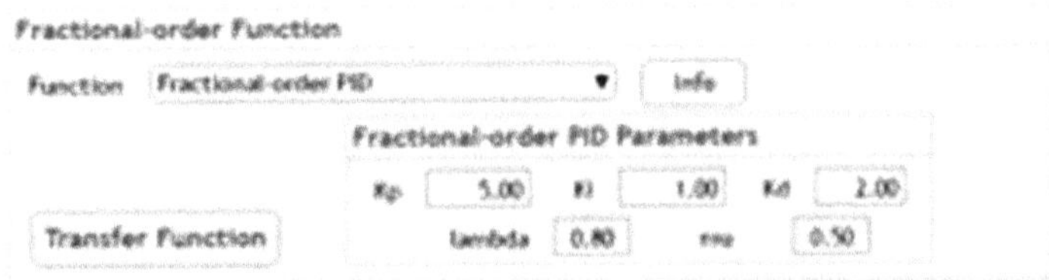

Approximation Method

In this second module, the GUI will allow the user to select the following four types of approximation methods namely Oustaloup, Refined Oustaloup, Matsuda and Curve fitting. Here also, a brief description of the approximation method is available from "info" button which is similar to the "info" button available in the previous module. An example of the description for refined Oustaloup approximation is given in Figure 5. In the figure, the definition of the refined Oustaloup approximation for s^{α} in the desired frequency range and order of approximation *N* is presented. As mentioned earlier in Section 2, *N* should be an even number for Matsuda approximation, otherwise, the approximation will yield into an improper transfer function. This will be notified to the user as shown in Figure 6 if the selected function is `Matsuda'.

Upon selection of an approximation method, the toolbox allows the user to select the order of approximation *N* and desired frequency range as shown in the second module of Figure 1. Once defined, the "Calculate" button will calculate the approximation of the selected function and will pop-up the "Results" module.

Figure 5. Details of the Refined Oustaloup Approximation

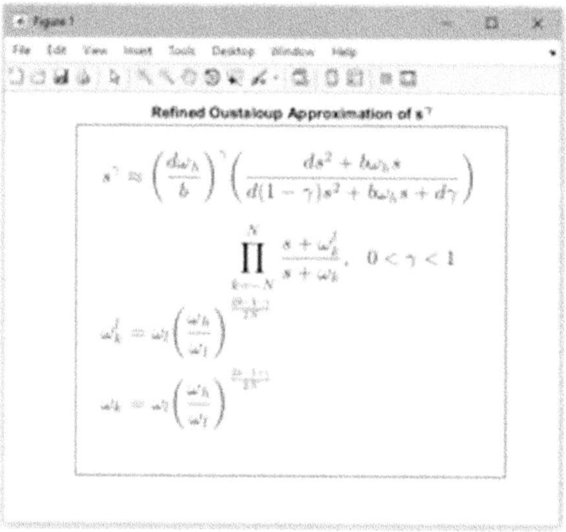

Figure 6. Warning and Error pop-up windows for Matsuda approximation

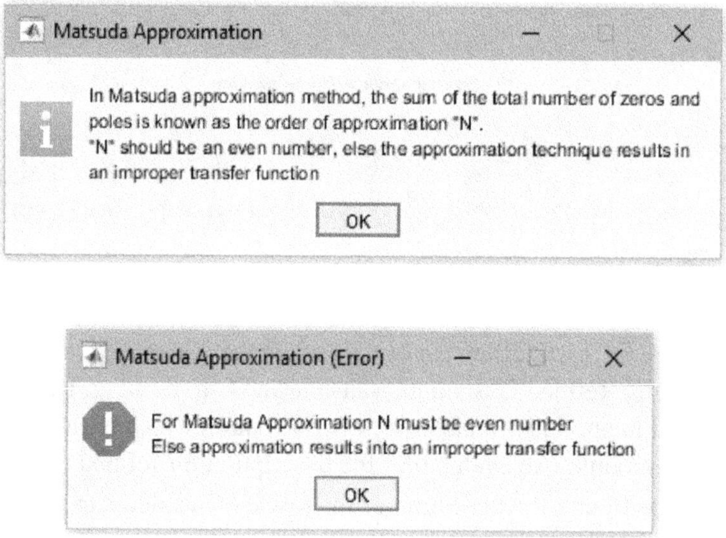

Results

The results section is the final module of the GUI. This module is further subdivided into the four sub-modules namely Approx. Transfer Function, Plots (time and frequency domain responses), Error (error between the exact and the approximated responses) and Stability. In the first sub-module, upon clicking the "Approx. Transfer Function", the approximated transfer function of the selected fractional-order system will be displayed in the command window.

In the second sub-module, all time and frequency domain plots like root locus, step, impulse, Bode, Nyquist and Nichols are available. Furthermore, in all plots, the approximated response is compared with the exact response by default. Meanwhile, numerical assessment such as mean absolute error (MAE), mean squared error (MSE) and root mean squared error (RMSE) defined in equations (10), (11) and (12) are available in the third sub-module. Lastly, the "Stability" analysis button of sub-module four allows the user to evaluate the stability of the approximated transfer function.

$$MAE = \frac{1}{n}\sum_{i=1}^{n}\left|y_i - x_i\right| \tag{10}$$

$$MSE = \frac{1}{n}\sum_{i=1}^{n}\left(y_i - x_i\right)^2 \tag{11}$$

$$RMSE = \frac{1}{n}\sum_{i=1}^{n}\sqrt{\left(y_i - x_i\right)^2} \tag{12}$$

where y_i is the approximated frequency response data, x_i is the exact frequency response data and *n* is the total size of the data. Here, the exact frequency response data of the fractional-order model is obtained by substituting $s = j\omega$ and then evaluating the function for different values of ω in the desired frequency range. On the other hand, the exact step response of the fractional-order model is obtained from the inverse Laplace transform of integer-order integrator $\frac{1}{s^n}$ as:

$$L^{-1}\left[\frac{1}{s^n}\right] = \frac{t^{n-1}}{(n-1)!}, n \in N \tag{13}$$

As in the case of equation (13), the inverse Laplace transform of the fractional integrator $\frac{1}{s^\alpha}$ is derived as:

$$L^{-1}\left[\frac{1}{s^\alpha}\right] = \frac{t^{\alpha-1}}{(\alpha-1)!}, 0 < \alpha < 1 \tag{14}$$

where $\Gamma(\alpha) = (\alpha - 1)!$. From (14), the step response is computed as

$$L^{-1}\left[\frac{1}{s^{\alpha+1}}\right] = \frac{t^\alpha}{\Gamma(\alpha+1)} \tag{15}$$

SIMULATION STUDY

In this section, the validation of MATLAB GUI for the approximation of fractional-order transfer function followed by control of pH neutralization process using the toolbox is presented.

Fractional-Order Transfer Function

To evaluate the GUI toolbox, consider the fractional-order transfer function model given as follows:

$$G(s) = \frac{s+1}{10s^{3.2} + 185s^{2.5} + 288s^{0.7} + 1} \tag{16}$$

The model will be approximated using the developed GUI. From *G(s)* in equation (16) the coefficients and powers of numerator and denominators are deduced as Coeff. of Num = [1 1], Coeff. of Den = [10 185 288 1], Pow of Num = [1 0] and Pow of Den = [3.2 2.5 0.7 0]. From the parameters, it can be seen that the orders of coefficients and powers should be same else the GUI will pop up and error showing that the "Order of coefficients and powers of Num/Den is not same". Therefore, these parameters of the model will be entered in the GUI as shown in Figure 1. In all the cases, the order of approximation and the desired frequency range are chosen as 5 and (10^{-3}, 10^{3}). However, for Matsuda, the order of approximation is chosen as 10 which is equal to the sum of 5 zeros and 5 poles for effective comparison with other approaches.

The bode plots of the approximated transfer functions of the model using Oustaloup, refined Oustaloup, Matsuda and curve fitting are given in Figures 7, 8, 9 and 10 respectively. The numerical analysis of the figures for all the three errors (MSE, MAE and RMSE) is given in Table 2. The obtained results show that using the toolbox the approximation and analysis can be done easily and effectively. Furthermore, the stability of the approximated transfer function from the table also shows that all the approximated transfer functions are stable.

Control of pH Neutralization Process

To further demonstrate the function of the toolbox, a simulation study is conducted for control of pH neutralization process using fractional-order PID controller. The schematic diagram of the pH neutralization process is shown in Figure 11. The plant consists of a continuous stirred tank reactor (CSTR) VE110 in which the base is added to the acid to regulate the pH. To control the flow-rate of acid and base, two flow control valves FCV120 and FCV121 are used. Also, two flow transmitters FT120 and FT121 are used to measure the flow-rate of the acid and base respectively. The sensor AT is used to measure the pH value of the solution while the agitator AG is used for mixing the two solutions homogeneously.

The model of the CSTR governed by the incoming flow-rate of acid F_1 and base F_2 and the outgoing flow-rate F_1+F_2 is given as:

$$V\frac{d}{dt}C[H^+] = F_1C_1 - (F_1 + F_2)C[H^+] \tag{17}$$

Figure 7. Bode plot of G(s) using Oustaloup approximation

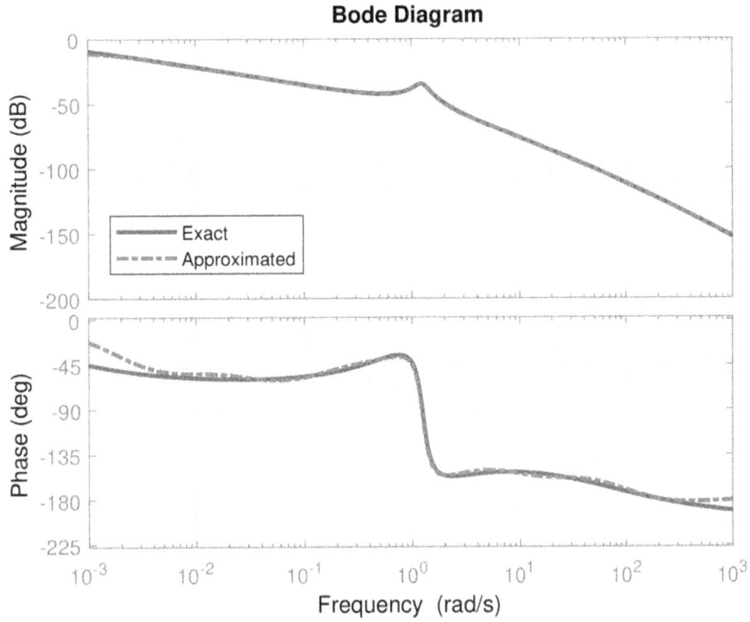

Figure 8. Bode plot of G(s) using Refined Oustaloup approximation

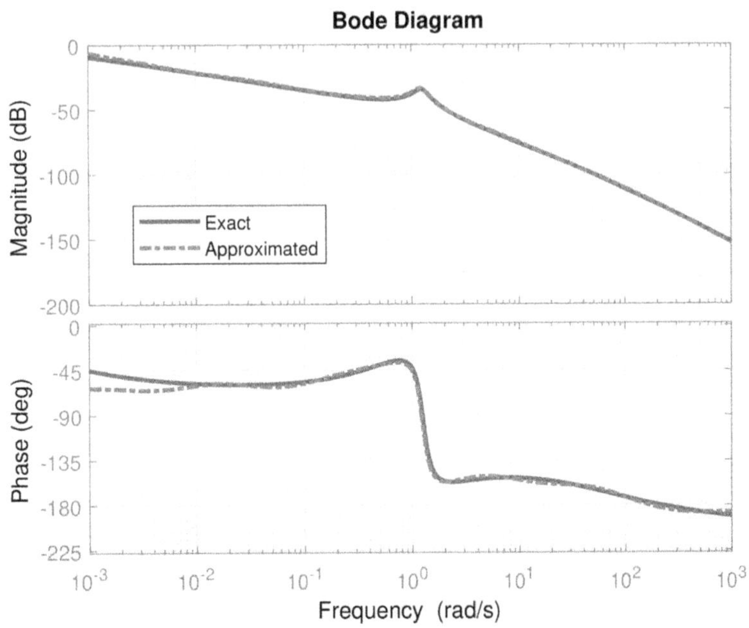

Figure 9. Bode plot of G(s) using Matsuda approximation

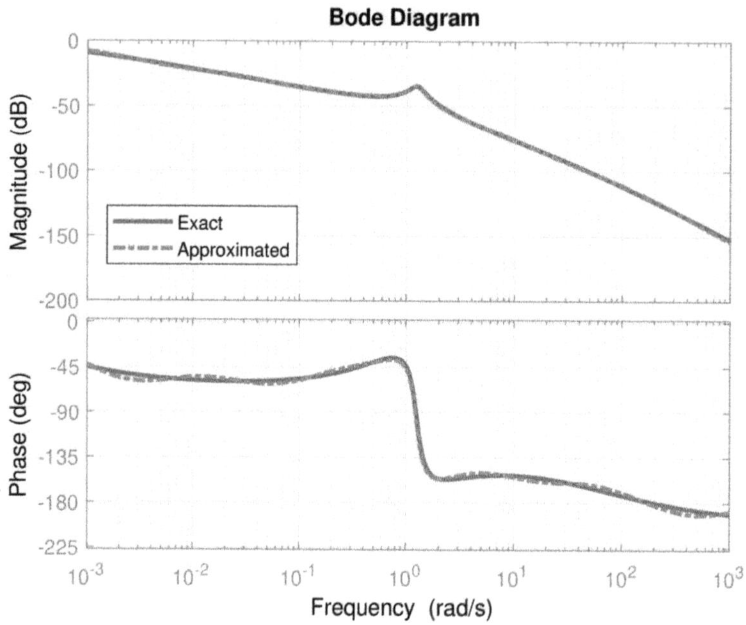

Figure 10. Bode plot of G(s) using curve fitting

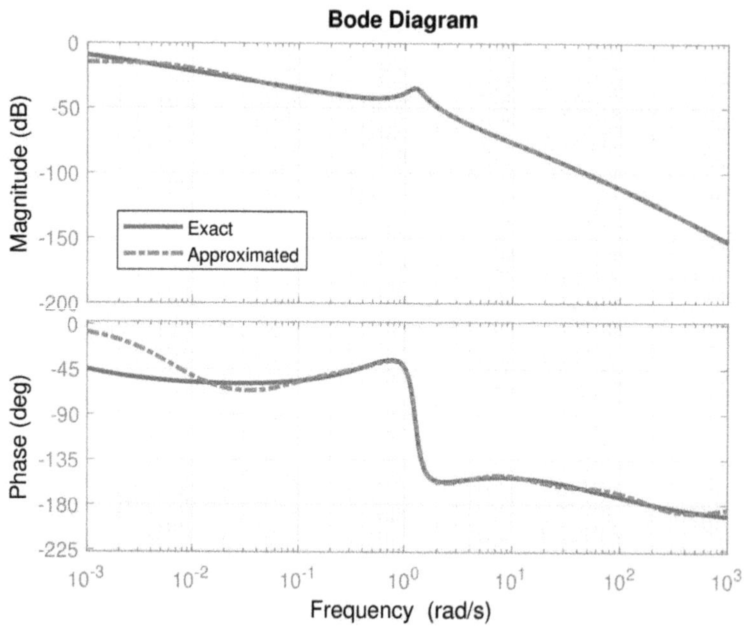

Table 2. Comparison of errors for Magnitude and Phase Response of G(s)

Technique	MSE		MAE		RMSE		Stable
	Magnitude	Phase	Magnitude	Phase	Magnitude	Phase	
Oustaloup	0.0829	38.3409	0.2227	5.1760	0.2879	6.1920	Yes
Refined Oustaloup	0.1159	5.3705	0.2985	1.9960	0.3403	2.3174	Yes
Matsuda	0.5236	9.4788	0.6405	2.6449	0.7236	3.0788	Yes
Curve Fitting	0.2293	10.7641	0.4199	2.6240	0.4789	3.2809	Yes

Figure 11. Schematic diagram of pH neutralization process

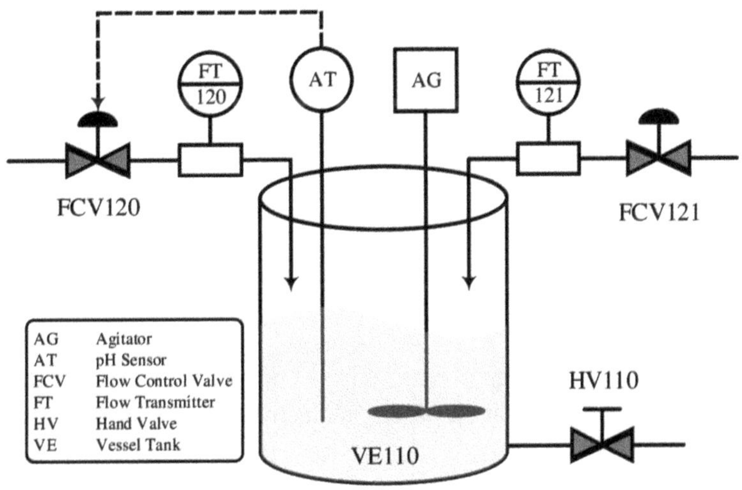

$$V\frac{d}{dt}C[OH^-] = F_2C_2 - (F_1 + F_2)C[OH^-] \tag{18}$$

where V is the volume of the tank, C_1 and C_2 are the concentrations of acid and base and $C[H^+]$ and $C[OH^-]$ are the concentrations of H^+ and OH^- components in the reactor solution respectively. Consequently, consider the titration process of acetic acid (CH_3COOH) neutralized with sodium hydroxide base (NaOH), the reactions of the process are given as follows:

$$CH_3COOH + NaOH \to CH_3COONa + H_2O \tag{19}$$

$$CH_3COOH \to H^+ + CH_3COO^- \tag{20}$$

$$NaOH \to Na^+ + OH^- \tag{21}$$

$$H^{+} + OH^{-} \rightleftharpoons H_2O \tag{22}$$

From equation (20), the concentration of non-reactant component C[H+] and the acid dissociation constant (K_a) are defined as

$$C[H^{+}] = [CH_3COOH] + [CH_3COO^{-}] \tag{23}$$

$$K_a = \frac{[H^{+}][CH_3COO^{-}]}{[CH_3COOH]} \tag{24}$$

Similarly, from equation (22), the concentration of non-reactant component C[OH-] and the ionic product of water (K_w) are defined as

$$C[OH^{-}] = [Na^{+}] \tag{25}$$

$$K_w = [H^{+}][OH^{-}] \tag{26}$$

Furthermore, the charge balance equation based on electroneutrality principle is given as

$$[H^{+}] + [Na^{+}] = [OH^{-}] + [CH_3COO^{-}] \tag{27}$$

Thus, substituting equation (23) through (26) in (27), a polynomial equation in [H+] is derived as

$$[H^{+}]^3 + [H^{+}]^2(K_a + C[OH^{-}]) + [H^{+}](K_a C[OH^{-}] - K_w - K_a C[H^{+}]) - K_a K_w = 0 \tag{28}$$

The result of equation (28) will be substituted in the following equation to calculate the pH value of the solution in the CSTR.

$$pH = -\log_{10}[H^{+}] \tag{29}$$

The model parameters used for the simulation are: V=50L, F_1=10m^3/h, C_1=C_2=0.5, K_a=1.86*10^{-5} and K_w=1*10^{-14}. The titration curve for the pH process is shown in Figure 12. From the figure, the process is highly nonlinear and very sensitive.

For the control of this neutralization process, the FOPID controller is designed as shown in equation (3). Furthermore, the parameters of the controller are tuned using accelerated particle swarm optimization based on the design in (Kishore Bingi et al., 2017). The optimization parameters of the algorithm

Figure 12. Titration curve

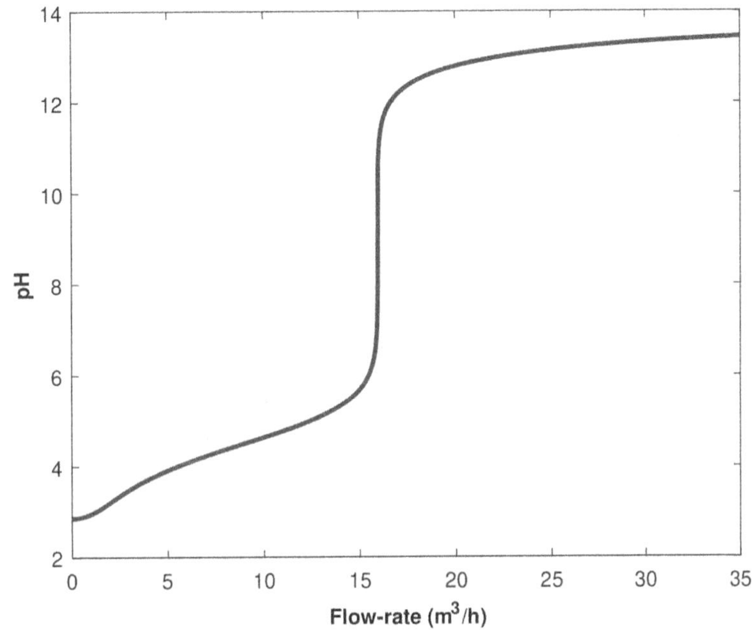

are given in Table 3. The objective function used for the optimization is integral time absolute error (ITAE) given as follows:

$$ITAE = \int_0^T t \mid e(t) \mid dt \tag{30}$$

Therefore, the transfer function of FOPID controller designed for the pH neutralization process is given as

Table 3. Optimization parameters of APSO algorithm

Parameter	Value
Dimensional search space	5
Population size	50
Bird Step	50
Randomness parameter	0.5
Control variable	0.7
Acceleration constant	0.7

$$C(s) = 35.9286 + \frac{20.0159}{s^{0.9164}} + 1.4016s^{0.9570} \tag{31}$$

The designed FOPID controller will be approximated using the developed GUI for simulating the pH neutralization process. In all the cases, the order of approximation and the desired frequency range are chosen as 5 and (10^{-3}, 10^{3}). The choice for the range of frequency and the order of approximation are based on reported works in (Chen & Atherton, 2007; Deniz et al., 2016; Xue, 2017). For the Matsuda approach, the order of approximation is chosen as 10 which is equal to the sum of 5 poles and 5 zeros for effective comparison with other methods.

The Bode plots of the approximated controllers as obtained from the toolbox using all the four approaches are shown in Figures 13, 14, 15 and 16. Likewise, through the toolbox, the numerical analysis based on MSE, MAE and RMSE is shown in Table 4. From the results, the magnitude response of the curve fitting approach outperformed the other approaches for all the performance indexes. However, for the phase response, the Matsuda approximation is better with respective MSE and RMSE values of 0.7903 and 0.8890.

To further see the functionality of the toolbox, the Nichols and Nyquist plots of C(s) using the Matsuda and curve fitting approaches shown in Figures 17, 18 and 19, 20. Whereas in the case of bode plot, the effectiveness of the Matsuda and curve fitting approaches have been confirmed. This was made possible and easily through the toolbox.

Figure 13. Bode plot of C(s) using Oustaloup approximation

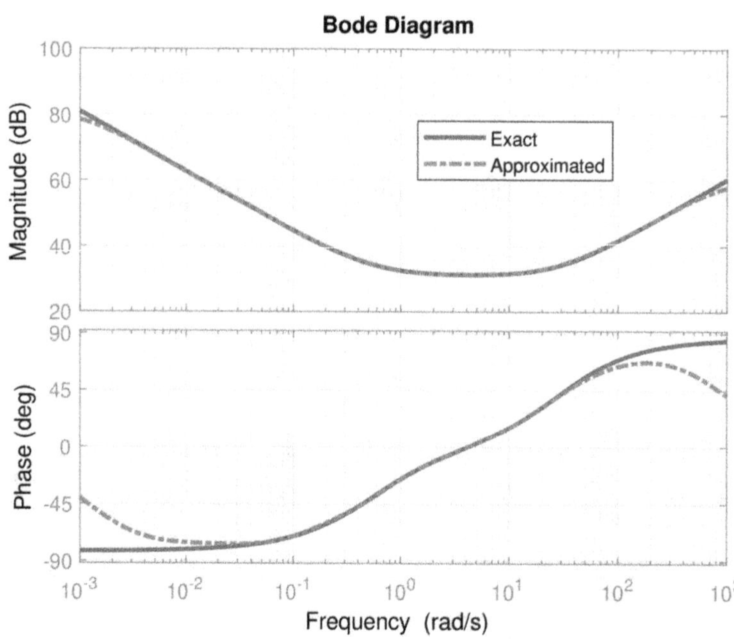

Figure 14. Bode plot of C(s) using Refined Oustaloup approximation

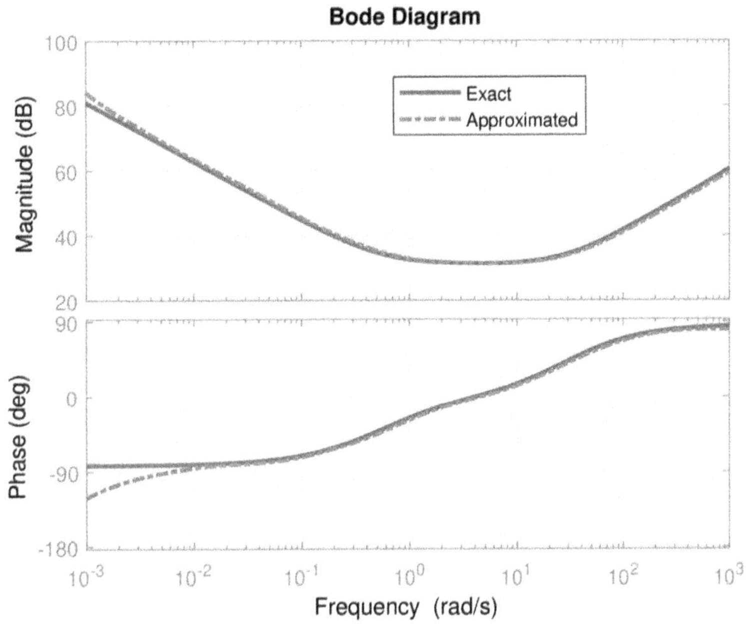

Figure 15. Bode plot of C(s) using Matsuda approximation

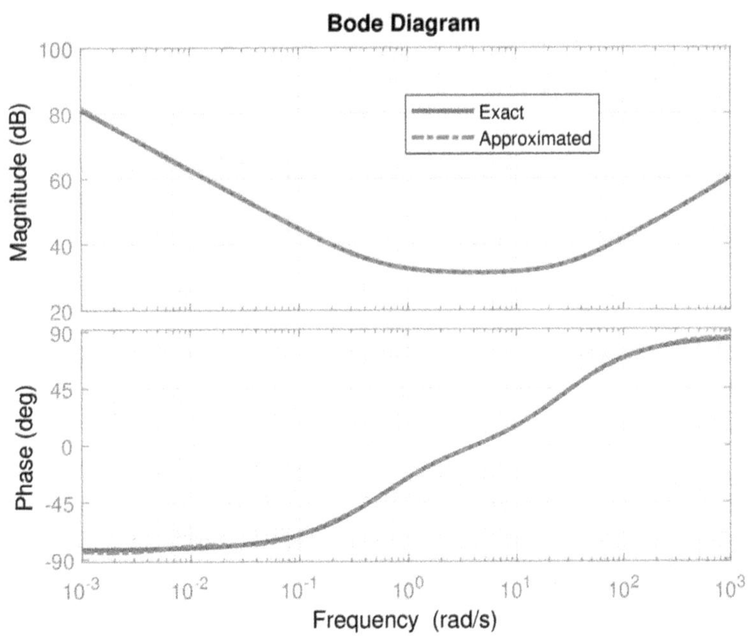

Figure 16. Bode plot of C(s) using curve fitting

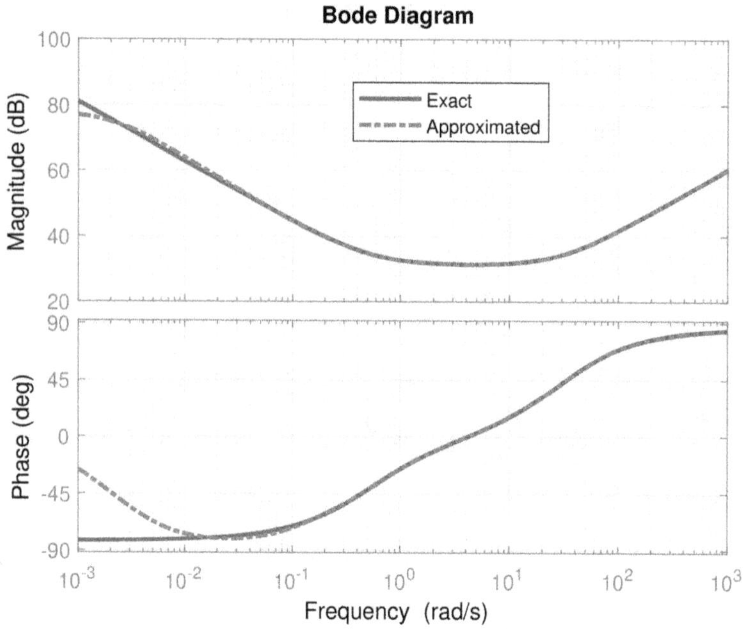

Table 4. Comparison of errors for Magnitude and Phase Response of C(s)

Technique	MSE		MAE		RMSE		Stable
	Magnitude	Phase	Magnitude	Phase	Magnitude	Phase	
Oustaloup	1.2611	19.2644	0.8166	23.5561	1.1230	26.8191	Yes
Refined Oustaloup	0.7732	8.3027	0.8637	2.5313	0.8793	2.8814	Yes
Matsuda	0.0219	0.7903	0.1234	0.8250	0.1481	0.8890	Yes
Curve Fitting	0.0169	3.1823	0.0259	0.2029	0.1301	1.7839	Yes

The comparison of the response of the pH neutralization process with controllers as obtained from the toolbox is shown in Figure 21. The regions of interest A, B, C and D of the figure are further highlighted in Figure 22. The numerical assessment of the response in terms of rise time (t_r), settling time (t_s) and overshoot (%OS) is given in Table 5. Similarly, the comparison of the response of the pH neutralization process for variation in set-point is shown in Figure 23. Furthermore, the regions of interest A, B, C and D of the figure are further highlighted in Figure 24. The results showed that the fractional-order controllers can be approximated using the proposed toolbox.

Figure 17. Nichols plot of C(s) using Matsuda approximation

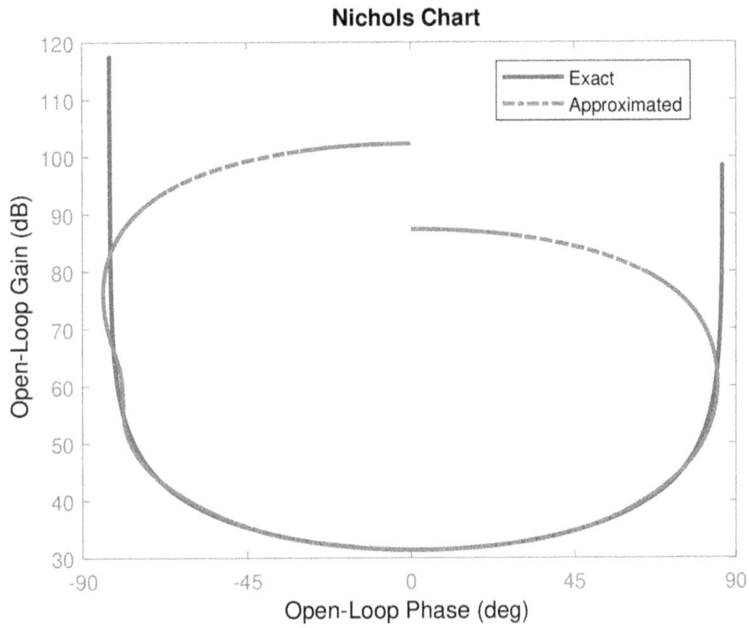

Figure 18. Nichols plot of C(s) using Curve Fitting approximation

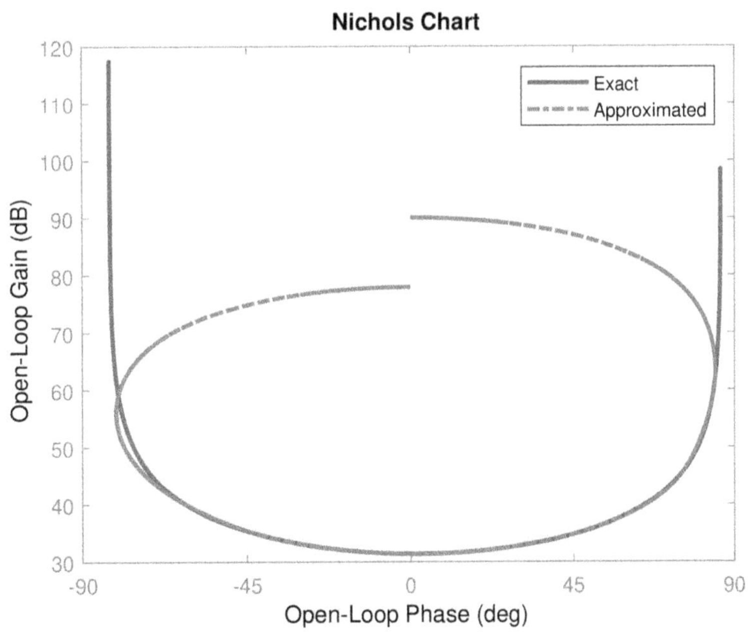

Figure 19. Nyquist plot of C(s) using Matsuda approximation

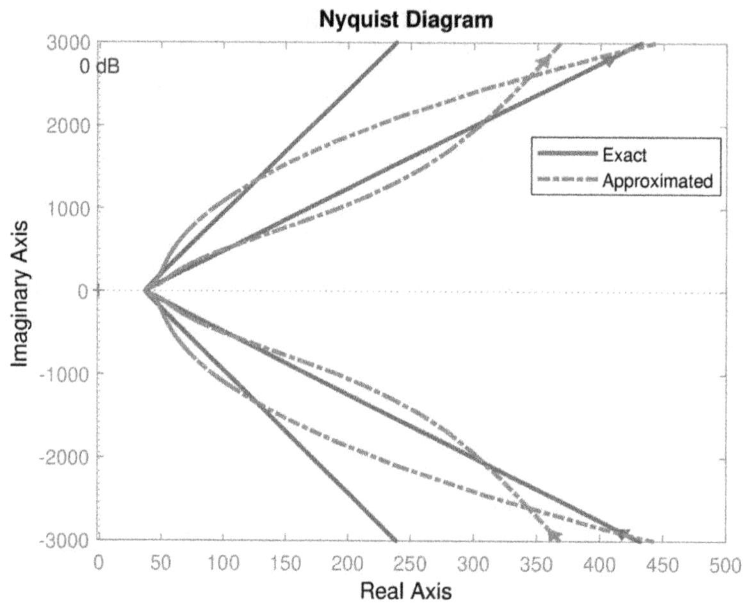

Figure 20. Nyquist plot of C(s) using Curve Fitting approximation

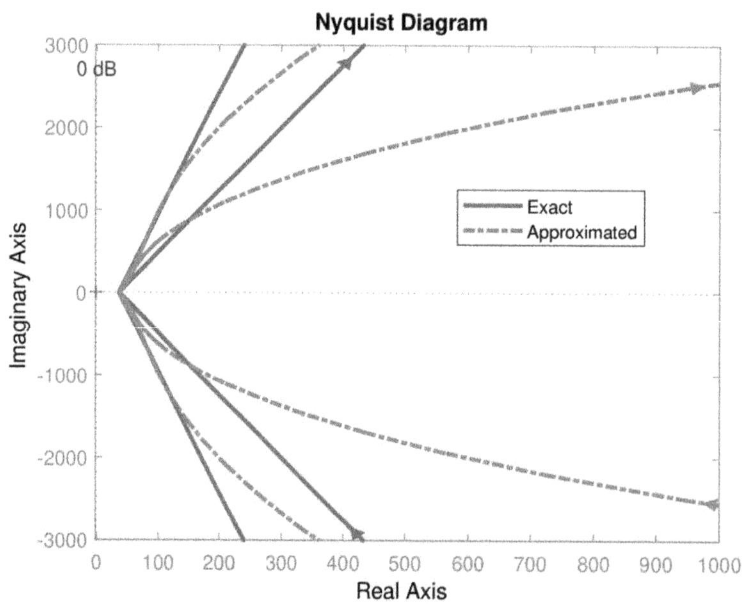

Table 5. Performance analysis of various fractional order PID controllers

Controller	Rise time (s)	Settling time (s)	%OS
PID	21.1272	98.6257	2.6755
$PI^{\lambda}D^{\mu}$ - Oustaloup	20.0930	85.2225	2.1675
$PI^{\lambda}D^{\mu}$ - Ref Oustaloup	19.7167	108.3747	2.4538
$PI^{\lambda}D^{\mu}$ - Matsuda	20.2031	77.4872	2.6550
$PI^{\lambda}D^{\mu}$ - Curve Fitting	20.2538	73.3282	2.4923

Figure 21. Performance over set-point tracking and disturbance rejection

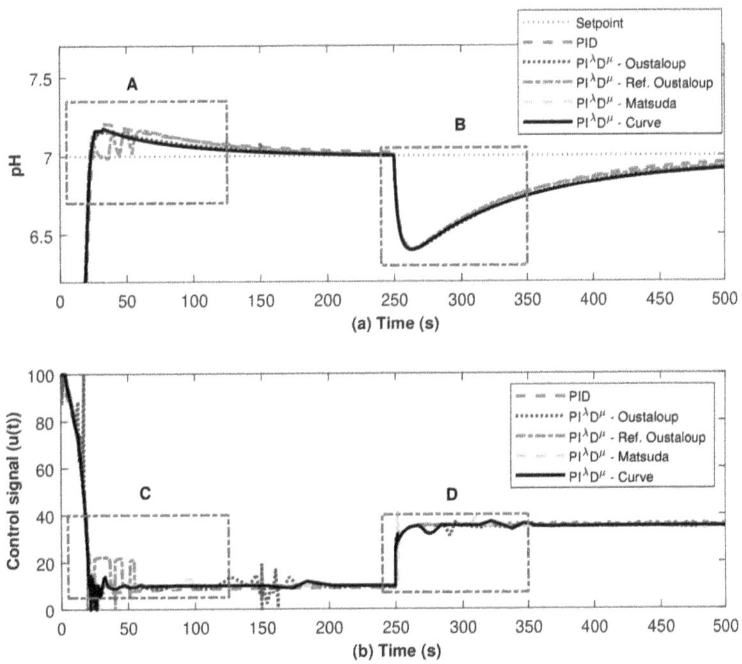

CONCLUSION

In this paper, a MATLAB based GUI toolbox is proposed for the approximation of fractional-order differentiator, integrator, PID controller and transfer function. The toolbox consists of four widely used approximation techniques i.e., Oustaloup, refined Oustaloup, Matsuda and Curve fitting. It also allows for graphical and numerical analysis in both time and frequency domains. The graphical analysis includes root locus, Bode, Nyquist, Nichols, Step and impulse responses while the numerical analysis includes MAE, MSE, RMSE of magnitude and phase response of Bode plot. The toolbox has been tested for the approximation of fractional-order PID controllers designed for pH neutralization process. The results obtained shows that using the toolbox the approximation and analysis can be done easily.

As part of future studies, MATLAB Simulink blocks will be developed for real-time implementation and evaluation of fractional-order models.

Figure 22. Zoomed-in view of regions A, B, C and D of Figure 20

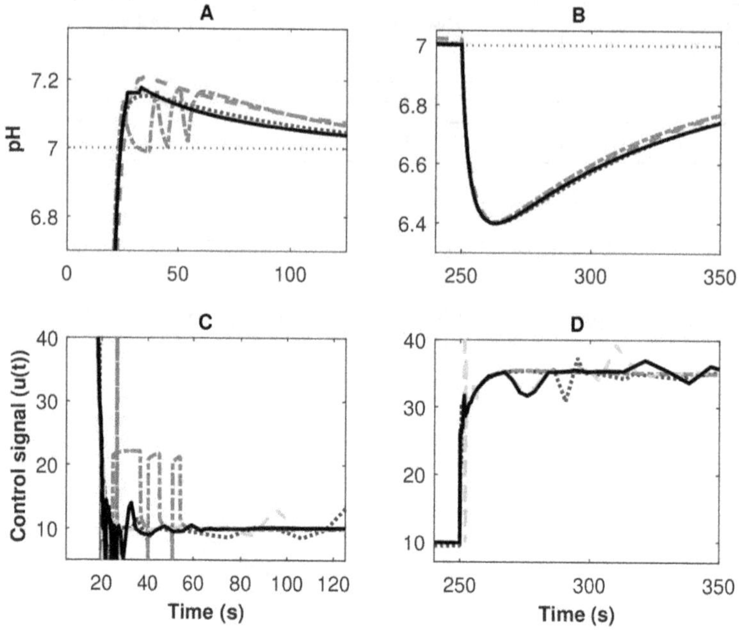

Figure 23. Performance over variable set-point tracking

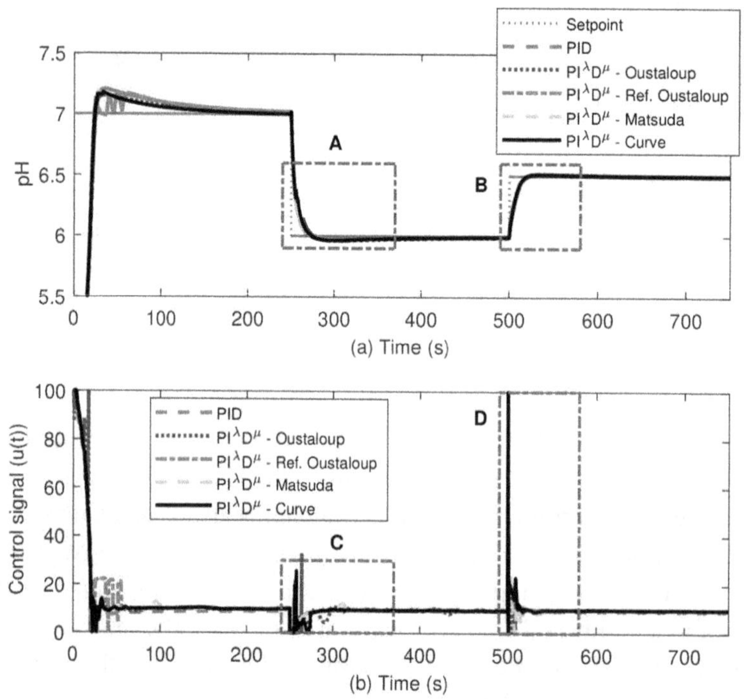

Figure 24. Zoomed-in view of regions A, B, C and D of Figure 22

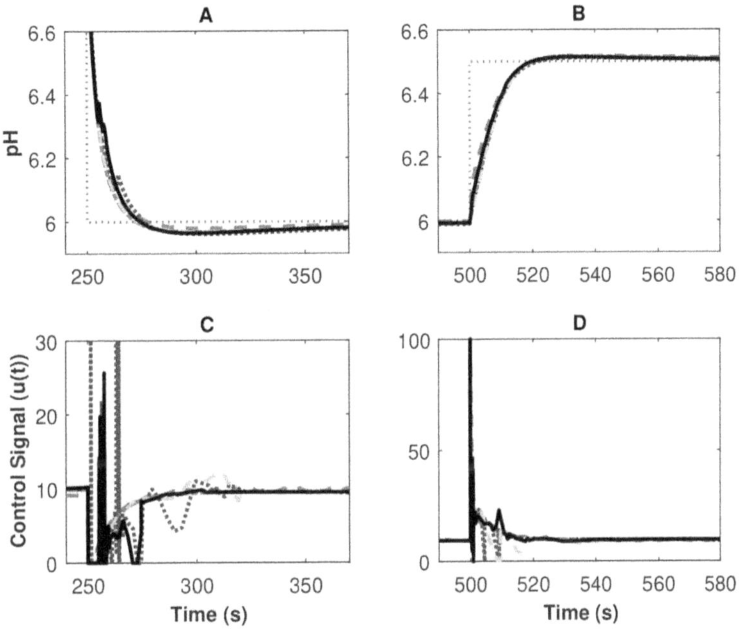

ACKNOWLEDGMENT

This work was supported by Yayasan Universiti Teknologi PETRONAS Fundamental Research Grant No. 0153AA-H16.

REFERENCES

Bingi, K., Ibrahim, R., Karsiti, M. N., & Hassan, S. M. (2017). Fractional-order filter design for set-point weighted PID controlled unstable systems. *International Journal of Mechanical and Mechatronics Engineering*, *17*(5), 173–179.

Bingi, K., Ibrahim, R., Karsiti, M. N., & Hassan, S. M. (2017). Fractional Order Set-point Weighted PID Controller for pH Neutralization Process Using Accelerated PSO Algorithm. *Arabian Journal for Science and Engineering*, 1–15.

Chen, Y., & Atherton, D. P. (2007). *Linear feedback control: analysis and design with MATLAB* (Vol. 14). Siam.

Chen, Y., Petras, I., & Xue, D. (2009). *Fractional order control-a tutorial.* Paper presented at the American Control Conference, 2009. ACC'09.

de Oliveira & Mata. (2005). *Ninteger v. 2.3 Fractional control toolbox for MATLAB.* Lisboa, Universidade Technical.

Deniz, F. N., Alagoz, B. B., Tan, N., & Atherton, D. P. (2016). An integer order approximation method based on stability boundary locus for fractional order derivative/integrator operators. *ISA Transactions*, *62*, 154–163. doi:10.1016/j.isatra.2016.01.020 PMID:26876378

Djouambi, A., Charef, A., & Besançon, A. (2007). Optimal approximation, simulation and analog realization of the fundamental fractional order transfer function. *International Journal of Applied Mathematics and Computer Science*, *17*(4), 455–462. doi:10.2478/v10006-007-0037-9

Dormido, S., Pisoni, E., & Visioli, A. (2010). An interactive tool for loop-shaping design of fractional-order PID controllers. *Proceedings of the 4th IFAC Workshop on Fractional Differentiation and Its Applications (FDA'10).*

Dormido, S., Pisoni, E., & Visioli, A. (2012). Interactive tools for designing fractional-order PID controllers. *International Journal of Innovative Computing, Information, & Control*, *8*(7), 4579–4590.

Du, B., Wei, Y., Liang, S., & Wang, Y. (2017). Rational approximation of fractional order systems by vector fitting method. *International Journal of Control, Automation, and Systems*, *15*(1), 186–195. doi:10.100712555-015-0351-1

Dzielinski, A., & Sierociuk, D. (2008). Simulation and experimental tools for fractional order control education. *Proc. IFAC*. 10.3182/20080706-5-KR-1001.01975

Kar, B., & Roy, P. (2018). A Comparative Study Between Cascaded FOPI–FOPD and IOPI–IOPD Controllers Applied to a Level Control Problem in a Coupled Tank System. *Journal of Control, Automation and Electrical Systems*, *29*(3), 340–349. doi:10.100740313-018-0373-z

Krishna, B. T. (2011). Studies on fractional order differentiators and integrators: A survey. *Signal Processing*, *91*(3), 386–426. doi:10.1016/j.sigpro.2010.06.022

Lachhab, N., Svaricek, F., Wobbe, F., & Rabba, H. (2013). *Fractional order PID controller (FOPID)-toolbox.* Paper presented at the Control Conference (ECC), 2013 European.

Lanusse, P., Malti, R., & Melchior, P. (2013). CRONE control system design toolbox for the control engineering community: tutorial and case study. *Phil. Trans. R. Soc. A, 371*(1990), 20120149.

Li, Z., Liu, L., Dehghan, S., Chen, Y. Q., & Xue, D. (2017). A review and evaluation of numerical tools for fractional calculus and fractional order controls. *International Journal of Control*, *90*(6), 1165–1181. doi:10.1080/00207179.2015.1124290

Machado, J., Kiryakova, V., & Mainardi, F. (2011). Recent history of fractional calculus. *Communications in Nonlinear Science and Numerical Simulation*, *16*(3), 1140–1153. doi:10.1016/j.cnsns.2010.05.027

Malti, R., Melchior, P., Lanusse, P., & Oustaloup, A. (2011). *Towards an object oriented CRONE toolbox for fractional differential systems.* Paper presented at the 18th IFAC world congress. 10.3182/20110828-6-IT-1002.02443

Malti, R., Melchior, P., Lanusse, P., & Oustaloup, A. (2012). Object-oriented crone toolbox for fractional differential signal processing. *Signal, Image and Video Processing*, *6*(3), 393–400. doi:10.100711760-012-0323-3

Malti, R., & Victor, S. (2015). *CRONE Toolbox for system identification using fractional differentiation models.* Paper presented at the 17th IFAC Symposium on System Identification (SYSID 2015). 10.1016/j.ifacol.2015.12.223

Marinov, T. M., Ramirez, N., & Santamaria, F. (2013). Fractional integration toolbox. *Fractional Calculus & Applied Analysis, 16*(3), 670–681. doi:10.247813540-013-0042-7 PMID:24812536

Matušů, R. (2011). Application of fractional order calculus to control theory. *International Journal of Mathematical Models and Methods in Applied Sciences, 5*(7), 1162-1169.

Monje, C. A., Chen, Y., Vinagre, B. M., Xue, D., & Feliu-Batlle, V. (2010). Fractional-order systems and controls: fundamentals and applications. Springer Science & Business Media.

Oustaloup, A., Melchior, P., Lanusse, P., Cois, O., & Dancla, F. (2000). *The CRONE toolbox for Matlab.* Paper presented at the Computer-Aided Control System Design, 2000. CACSD 2000. IEEE International Symposium on. 10.1109/CACSD.2000.900210

Petrás, I. (2011). *Fractional derivatives, fractional integrals, and fractional differential equations in Matlab. In Engineering education and research using MATLAB.* InTech.

Petráš, I. (2011). *Fractional-order nonlinear systems: modeling, analysis and simulation.* Springer Science & Business Media. doi:10.1007/978-3-642-18101-6

Pisoni, E., Visioli, A., & Dormido, S. (2009). *An interactive tool for fractional order PID controllers.* Paper presented at the Industrial Electronics, 2009. IECON'09. 35th Annual Conference of IEEE. 10.1109/IECON.2009.5414720

Shah, P., & Agashe, S. (2016). Review of fractional PID controller. *Mechatronics, 38*, 29–41. doi:10.1016/j.mechatronics.2016.06.005

Sierociuk, D. (2005). *Fractional order discrete state-space system simulink toolkit user guide.* Retrieved from http://www.ee.pw.edu.pl/~dsieroci/fsst/fsst.htm

Tepljakov, A. (2017a). *FOMCON: Fractional-Order Modeling and Control Toolbox. In Fractional-order Modeling and Control of Dynamic Systems* (pp. 107–129). Springer. doi:10.1007/978-3-319-52950-9_6

Tepljakov, A. (2017b). *Fractional-order modeling and control of dynamic systems.* Springer. doi:10.1007/978-3-319-52950-9

Tepljakov, A., Petlenkov, E., & Belikov, J. (2011). FOMCON: A MATLAB toolbox for fractional-order system identification and control. *International Journal of Microelectronics and Computer Science, 2*(2), 51–62.

Tepljakov, A., Petlenkov, E., Belikov, J., & Finajev, J. (2013). *Fractional-order controller design and digital implementation using FOMCON toolbox for MATLAB.* Paper presented at the Computer Aided Control System Design (CACSD), 2013 IEEE Conference on. 10.1109/CACSD.2013.6663486

Valério, D., & Da Costa, J. S. (2004). Ninteger: a non-integer control toolbox for MatLab. *Proceedings of the Fractional Differentiation and its Applications.*

Valério, D., Trujillo, J. J., Rivero, M., Machado, J. A. T., & Baleanu, D. (2013). Fractional calculus: A survey of useful formulas. *The European Physical Journal. Special Topics*, *222*(8), 1827–1846. doi:10.1140/epjst/e2013-01967-y

Vinagre, B. M., Podlubny, I., Hernandez, A., & Feliu, V. (2000). Some approximations of fractional order operators used in control theory and applications. *Fractional Calculus & Applied Analysis*, *3*(3), 231–248.

Xue, D. (2017). *Fractional-order Control Systems: Fundamentals and Numerical Implementations* (Vol. 1). Walter de Gruyter GmbH & Co KG. doi:10.1515/9783110497977

Yousfi, N., Melchior, P., Rekik, C., Derbel, N., & Oustaloup, A. (2012). Design of Centralized CRONE Controller Combined with MIMOQFT Approach Applied to Non Square Multivariable Systems. *International Journal of Computer Applications, 45*(16).

Chapter 13
Lean Six Sigma Implementation Framework Using Resource-Based Theory Approach:
An Integrative Literature Review

Michael Sony
Namibia University of Science and Technology, Namibia

ABSTRACT

Resource-based theory (RBT) is one of the most popular paradigms in operations management. Lean Six Sigma (LSS) is the most widely used business improvement initiative for the last two decades. The previous LSS frameworks were concentrated on the LSS processes, and none of them proposed a link between organizational strategy, organizational resources and capabilities for gaining competitive advantage. The purpose of this chapter is to construct an RBT-based LSS framework. By reviewing the previous literature on RBT, LSS and organizational strategy, this chapter builds an implementation framework. This framework considers the firm resources, firm capability, competitive advantage, and LSS integration as the foundation of a firm's strategy. The research suggests that application of RBT can add richness to LSS implementation within an organization, which will bring practical implications for forming the resource-based strategy in the organizations leading to competitive advantage. This is the first chapter to suggest an integration framework of RBT with LSS for strategy implications.

INTRODUCTION

Resource-based Theory (RBT) is a popular paradigm in strategy management, however, its popularity has also increased in Operations Management(Grant, 1999; Hitt, Xu, et al., 2016). The fundamental proposition in RBT is that firms try to differentiate from its rival to attain a competitive advantage. Thus, a firm can differentiate from its rivals by making the best use of its resources and capabilities to gain competitive advantage. Lean Six Sigma (LSS) is onc of the most widely used methodology which relies on the collaborative efforts of group of people and between people and resources in the organizations to

DOI: 10.4018/978-1-5225-8223-6.ch013

improve the organizations performance by systematically reducing waste and variations(Naslund, 2008; Asefeso, 2014; de Freitas et al., 2017; Laureani and Antony, 2017). Many researchers have devoted considerable attention to testing LSS implementation models(Kumar et al., 2006; Hilton and Sohal, 2012; Karunakaran, 2016; Hill et al., 2018). Previous studies on LSS frameworks were basically case studies. The case studies of LSS frameworks primarily concentrated on auto industry(Kumar et al., 2006; Vinodh et al., 2011), electronics(Chen and Lyu, 2009), manufacturing(Gnanaraj et al., 2012; Vinodh et al., 2014), Aero maintenance(Thomas et al., 2015; Karunakaran, 2016), Maintenance Repair and Overhaul (Hill et al., 2018) etc. These industry-specific LSS frameworks were concentrated on LSS processes, and none of them proposed a link between organizational strategy and organizational resources and capabilities for the implementation of LSS. Besides, the previous LSS frameworks were also specific to the industry suggesting various practices to implement LSS. However, organizations still need to identify and or/ acquire resources which are needed in its portfolio for developing capabilities for implementing these practices in an effective manner(Hitt, Xu, et al., 2016). These capabilities may be acquired in a different manner depending on the strategy employed by the organizations(Hitt, Carnes, et al., 2016). To cite an example organization who aspire to be a market leader will deploy LSS strategy successfully in product or service R & D and another organization who wants to extract more out of manufacturing process will deploy LSS into manufacturing processes. The firm's resources and capabilities in these organizations are therefore likely to be different depending on the organization strategy(Grant, 1999) even though they are deploying LSS. If LSS or TQM is implemented in isolation without linking to organization strategy, they may attain operational effectiveness but may be insufficient to achieve a competitive advantage. The organizations have to coordinate across all of their activities to produce superior value to the customers and achieve competitive advantage(Porter, 1996; Hitt, Carnes, et al., 2016). An important point to consider is the imitation of LSS specific practices or capabilities may be easily carried out by the rivals. However, what will be difficult to imitate by the rivals will be when the activities are coordinated or synchronized in a strategic manner. When these capabilities are coordinated it means rivals have to imitate a series of capabilities which are needed to perform the practices, opposed to specific practices(Peteraf, 1993). Thus, there is a need for an LSS framework which will link organizational strategy and organizational resources and capabilities for its implementation. Recently, there has been a resurgence of interest in operations management. Hitt et al., (2016) in the Journal of Operations Management highlighted the importance of RBT in Operations Management and further explicated applying RBT to all operation strategy for gaining deep understanding from these perspectives. There has been no research done yet on RBT integration during the implementation of LSS. Thus, in this study, we intend to ask the research question on *1) How to integrate RBT while implementing LSS for gaining a competitive advantage? 2) How should future research proceed given our findings?* The integrative review method is the only approach that allows for the combination of varied methodologies (for example, experimental and non-experimental research), and has the potential to play a greater role in evidence-based studies(Whittemore and Knafl, 2005). An integrative literature review summarizes the past literature and draws an overall conclusion from the body of literature on the said topic(Beyea and Nicoll, 1998). LSS and RBT being a matured topic and as our purpose was to summarize the previous studies and build a framework to integrate RBT with LSS methodologies. Therefore, an integrative literature review is conducted as the topic under consideration. The chapter is organized as follows the next section in this chapter elucidates the background theory, which will be followed by the methodology and therefore the proposed RBT based LSS framework, discussion, conclusion, and scope for future research.

BACKGROUND THEORY

RBT is an important paradigm in strategic management, however, it has also become popular in operation management(Hitt, Carnes, et al., 2016; Hitt, Xu, et al., 2016). RBT assumes that firms within an industry are heterogenous based on their difference in resources. Strategic management scholars used RBT to understand how a firm creates an advantage over industry rivals with strategy. The proposed link between competition among product market position and competition among resources position was the first starting point of RBT as a strategy (Wernerfelt, 1984). The firm's sustainable competitive advantage depends on the ability to continuously recombine its assets stocks and apply them to new markets opportunities. Therefore, a thinking arose that firm's important resources are accumulated rather than acquired through strategic factor markets. Firms with similar resources may vary in their performance in the market. Similarly, firms with similar investments over a period of time may show different outcomes over the period of time(Dierickx and Cool, 1989). From such a viewpoint it transpires that firms need valuable and rare resources to gain competitive advantage. However, to sustain the competitive advantage the resources should be not be imitated or substituted by other firms' resources(Barney, 1991). In addition, it also called for managing or orchestrating the firm's resources to create capabilities which will be further difficult to imitate. Further, these capabilities can be leveraged with appropriate strategies(Sirmon et al., 2011) to creating a competitive advantage. The field of operations management recently has begun to emphasize strategic and macro issues, leading to importing various theories from other disciplines(Hitt, Carnes, et al., 2016). RBT was important to explain various issues in operations management(Pilkington and Meredith, 2009). Operations strategy, therefore, establishes a connection between operations and corporate strategy. An organization by proper strategic positioning or aligning of operations capabilities can significantly impact competitive strength and business performance (Anderson et al., 1989). Operations can act as a cushion against rival attacks. If operations are well integrated with firms employees and processes, it can be very difficult to imitate(Hayes and Upton, 1998). In operations, there have been many approaches or methods for improving the product, service or process performance. LSS is one such popular methodology in the last two decades(Laureani and Antony, 2017). LSS is the integration of Lean with Six Sigma. LSS is defined as a business strategy and methodology that increases process performance, resulting in greater customer satisfaction and results(Snee, 2010). The difference between Lean and Six Sigma was that Six Sigma took longer training time, larger investment, reduce defects, increase capability and stability. Lean on the other hand reduce inefficiency within the system and Six Sigma improves the effectiveness(Antony, 2011). During the last 15 years LSS is implemented throughout the world from the manufacturing to services such as Police, higher education, financial sector, healthcare and so on(Antony et al., 2018). While success stories on LSS usually speak for themselves and do not need any shreds of evidence due to the large plethora of studies. However, the failures were due to various reasons which were specific to the organization and the manner it was implemented(Albliwi et al., 2014; McLean et al., 2017). Initially organizations felt that LSS was a large toolkit of both Lean and Six Sigma tools, however, later they realized the importance of LSS as an organizational strategy(Antony et al., 2018), which is used in a strategic sense can benefit the organization in the long run. Thus, integration of LSS and RBT could be beneficial for organizations.

FOCUS OF THE CHAPTER

The previous studies of LSS suggest that its use as a strategic tool could benefit the organization by creating a competitive advantage. The integration of RBT with LSS can add value in two dimensions. First, LSS is considered a business strategy and methodology that reduce waste and variation with the operations resources and capabilities leading to competitive positioning of the firm(Antony et al., 2017). In RBT the resource especially orchestration complements various business strategy which is similar to the LSS thinking by with a focus on using existing resources optimally and acquiring or unbundling of strategic resources to create competitive advantage(Grant, 1999; Hitt, Carnes, et al., 2016; Hitt, Xu, et al., 2016). Secondly, for LSS to be sustainable, it requires a synergistic process of integrating and aligning all business operations in a strategic manner from the supplier to the consumer(Albliwi et al., 2014). RBT with a special focus on resource orchestration will help in optimizing, acquiring, bundling, leveraging resource and capability for creating competitive advantage(Wernerfelt, 1984; Hitt, Xu, et al., 2016). Thus, RBT based LSS framework may also help in using resources and capability to acquire a sustainable competitive advantage for the firm. The purpose of this chapter is first 1) to summarise & analyse the literature on LSS and RBT 2) to build an integrative framework 3) to suggest how future research should be directed given the findings.

INTEGRATIVE LITERATURE REVIEW METHODOLOGY

The integrative literature review is a specific review methodology that summarizes the past empirical and theoretical studies to build a comprehensive framework for understanding and implementing the phenomenon under consideration(Whittemore and Knafl, 2005). It has the potential to influence the research, practice, policy and strategy in the domain area under study. In other words, a well done integrative literature review can offer a lot of insights for both academicians and practitioners(Burke and Hutchins, 2007). The integrative review is also known for presenting varied perspectives on the phenomenon under consideration(Evans and Pearson, 2001). Therefore, this integrative review is intended to build on the work of earlier literature reviews in both RBT and LSS to construct a new body of knowledge through an integrative framework. Our goal is to add to the knowledge of LSS and RBT by completing a broad sweep of the literature which is focused on the field of both the domains. In order to conduct the literature review, we use guidelines given for literature review by Torraco(2005). The literature review is conducted with the aim to answer the following research questions.

1. How to integrate RBT while implementing LSS for gaining a competitive advantage?
2. How should future research proceed given our findings?

DATA SOURCES

A methodological review process was undertaken. The first step was intended to search electronic databases. The search criteria employed for this research was Lean Six Sigma, LSS, LSS success, LSS failures, Resource based theory, RBT in operations management, RBT and Lean, RBT and Six Sigma, RBT and LSS. RBT and continuous improvement, RBT and business improvement etc. The scope was restricted to last 25 years i.e. 1992 to 2018. The databases which was included in the study include was Academic Source Premier (EBSCO), Google Scholar, Business Source Premier (EBSCO), Emerald, IEEE Xplore Digital Library, JSTOR, ProQuest Dissertations and Theses, Science Direct, Taylor & Francis and World Public Library, Scopus, emerald, Inderscience, Wiley. Though some authors have concluded that conference proceedings should be excluded (Scott-Findlay and Estabrooks, 2006), nevertheless the conference proceedings offer some insights in an emerging research area (Flick, 2015) such as RBT and LSS. Therefore, conference proceedings by reputed publishers were included.

INCLUSION AND EXCLUSION

The screening process for the literature review followed a goal of finding articles which were focused on Lean Management criticisms. The articles included in this study were qualitative, quantitative, Editorial opinion, theoretical studies and conference papers were included in this review. The screening criteria were that articles had to be published in English between 1992 and May 2018 and other criteria were that it should be focused specifically on LSS and RBT. The articles were excluded if they did not focus on LSS and RBT or if the research design was poor or argument which was presented was not clear.

SCREENING

In order to screen the articles first a broad search of articles was conducted. The second step was to read the title and abstract to sort the articles that met the screening criteria. This step also helped to remove the duplicate articles. The third step was to read the full articles which were then used to meet the inclusion / exclusion criteria. The last step consisted of reading the reference list of articles which helped to further improve the search criteria.

DATA ANALYSIS

To conduct an integrative literature review, there is no well-set standard (Burke and Hutchins, 2007; Conn et al., 2003; Smith et al., 2009; Whittemore, 2005). As the primary goal of this research was to decide the integration mechanism of LSS and RBT. This was intended to be carried out by determining the patterns, directions, similarities, and differences within the sample. To bring in methodological approach a framework was used (Smith et al., 2009; Whittemore, 2005). The methodology adopted was

1. The articles which were retained was read three times. It was done so that themes and categories could emerge from the analysis and categorization
2. In order to decide the quality of the articles, the reasoning used by each author was used as guide post in relation to LSS and RBT.
3. For determining the quality of the theoretical articles, it was done by how well the authors have described the LSS and RBT. Also, the quality of articles relation to scholarship was also considered.
4. Kitson (2006) stressed that the ability to communicate without bias and clearly determined the scholarship of the articles, therefore these criteria was used as a guidepost in scholarship determination.

5. Some further guidelines were also considered such as the research design, sample characteristics or description, measurement, statistical analysis and relevance to knowledge development.

Thirty-eight articles met the inclusion and exclusion criteria. In order to thematically analyse the articles, the following guidelines were used. The articles after they were read were summarised in writing. The motto behind this was to classify the articles in a thematic manner. To do so patterns and categories were formed, which was subsequently used to group these similar categories in a major theme. The entire sample was then critically analysed to gain an understanding of the state of overall knowledge in relation to LSS and RBT. After studying all the papers, 38 articles that were relevant. The articles were analysed, and following themes of criticisms emerged from the analysis.

SOLUTIONS AND RECOMMENDATIONS

The final list of all 38 literature was analysed in a thematic manner to construct the LSS and RBT integration framework. Organizational strategy is a match between the organizational internal resources, skills etc. and the opportunities and the risks created by the external environment(Grant, 1999). While implementing LSS initiatives from a strategic point of view, the role of organization resources should form the foundation of the firm's strategies. From the corporate strategic point of view, firms operate in industrial and geographical boundaries where there are economies of scope and reduced transaction cost(Hitt et al., 1997). When the same is looked upon from a business strategy point of view for a firm to have a competitive advantage, factors like competitive imitation, the appropriability of return to innovative products, services or processes, information which can create profitability between firms and the means by which resources accumulation or improvement can result in competitive advantage, should be analysed for the relationship between resources, competition and profitability for firms(Grant, 1999). The customer needs are constantly changing and volatile, technologies for delivering them are also changing, the market is volatile, under such circumstances an external orientation does not lead to a safe strategy(Kotter, 2012). LSS improves the capability of the resources of the organization(Jeyaraman and Kee Teo, 2010), hence the foundation of the strategy in terms of what the business is capable of doing may offer a sustainable basis for a long term strategy than an orientation to meeting the needs of the marketplace. An organization implementing LSS can earn profits on two factors a) The attractiveness of the industry where it is located and b) ability to establish a competitive advantage over its rivals. With the advent of fourth industrial revolution and market changes, there is the international competition, technological change, big data revolution and diversification of firms across the boundaries of industries have intensified the competition and hence differences of profitability within the industry are much more important than across the industry(Pérez et al., 2015). Thus, the ability to establish a competitive advantage across rivals will be a more important factor for inter-firm profit differentials(Peteraf, 1993; Nason and Wiklund, 2018). The resource-based view of the firm explains the competitive advantage better than the strategic positioning of the firm in the choice between cost and differentiation advantage or between broad and narrow market scope(Hitt, Carnes, et al., 2016). Therefore, the business strategy should be viewed as a mission for return for resources which confer the competitive advantage over the real costs of the resources(Grant, 1999). The return for resources may be maximised by eliminating waste and variation within the system in a sustainable manner.

FIRM RESOURCES

The resources are key inputs to the production process. It could be equipment's, tools, skills, processes, employees, patents and so on. On their own only few resources are productive. Identification of resources becomes a major challenge, as most of the management information system provides a fragmented and incomplete picture of the firm. Also, the financial balance sheets do not consider intangible resources and people-based skills and abilities(Grant, 1999; McGee, 2015). A solution for identification of resources is to classify the resources into a) financial b) physical c) human d) technological e) reputation and f) organizational resources(Miles et al., 1978; Helfat, 2017). Maximising the resource over time is the primary goal. LSS can be used to maximise the productivity of all tangible and intangible resources of the firm(Antony et al., 2017). Thus, by deploying LSS one can economize the use of resources in the best possible method because waste and variation are tackled in the best manner(Albliwi et al., 2014). Another facet of LSS is that it can be used so that existing resources can be used more intensely and in a more profitable manner(Naslund, 2008). The return an organization gains by making the existing resources into something more useful can put the organization on a competitive map.

Research Proposition 1: *LSS can be used to maximize the productivity of all tangible and intangible resources leading to maximizing of firm resources over a period of time.*

Research Proposition 2: *By using LSS existing resources can be used intensely and profitably, which can put the firm on a competitive map.*

FIRM CAPABILITY

It is the capacity of a team or group of resources within an organization to perform some task or activity together(Grant, 1999). Capabilities are the main source of competitive advantage. It is usually identified as a classification of functional activity. A capability cannot be created by simply assembling a group of resources(Dyer and Singh, 1998). To create firm capability a firm requires a complex coordination among people, and between people and resources. Therefore, such a coordination can be acquired through learning and repetition(Kavin and Narasimhan, 2018). A capability is nothing but a routine or set of routines that interact with a predetermined objective. Organization routine in simple ways is a set of regular and predictable patterns of activity which are made up of the sequence of coordinated actions by individuals(Grant, 1999). LSS can help in reducing the variation and waste in the product, service or processes(Antony et al., 2017). These are done through coordinated actions among people and between people and resources. LSS can help to optimize the relationship between resource and capability. The capability some organization possesses depends on the resources it has and also it depends on the proper use of it. LSS can also help the organization to effectively coordinate and cooperate with teams or groups(Asefeso, 2014) to effectively use the resources to attain the capability to perform a task or activity within the organization. LSS can, therefore, be used effectively as a trade- off, between efficiency and flexibility in the organizational routines in various functional areas.

Research Proposition 3: *LSS can be used to improve the coordination among people and between people and resources so that firm's capability improves.*

COMPETITIVE ADVANTAGE

The competitive advantage can arise due to the returns available from the firm's resources and capability(Grant, 1999; Hitt, Xu, et al., 2016). These competitive factors can be eroded in the long run, if there is depreciation of the resources and capability of the firm. Likewise, it can also be eroded if there is an imitation by the rival(Peteraf, 1993). The rate of erosion depends on the characteristics of resources and capabilities(Grant, 1999; Belton, 2017). LSS if implemented can maintain the resources and capabilities within the firm, by continuously reducing the waste and arresting variations within products, services and processes(Yadav et al., 2017). Thus, the competitive advantage of the firms implementing LSS will depend upon the resources and capabilities. If the firm is competing in an efficient market, then the rivals will immediately catch up and firm competitive position will erode(Hitt et al., 1997; Grant, 1999). This can happen if the LSS is not deployed across various levels or in all departments within in the organization(Sony and Naik, 2012). In not so competitive markets a firm can be a bit lenient in efficiency while implementing LSS in the short run. This could be due to lack of competition. However, in the long run, efficient rivals will take over. Thus, LSS should be a tool for building a competitive advantage of the firms.

Research Proposition 4: *The competitive advantage of firms implementing LSS will depend on the resources and capabilities and imitation of the same by the rivals.*

FORMULATION OF STRATEGY

Resources, capability and competitive advantage translate into strategy. Strategies are designed so that so that a firm can make the best use of resources, capabilities and sustain the competitive advantage. The firm's strategic plans should on a timeline revolve around the ability of a firm to make the best use of the resources and capability in order to sustain the competitive advantage(Grant, 1999; Bhatt and Grover, 2005; Liu and Atuahene-Gima, 2018). LSS will help the firms to make the resource and capability free from waste or non-value added activities and also reduce the variation(Asefeso, 2014). When the firms, resources, and capabilities are easy to imitate then strategies for sustaining competitive advantage warrants catering small markets(Grant, 1999; Shay and Rothaermel, 1999). LSS will help here to maintain the competitive advantage in small markets. Even though the resources and capabilities can be imitated, by deploying LSS throughout the organization, the learning effect of continuous improvement(Sony and Naik, 2011) will be difficult to imitate in the long run. Also, the innovative segment within LSS will help to improve or design products, process or services(Laureani and Antony, 2011) in a small window of opportunity where the firm will also protect the existing products and services, and also venture out to design new products and services. Such a strategy will help, because, the rate of erosion of competitive advantage can be arrested or slowed down by LSS.

Research Proposition 5: *Firm strategies should focus on making the best use of resources and capabilities by using LSS to sustain the competitive advantage.*

Research Proposition 6: *When the firm's resources and capabilities can be easily imitated by the rivals, LSS should be deployed to reduce the rate of erosion of competitive advantage by the learning effect of continuous improvement of reduction in waste and variation.*

IDENTIFYING THE RESOURCE GAP

The analysis so far has been an analysis of existing resources and capabilities so as to maximise the return for an organization. The gap between the existing resources and capabilities and future needed resource and capability can be strategically analysed. This part of the analysis is significant so that the strategic objectives can be met. LSS can help in the development of the firm's resource and capability base(Laureani and Antony, 2017). This can be done in two ways. One is improving the existing resource and capability. The second is augmenting the resources and capability to sustain or maintain or diversify the competitive position. LSS definitely will help in improving the existing resources and capabilities by reducing the waste and variation which will lead to maximizing the returns(Antony et al., 2016). Thus, by utilising existing resources and capabilities various strategic positioning can be sustained. As regards to the second option for augmenting existing resources and capabilities, LSS could be deployed so that the augmented system will be designed using LSS principles(Kasemsap, 2018) which result in the optimum utilization of resources and capabilities, leading to maximization of returns.

RBT BASED LEAN SIX SIGMA IMPLEMENTATION FRAMEWORK

For improving the resources and capabilities for competitive advantage, simultaneous implementation of Lean and Six Sigma principles is advocated. However, an order of precedence of Lean and Six Sigma is important not only from an academic point of view but also from a practical deployment point of view. The connectivity between Six Sigma and Lean is linear(Jeyaraman and Kee Teo, 2010; Albliwi et al., 2014; Asefeso, 2014). The concept here is Six Sigma being applied first so that variation is reduced and subsequently Lean is applied(George and George, 2003). From a resource and capability perspective, the variation in resources and capabilities must reduce first and then subsequently Lean applied to remove Mura, Muri. and Muda(Pieńkowski, 2014; Timans et al., 2016; Hill et al., 2018). Thus, in resource-based view, Lean is a subset of Six Sigma for LSS. The RBT based framework is given in Figure 1.

Define Phase

In this phase, the definition of the problem or improvement opportunity about the resource or capability or both is defined(Deshmukh and Chavan, 2012). The problem must be clearly defined in terms of organization competitive advantage point of view. This is important in the strategic sense because once the problem is defined in terms of both the resource or capability and competitive advantage point of view, the organization can clearly strategize it in terms of overall mission and vision of the organization. This new framework of LSS also suggests that this definition must be in line with the Lean principles. The definition of the goals for both resource and capability must be analysed first from Muda (waste) point of view or towards the banishment of eight wastes. The unevenness of resources or capability needs also to be defined from Mura(unevenness) and at last from Muri (overburden) to get a sustainable perspective.

Research Proposition 7: *Firms while defining a process improvement or goals for resources and capabilities for competitive advantage using LSS framework must be holistically analysed in consonance with Mura, Muri, and Muda for either improving existing resource and capability or acquiring new resource and capability.*

Measure Phase

In the measure phase, the performance of the resources and capability is measured(Ramberg, 2000) with the objective or goal in the define phase. The performance of the resources or capability also must be measured with respect to Mura, Muri and Muda. The detailed data collection plan must be made to include all the details of measurement. It must provide a significant and reliable measurement plan. This will indicate how the current resource and capability is measuring. Thus, the measurement system should be a holistic measurement system considering Mura, Muri and Muda along with the objective or goal of the problem defined in the previous phase.

Research Proposition 8: *While measuring the performance of resources and capabilities the effect on Mura, Muri and Muda have also to be measured to capture the measurement.*

Analyse Phase

In this phase, the analyses of the resources and capability are carried out based on the data collected in the previous phase. The causes of variations in resources or capability(Kaushik and Khanduja, 2009) are carried out and further analysed with respect to Mura, Muri, and Muda. An analysis which advocates the removal of too much Muda from resources and capabilities may result in Muri(Hampson, 1999). Also, the Mura if is not addressed properly in resources and capability, it may also result in Muri(Bhasin and Burcher, 2006). Thus, analysis cannot proceed with the concentration of just one aspect of Mura, Muri and Muda but in a holistic manner which takes care of the entire system holistically.

Research Proposition 9: *Analyses of resources and capabilities must be carried out through LSS by considering the aspects of Muda, Muri and Mura in totality.*

Improvement Phase

In this phase, the improvement measures of the resources and capability are suggested by addressing the root causes(Gee, 2018). While suggesting the improvement measures for resources and capabilities Mura, Muri and Muda must be considered. An improvement measure may be viable with respect to Muda but may not be feasible with respect to Mura and Muri(Bhasin and Burcher, 2006; Sony, 2018). An organization while improving the resources and capabilities, has to blend a holistic perception for analysis which takes into account Muda, Muri and Mura

Research Proposition 10: *The improvement measures for improving the performance of resources and capabilities, has to be analyzed with respect to Muri, Mura, and Muda.*

Control Phase

In this phase, it is made sure that future performance of the resources and capabilities are maintained (Gee, 2018)with respect to Muda, Mura and Muri as envisaged in the improvement phase. This is an important phase in the improvement of resources and capabilities as sustainability of improvement measures are finalized in this phase.

Figure 1.

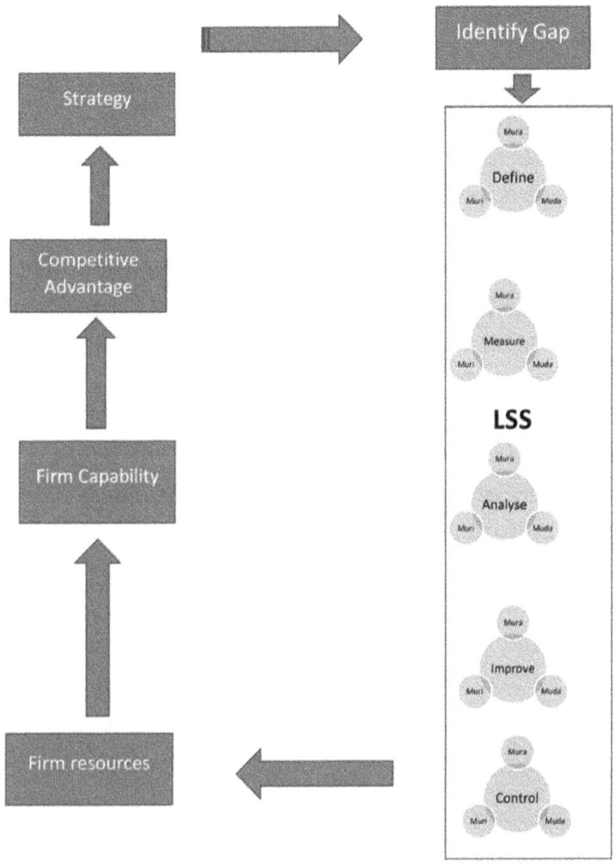

Research Proposition 11: *The sustainability of improvement measure for the resource and capability has to be controlled or monitored not in isolation but in consonance with respect to Muri, Mura, and Muda.*

DISCUSSION

Resources and capabilities are important inputs for an organization to achieve competitive advantage leading to the improved image of the firm and financial profits(Alexy et al., 2018). Thus, resource and capabilities have to be meticulously strategized either for improving the existing resources and capabilities or acquiring new resources and capabilities(Grant, 1999; Hitt, Carnes, et al., 2016). LSS can be used as a strategic tool for managing existing resources and capabilities(Byrne et al., 2007). While implementing LSS in an organization, the organization resources should form the foundation on which the strategies are built. In order to maximise the productivity of tangible and intangible resources, LSS would be an important methodology. The phenomenon of waste reduction and the reduction of variation in a holistic sense will result in using existing resources intensely and profitably in the long run(Laureani and Antony, 2017). A firm's capability is the capacity of a team or a group of resources to carry out a particular task or activity. A capability is an important tool for building a competitive advantage(Grant, 1999). By using LSS a firm can improve the coordination between the people(Andersson et al., 2006) and between people and resource(Asefeso, 2014) in order to improve the firm's capability. This coordination is improved by reducing the non- value adding activities and reducing the variation within the context of Mura, Muri, and Muda. An effective use of the firm's resources and capabilities will result in a competitive advantage for the firms. The competitive advantage of the firm will sustain depending upon the efficient use of resources and capabilities and also imitation by rivals(Grant, 1999; Hitt, Carnes, et al., 2016). LSS will help in the efficient use of resources and capabilities by reducing waste and variation. Also, the tendency to imitate by the rivals can be reduced by deploying LSS, as it will result in a long-term learning effect in using the continuous improvement strategy (Sony and Naik, 2012). The long-term organizational learning of continuous improvement and variation reduction by considering Muri, Mura and Muda can result in a sustainable competitive advantage which will be difficult to imitate in the long run. Every organization should devise appropriate LSS strategy while improving existing resource and capability or acquiring new resource or capability to sustain or lead to a competitive advantage. Likewise, in markets where it is not possible in the long run to maintain competitive advantage like in efficient markets, LSS can be deployed so that rate of erosion of competitive advantage can be prolonged by the continuous improvement of products, services, and processes. This is possible because the rate of erosion of competitive advantage(Sarason and Tegarden, 2003) will be offset by the continuous improvement in products, services and processes(Elmuti and Kathawala, 1997). In addition, the search for new opportunities can be expanded by deploying LSS strategically to develop innovative products, services or processes(Byrne et al., 2007) leading to acquiring and maintenance of competitive advantage.

CONCLUSION

Organizational Strategy is a match between the organizational internal resources, skills, capabilities etc. and the opportunities and the risks created by the external environment(Grant, 1999). Lean Six Sigma is one of the most widely used quality management strategy used by organizations in the last two decades. Most of the research has been around LSS as a quality management strategy and the external environment. Therefore, the link between LSS strategy and firms' internal resources and capabilities have received a neglect. Recently, there has been a resurgence in the role of the firm's internal resource as a foundation of strategy. This paper proposes the LSS implementation framework that considers resource-based view, for building an organization strategy. Eleven research propositions are developed so that future research can be directed in this area. The limitation of this study is based on the search criteria and the database used. Though most care is taken to include articles based on inclusion and exclusion criteria. The data sources are based up to 2018, and future studies should be conducted to further strengthen the research. Though the integrative review method has been critiqued for its potential for bias and lack of rigour, to avoid it this study uses Torraco (2005) methodology. Empirical studies should be conducted using this framework to further strengthen the proposed framework.

FUTURE RESEARCH AGENDA

The RBT based Lean Six Sigma Implementation framework work is built based on the previous literature. Future research should first test the framework using the case studies. This is important because many variables are involved while study the resource-based view, hence case study will help to describe the knowledge in a better manner. The research propositions may be further tested through quantitative research to gain further insights into the resource-based framework. The applicability of the framework may also be studied with respect to manufacturing and service industries. This is important as the resources and capabilities required in the manufacturing and service sectors are different and hence it will help to implement the resource-based framework in these sectors. The research should first explore case studies by linking LSS with RBT. Case studies are suggested initially because of the ability to study many variables. Though generalizability is a problem when it comes to case studies, however, by studying the case in a longitudinal manner will help us to understand the time varying component of organization internal strategies of LSS and competitive advantage. Another interesting area to study would be the applicability of LSS and RBT framework in different market structures to create a competitive advantage. Future research should also link LSS, RBT and Sustainability. In sustainability it would interesting to link the triple bottom line approach (Slaper and Hall, 2011) with LSS and RBT because this will help organizations to be sustainable as well as competitive a new dimension to survive in this modern world. Another perspective could be the integration of Industry 4.0, LSS and RBT, will lead to a new perspective for competitive advantage.

MANAGERIAL IMPLICATIONS

This chapter will help the managers working in the organizations to analyze the existing LSS programs or new LSS programs to be implemented so as to create and sustain competitive advantage through the collection, integration and organization of rare, valuable and inimitable LSS based resources. Managers can use this study to identify the firm resources that exist within the organization which can create a competitive advantage. It is important to identify these resources because most organizations confuse between the resources and capabilities. tangible and intangible resources can be bundled to create capabilities for an organization., therefore identifying organization capabilities forms a next important element. After identifying the capabilities, the competitive advantage for the firm can be ascertained, leading to the formulation of strategy. This makes the organization to understand the existing state and future state. A unique LSS methodology suggested in this chapter can be used to create a competitive advantage. Being a cyclical process, organizations must revisit these steps to sustain the competitive advantage gained by the organizations. While acquiring new resources the proposed LSS methodology can be used to strategically acquire new resources.

REFERENCES

Albliwi, S., Antony, J., Abdul Halim Lim, S., & van der Wiele, T. (2014). Critical failure factors of Lean Six Sigma: A systematic literature review. *International Journal of Quality & Reliability Management Vol., 31*(9), 1012–1030. doi:10.1108/IJQRM-09-2013-0147

Alexy, O., West, J., Klapper, H., & Reitzig, M. (2018). Surrendering control to gain advantage: Reconciling openness and the resource-based view of the firm. *Strategic Management Journal, 39*(6), 1704–1727. doi:10.1002mj.2706

Anderson, J.C., Cleveland, G., & Schroeder, R.G. (1989). *Operations Strategy a Literature Review*. Academic Press.

Andersson, R., Eriksson, H., & Torstensson, H. (2006). Similarities and differences between TQM, six sigma and lean. *The TQM Magazine, 18*(3), 282–296. doi:10.1108/09544780610660004

Antony, J. (2011). Six Sigma vs Lean: Some perspectives from leading academics and practitioners. *International Journal of Productivity and Performance Management, 60*(2), 185–190. doi:10.1108/17410401111101494

Antony, J., Gupta, S., Sunder, M. V., & Gijo, E. V. (2018). Ten commandments of Lean Six Sigma: A practitioners' perspective. *International Journal of Productivity and Performance Management, 67*(6), 1033–1044. doi:10.1108/IJPPM-07-2017-0170

Antony, J., Setijono, D., & Dahlgaard, J. J. (2016). Lean Six Sigma and Innovation–an exploratory study among UK organisations. *Total Quality Management & Business Excellence, 27*(1–2), 124–140. doi:10.1080/14783363.2014.959255

Antony, J., Snee, R., & Hoerl, R. (2017). Lean Six Sigma: Yesterday, today and tomorrow. *International Journal of Quality & Reliability Management, 34*(7), 1073–1093. doi:10.1108/IJQRM-03-2016-0035

Asefeso, A. (2014). Lean Six Sigma: Cost Reduction Strategies (2nd ed.). Academic Press.

Barney, J. (1991). Firm resources and sustained competitive advantage. *Journal of Management, 17*(1), 99–120. doi:10.1177/014920639101700108

Belton, P. (2017). *Competitive Strategy: Creating and Sustaining Superior Performance*. Academic Press.

Beyea, S. C., & Nicoll, L. H. (1998). Writing an integrative review. *AORN Journal, 67*(4), 877–880. doi:10.1016/S0001-2092(06)62653-7

Bhasin, S., & Burcher, P. (2006). Lean viewed as a philosophy. *Journal of Manufacturing Technology Management, 17*(1), 56–72. doi:10.1108/17410380610639506

Bhatt, G. D., & Grover, V. (2005). Types of information technology capabilities and their role in competitive advantage: An empirical study. *Journal of Management Information Systems, 22*(2), 253–277. doi:10.1080/07421222.2005.11045844

Burke, L. A., & Hutchins, H. M. (2007). Training transfer: An integrative literature review. *Human Resource Development Review, 6*(3), 263–296. doi:10.1177/1534484307303035

Byrne, G., Lubowe, D., & Blitz, A. (2007). Using a Lean Six Sigma approach to drive innovation. *Strategy and Leadership*, *35*(2), 5–10. doi:10.1108/10878570710734480

Chen, M., & Lyu, J. (2009). A Lean Six-Sigma approach to touch panel quality improvement. *Production Planning and Control*, *20*(5), 445–454. doi:10.1080/09537280902946343

de Freitas, J. G., Costa, H. G., & Ferraz, F. T. (2017). Impacts of Lean Six Sigma over organizational sustainability: A survey study. *Journal of Cleaner Production*, *156*, 262–275. doi:10.1016/j.jclepro.2017.04.054

Deshmukh, S. V., & Chavan, A. (2012). Six Sigma and SMEs: A critical review of literature. *International Journal of Lean Six Sigma*, *3*(2), 157–167. doi:10.1108/20401461211243720

Dierickx, I., & Cool, K. (1989). Asset stock accumulation and sustainability of competitive advantage. *Management Science*, *35*(12), 1504–1511. doi:10.1287/mnsc.35.12.1504

Dyer, J. H., & Singh, H. (1998). The relational view: Cooperative strategy and sources of interorganizational competitive advantage. *Academy of Management Review*, *23*(4), 660–679. doi:10.5465/amr.1998.1255632

Elmuti, D., & Kathawala, Y. (1997). An overview of benchmarking process: A tool for continuous improvement and competitive advantage. *Benchmarking for Quality Management & Technology*, *4*(4), 229–243. doi:10.1108/14635779710195087

Evans, D., & Pearson, A. (2001). Systematic reviews: Gatekeepers of nursing knowledge. *Journal of Clinical Nursing*, *10*(5), 593–599. doi:10.1046/j.1365-2702.2001.00517.x

Gee, J. (2018). *The New Work Order*. Academic Press.

George, M.L., & George, M. (2003). *Lean Six Sigma for Service*. Academic Press.

Gnanaraj, S. M., Devadasan, S. R., Murugesh, R., & Sreenivasa, C. G. (2012). Sensitisation of SMEs towards the implementation of Lean Six Sigma–an initialisation in a cylinder frames manufacturing Indian SME. *Production Planning and Control*, *23*(8), 599–608. doi:10.1080/09537287.2011.572091

Grant, R. M. (1999). *The resource-based theory of competitive advantage: implications for strategy formulation*. Knowledge and Strategy.

Hampson, I. (1999). Lean Production and the Toyota Production System Or, the Case of the Forgoften Production Concepts. *Economic and Industrial Democracy*, *20*(3), 369–391. doi:10.1177/0143831X99203003

Hayes, R. H., & Upton, D. M. (1998). Operations-based strategy. *California Management Review*, *40*(4), 8–25. doi:10.2307/41165962

Helfat, C. E. (2017). *Stylized facts regarding the evolution of organizational resources and capabilities*. The SMS Blackwell Handbook of Organizational Capabilities.

Hill, J., Thomas, A. J., Mason-Jones, R. K., & El-Kateb, S. (2018). The implementation of a Lean Six Sigma framework to enhance operational performance in an MRO facility. *Production & Manufacturing Research*, *6*(1), 26–48. doi:10.1080/21693277.2017.1417179

Hilton, R. J., & Sohal, A. (2012). A conceptual model for the successful deployment of Lean Six Sigma. *International Journal of Quality & Reliability Management*, *29*(1), 54–70. doi:10.1108/02656711211190873

Hitt, M. A., Carnes, C. M., & Xu, K. (2016). A current view of resource based theory in operations management: A response to Bromiley and Rau. *Journal of Operations Management, 41*(10), 107–109. doi:10.1016/j.jom.2015.11.004

Hitt, M. A., Hoskisson, R. E., & Kim, H. (1997). International diversification: Effects on innovation and firm performance in product-diversified firms. *Academy of Management Journal, 40*(4), 767–798.

Hitt, M. A., Xu, K., & Carnes, C. M. (2016). Resource based theory in operations management research. *Journal of Operations Management, 41*, 77–94. doi:10.1016/j.jom.2015.11.002

Jeyaraman, K., & Kee Teo, L. (2010). A conceptual framework for critical success factors of lean Six Sigma: Implementation on the performance of electronic manufacturing service industry. *International Journal of Lean Six Sigma, 1*(3), 191–215. doi:10.1108/20401461011075008

Karunakaran, S. (2016). Innovative application of LSS in aircraft maintenance environment. *International Journal of Lean Six Sigma, 7*(1), 85–108. doi:10.1108/IJLSS-01-2015-0001

Kasemsap, K. (2018). *Applying Lean Production and Six Sigma in global operations. In Operations and Service Management: Concepts* (pp. 582–612). Methodologies, Tools, and Applications.

Kaushik, P., & Khanduja, D. (2009). Application of Six Sigma DMAIC methodology in thermal power plants: A case study. *Total Quality Management, 20*(2), 197–207. doi:10.1080/14783360802622995

Kavin, L., & Narasimhan, R. (2018). An Investigation of Contextual Influences on Innovation in Complex Projects. *Innovation and Supply Chain Management: Relationship, Collaboration and Strategies*, 51–77.

Kitson, A. (2006). The relevance of scholarship for nursing research and practice. *Journal of Advanced Nursing, 55*(5), 541–543. doi:10.1111/j.1365-2648.2006.04004_1.x

Kotter, J.P. (2012). *Leading Change*. Academic Press.

Kumar, M., Antony, J., Singh, R. K., Tiwari, M. K., & Perry, D. (2006). Implementing the Lean Sigma framework in an Indian SME: A case study. *Production Planning and Control, 17*(4), 407–423. doi:10.1080/09537280500483350

Laureani, A., & Antony, J. (2011). Standards for lean six sigma certification. *International Journal of Productivity and Performance Management, 61*(1), 110–120. doi:10.1108/17410401211188560

Laureani, A., & Antony, J. (2017). Leadership and Lean Six Sigma: A systematic literature review. *Total Quality Management & Business Excellence*, 1–29. doi:10.1080/14783363.2017.1288565

Liu, W., & Atuahene-Gima, K. (2018). Enhancing product innovation performance in a dysfunctional competitive environment: The roles of competitive strategies and market-based assets. *Industrial Marketing Management, 73*, 7–20. doi:10.1016/j.indmarman.2018.01.006

McGee, J. (2015). *Resource-Based View*. Wiley Encyclopedia of Management.

McLean, R. S., Antony, J., & Dahlgaard, J. J. (2017). Failure of Continuous Improvement initiatives in manufacturing environments: A systematic review of the evidence. *Total Quality Management & Business Excellence, 28*(3–4), 219–237. doi:10.1080/14783363.2015.1063414

Miles, R. E., Snow, C. C., Meyer, A. D., & Coleman, H. J. Jr. (1978). Organizational strategy, structure, and process. *Academy of Management Review*, *3*(3), 546–562. doi:10.5465/amr.1978.4305755

Naslund, D. (2008). Lean, six sigma and lean sigma: Fads or real process improvement methods? *Business Process Management Journal*, *14*(3), 269–287. doi:10.1108/14637150810876634

Nason, R. S., & Wiklund, J. (2018). An assessment of resource-based theorizing on firm growth and suggestions for the future. *Journal of Management*, *44*(1), 32–60. doi:10.1177/0149206315610635

Pérez, F., Irisarri, E., Orive, D., Marcos, M., & Estevez, E. (2015). A CPPS Architecture approach for Industry 4.0. *Emerging Technologies & Factory Automation (ETFA), 2015 IEEE 20th Conference on*, 1–4. 10.1109/ETFA.2015.7301606

Peteraf, M. A. (1993). The cornerstones of competitive advantage: A resource-based view. *Strategic Management Journal*, *14*(3), 179–191. doi:10.1002mj.4250140303

Pieńkowski, M. (2014). Waste measurement techniques for Lean companies. *International Journal of Lean Thinking*, *5*(1), 9–24.

Pilkington, A., & Meredith, J. (2009). The evolution of the intellectual structure of operations management—1980–2006: A citation/co-citation analysis. *Journal of Operations Management*, *27*(3), 185–202. doi:10.1016/j.jom.2008.08.001

Porter, M. E. (1996). *What is strategy*. Academic Press.

Ramberg, J. S. (2000). Six sigma: Fad or fundamental. *Quality Digest*, *6*(5), 30–31.

Sarason, Y., & Tegarden, L. F. (2003). The erosion of the competitive advantage of strategic planning: A configuration theory and resource based view. *Journal of Business and Management*, *9*(1), 1.

Shay, J. P., & Rothaermel, F. T. (1999). Dynamic competitive strategy: Towards a multi-perspective conceptual framework. *Long Range Planning*, *32*(6), 559–572. doi:10.1016/S0024-6301(99)00073-4

Sirmon, D. G., Hitt, M. A., Ireland, R. D., & Gilbert, B. A. (2011). Resource orchestration to create competitive advantage: Breadth, depth, and life cycle effects. *Journal of Management*, *37*(5), 1390–1412. doi:10.1177/0149206310385695

Slaper, T. F., & Hall, T. J. (2011). The triple bottom line: What is it and how does it work. *Indiana Business Review*, *86*(1), 4–8.

Snee, R. D. (2010). Lean Six Sigma–getting better all the time. *International Journal of Lean Six Sigma*, *1*(1), 9–29. doi:10.1108/20401461011033130

Sony, M. (2018). Industry 4.0 and lean management: A proposed integration model and research propositions. *Production & Manufacturing Research*, *6*(1), 416–432. doi:10.1080/21693277.2018.1540949

Sony, M., & Naik, S. (2011). Successful implementation of Six Sigma in services: An exploratory research in India Inc. *International Journal of Business Excellence*, *4*(4), 399–419. doi:10.1504/IJBEX.2011.041059

Sony, M., & Naik, S. (2012). Six Sigma, organizational learning and innovation: An integration and empirical examination. *International Journal of Quality & Reliability Management*, *29*(8), 915–936. doi:10.1108/02656711211258535

Thomas, A. J., Mason-Jones, R., Davies, A., & John, E. G. (2015). Reducing turn-round variability through the application of Six Sigma in aerospace MRO facilities. *Journal of Manufacturing Technology Management, 26*(3), 314–332. doi:10.1108/JMTM-05-2013-0052

Timans, W., Ahaus, K., van Solingen, R., Kumar, M., & Antony, J. (2016). Implementation of continuous improvement based on Lean Six Sigma in small-and medium-sized enterprises. *Total Quality Management & Business Excellence, 27*(3–4), 309–324. doi:10.1080/14783363.2014.980140

Vinodh, S., Gautham, S. G., & Ramiya, R. A. (2011). Implementing lean sigma framework in an Indian automotive valves manufacturing organisation: A case study. *Production Planning and Control, 22*(7), 708–722. doi:10.1080/09537287.2010.546980

Vinodh, S., Kumar, S. V., & Vimal, K. E. K. (2014). Implementing lean sigma in an Indian rotary switches manufacturing organisation. *Production Planning and Control, 25*(4), 288–302. doi:10.1080/09537287.2012.684726

Wernerfelt, B. (1984). A resource-based view of the firm. *Strategic Management Journal, 5*(2), 171–180. doi:10.1002mj.4250050207

Whittemore, R., & Knafl, K. (2005). The integrative review: Updated methodology. *Journal of Advanced Nursing, 52*(5), 546–553. doi:10.1111/j.1365-2648.2005.03621.x

Yadav, G., Seth, D., & Desai, T. N. (2017). Analysis of research trends and constructs in context to lean six sigma frameworks. *Journal of Manufacturing Technology Management, 28*(6), 794–821. doi:10.1108/JMTM-03-2017-0043

ADDITIONAL READING

Barney, J. B. (2017). The Evolutionary Roots of Resource-based Theory. The SMS Blackwell Handbook of Organizational Capabilities, 269-271.

Franchetti, M. J. (2015). *Lean Six Sigma for Engineers and Managers: With Applied Case Studies*. CRC Press. doi:10.1201/b18234

Furterer, S. L. (2016). *Lean Six Sigma in service: applications and case studies*. CRC press.

Geary, J., Hubbard, E., King, B., Hahn, D., Clark, K., & Sturtevant, O. J. (2018). Automated resource based scheduling system for cellular product manufacturing. *Cytotherapy, 20*(5), S74–S75. doi:10.1016/j.jcyt.2018.02.208

Jensen, J. A., Cobbs, J. B., & Turner, B. A. (2016). Evaluating sponsorship through the lens of the resource-based view: The potential for sustained competitive advantage. *Business Horizons, 59*(2), 163–173. doi:10.1016/j.bushor.2015.11.001

Lin, Y., & Wu, L. Y. (2014). Exploring the role of dynamic capabilities in firm performance under the resource-based view framework. *Journal of Business Research, 67*(3), 407–413. doi:10.1016/j.jbusres.2012.12.019

Chapter 14
Bio-Inspired Meta-Heuristic Multi-Objective Optimization of EDM Process

Goutam Kumar Bose
Haldia Institute of Technology, India

Pritam Pain
Haldia Institute of Technology, India

Supriyo Roy
https://orcid.org/0000-0001-8731-0068
Haldia Institute of Technology, India

ABSTRACT

Modern-day engineering is trending toward complex devices with high accuracy and precision while at the same time the workpiece materials are becoming harder and more complex alloys. Non-conventional machining helps to sustain industries in this challenging environment. Electric discharge machining is one such precision machining that can produce complex product with great accuracy. This machine can be utilized in the manufacturing of complicated shaped die for plastic molding, automobile parts, aerospace, and other applications. In solving real-world problems like engineering design, business planning, and network design, challenges are being faced due to highly non-linear data and limited resources like time and money. Optimization is the best choice to solve such practical problems efficiently. This chapter deals with the optimization of EDM process parameters using two different bio-inspired optimization algorithms, namely, artificial bee colony algorithm and whale optimization algorithm, for both single as well as multiple responses and compares the results.

DOI: 10.4018/978-1-5225-8223-6.ch014

INTRODUCTION

As the modern-day engineering is trending for the complex device with high accuracy and precision and at the same time the workpiece material is becoming harder and more complex alloy. It is very difficult to machining those products by only employing conventional machining process. Non-conventional machining helps to sustain those industries in this challenging environment. Electric Discharge Machining (EDM) is one such precision machining which can produce complex product with great accuracy. This machine can be utilised in the manufacturing of complicated shaped die for plastic moulding, automobile parts, aerospace and other applications. This machining process involves thermo chemical reaction in between toll and the workpiece. Both the cathode tool and the anode workpiece are submerged in the dielectric fluid and separated by a small gap, commonly known as spark gap. When the current flows, high number of electrons flows from tool to workpiece and as a result the thrust energy of the electron converted into heat energy and melt and vaporised the workpiece material instantly regardless of the hardness of the material. The temperature range during machining in between 8000^0C to 12000^0C.As the machining condition does not involve the hardness of the workpiece material, this EDM is very useful for machining hard electricalconductive material with high precision. The main influencing parameters for EDM are Pulse on Time (T_{ON}), Duty factor, Pulse off Time (T_{OFF}), sensitivity of the dielectric fluid, Spark Gap (SG), Gap Current (GI), flow of the dielectric fluid etc. AISI EN-24 is one such hard steel material which is very difficult to machining by conventional machining process.

In solving real-world problems like, engineering design, business planning, network design, even in case of holiday planning, challenges are being faced due to highly non-linear data and limited resources like, time and money. Optimization is the best choice to solve such practical problems efficiently. Optimization means to find out the best possible outcome on the basis of self-contradictory input parameters. Thus, optimization problems can be classified as constrained minimization problem in case of cost, business loss, energy consumption, surface roughness, machining time etc. and also as constrained maximization problem in case of profit, energy output, efficiency etc. During past years, several optimization algorithms have been developed by researchers to cope up the challenges of real-world problems. It is noteworthy, the recent development in computer simulation technology is essential to carry out those optimization algorithms successfully.

Inspired by the various intelligent activities spread over within nature, researchers have developed various nature-based optimization algorithms to solve non-linear problems. Some of these algorithms are based on intelligent behaviour of biological (living) characters are called bio-inspired algorithm. Within the bio-inspired algorithms, some are based on intelligent foraging behaviour of a group of biological characters known as Swarm Intelligence (SI-based) algorithm and some algorithms which are not based on swarm intelligent behaviour of biological character known as not SI-based algorithm. Some examples of SI-based algorithms are, Artificial Ant Colony (AAC), Artificial Bee Colony (ABC), Cuckoo Search (CS), Artificial Fish Swarm (AFS), Firefly Algorithm (FA) etc. Examples of not SI-based algorithms are Dolphin Echolocation (DE), Whale Optimization (WO), etc. EDM is one such complex machining process which involves various control parameters to optimize various different contradictory responses like Material Removal Rate (MRR), Surface Roughness (R_a), Overcut (OC) etc. It is very important to optimize this machining process within the given resources.

A brief literature review of the past research work is presented here.Das et al. (2014) has studied on the effect of different control parameters on MRR and Ra while machining EN31 steel by using EDM process.Mukherjee and Chakraborty (2012) had applied Biogeography Based Optimization (BBO) algorithm in order to optimize the responses and they finally compared BBO with genetic algorithm and artificial ant colony optimization algorithm and they have concluded that BBO is superior than other two algorithm. Baraskar et al. (2013) developed a design of experiments based on response surface methodology and optimised that design by using sorting genetic algorithm-II in order to achieve the pareto optimal set where contradictory responses are optimized. Gholipoor et al. (2015) have compared the effect of process parameters of near dry EDM with wet and dry EDM while drilling of SPK materials and find out that the high level of the discharge energy for wet EDM has most MRR, tool wear rate and surface roughness but in case of dry EDM the responses are very least, but while in case of near dry EMD process, MRR is high but surface roughness is low. Kumar and Kumar (2014) have optimized the process control parameters for the EDM process while the machining has been done by using cryogenic cooling of liquid nitrogen copper electrode and AlSiCp workpiece material and find out the optimal machining condition by employing grey relation analysis. Gostimirovic et al. (2012) investigated on the effect of discharge energy parameters on the MRR, surface roughness and recast layer and find out the most suitable parametric condition for EMD. Song & Chu (2015) compared the MRR for strip EDM and weir EDM and find out that the MRR for strip EDM is much higher than the wire EMD. Majumder, (2013) studied on the effect of process control parameters on AISI 316LN steel material while machining through EDM process and find out the feasible parametric combination by employing fuzzy model along with PSO. Kanyongo (2006) studied the relationship between home environment factor with reading achievement in Zimbabwe by using linear regression analysis through structural equation modelling and predicted the reading achievement. Mirjalili and Lewis (2016) developed a new metaheuristic optimization technique called Whale Optimization Algorithm (WOA) which mimic the characteristics of humpback whales. Kaveh and Ghazaan (2017) modified the WOA into Enhanced Whale Optimization Algorithm (EWOA) and applied on the frame structure in order to optimize the sizing problem and found an efficient result. Nasiri and Khiyabani (2018) updated the Whale Clustering Optimization Algorithm and compared this algorithm with other Particle Swarm Optimization (PSO) technique and found that this new developed algorithm is very much successful for data clustering. Kaur and Arora (2018) enhanced the WOA by using chaotic method and reduce the time requirement for the solution. Reddy et al. (2017) applied the WOA in order to optimize the Distributed Generator (DG) size and find out a better result compare to other optimization technique. Rajesh and Anand (2012) have studied the characteristics of the EDM process and applied Genetic Algorithm (GA) to optimize the MRR and surface finish (Ra) at same time considering the process control parameters such as working voltage, oil pressure, pulse on time, pulse off time and spark gap.

However, present research work deals with the optimization of EDM process parameters using two different bio-inspired optimization algorithms namely, Artificial Bee Colony algorithm and Whale optimization algorithm for both single as well as multiple responses and compares the results.

EXPERIMENTAL DETAILS

The current study is operated on Electric Discharge Machine (ACTSPARK SP1, China) die-sinking type with servo-head having uniform gap and positive polarity for copper tool electrode is used for testing. Commercial grade EDM-30 oil (specific gravity of 0.80 @ 25°C, Viscosity of 3.11 cSt. @ 38°C) is utilized as dielectric fluid with lateral flushing of pressure 0.2 kgf/cm^2. A square-shaped Cu tool (12x12 mm^2) is used for machining purpose. AISI EN24 tool steel work piece material is chosen for the experiment. EN24 tool steels are either nitride or carburized and are pre-hardened to (29-33) HRC. These hard alloy steels are capable of being machined into complex and large dies and moulds. The pulsed discharge current is applied in several steps in positive mode. The EDM setup comprises of dielectric reservoir, power supply and control unit, pump and flow system, operational tank with workpiece holding device, X-Y table compliant to the working table, tool holder with Z axis movement and the servo system to feed the tool part as shown in Figure 1.

The servo control unit is essential to maintain the pre-determined spark gap. It senses the gap voltage and relates it with the present value and the change in voltage is then used to switch the movement of servo motor to regulate the gap.

The MRR is expressed as the ratio of the volume of the work piece material removed during machining to the actual machining time. Surface smoothness of the machined surface is articulated as R_a is measured by using stylus type profilometer named Perthometer-M1 (MahrGmbh make). Overcut is expressed as half of the difference betweenthe area of the cavity produced to the tool frontal area. Area of cavity & frontal area of electrode can be calculated by measuring the respective length & width using Toolmaker's microscope.

Figure 1. Electric discharge machine setup

While executing those experiments, varying the levels of the factors in random rather than one at a time is proficient in terms of time and cost and also permits for the study of interactions between the factors. Based on former research works and initial study, four major control parameters are selected as input. Four input parameters are varied with three levels in sixteen experimental run. Those control parameters and their different boundary conditions are shown in Table 1 below. There are some other factors which has influence on performance like Flushing pressure, Duty cycle, Lift time etc., however, are kept uniform during experimental run. Table 2 display the response parameters with there maximum and minimum values.

ANALYSIS METHODOLOGY

Modelling of Responses

Experiments have been performed according to design of experiment (DOE). Sixteen experimental runs were carried out by varying Pulse on time (T_{ON}), Pulse off time (T_{OF}), Gap current (GI) and Spark gap (SG). Material removal rate (MRR, *mg/s*), centreline average roughness (R_a, *μm*) and overcut (OC, *mm*^{2}) were measured immediately after every machining operation. To obtain the objective functions for each response in terms of machining parameters, second-order polynomial regression equations have been generated for *MRR, Ra and OC* as a function of machining parameters using Minitab software.

The objective functions are given as follows:

$$\begin{aligned} MRR = 4.958 - 0.33634x_1 - 0.46422x_2 + 0.5641x_3 + 8.128x_4 \\ +0.019341x_1x_2 - 0.014695x_1x_3 + 0.000514x_2x_3 \end{aligned} \quad (1)$$

Table 1. Control parameters and boundary

Cutting Parameters	Symbol	Units	Boundary Condition	
			Lower	Upper
Pulse on time	T_{ON}	(μSec)	16	25
Pulse off time	T_{OFF}	(μSec)	12	21
Gap current	GI	(Amp)	7	10
Spark gap	SG	(mm)	0.16	0.19

Table 2. Response parameters and boundary

Responses Parameters	Symbol	Units	Boundary Condition	
			Lower	Upper
Material Removal Rate	MRR	(mg/s)	0.048	1.839
Surface Roughness	R_a	(μm)	4.4	9.3
Overcut	OC	(mm^2)	5.992	17.41

$$R_a = 35.2 - 0.81x_1 - 0.51x_2 - 4.66x_3 - 14.2x_4 \\ -0.0100x_1x_2 + 0.109x_1x_3 + 0.1350x_2x_3 \quad (2)$$

$$OC = -27 + 3.1x_1 - 4.1x_2 + 15.2x_3 - 139x_4 \\ +0.197x_1x_2 - 0.69x_1x_3 - 0.056x_2x_3 \quad (3)$$

where, x_1 = Pulse on time (µs), x_2 = Pulse off time (µs), x_3 = Gap current (A), x_4 = Spark gap (mm), MRR, R_a & OC carries usual meanings as above.

These objective functions have been optimized by following two nature based optimization algorithms obeying following constrains,

$$16 \leq x_1 \leq 25$$
$$12 \leq x_2 \leq 21$$
$$7 \leq x_3 \leq 10$$
$$0.16 \leq x_4 \leq 0.19$$

The boundary values have been selected based on extensive literature review (Bose & Pain, 2016).

Artificial Bee Colony Algorithm

Among all intelligent phenomena available in nature, Karaboga and Basturk were much fascinated by the intelligent foraging behaviour of honey bees. These researchers have modelled the foraging behaviour of honey bees in an optimization algorithm, well known as Artificial bee colony (ABC) algorithm (Karaboga, 2005; Karaboga & Basturk, 2007; Karaboga & Basturk, 2008). In this present study, this nature-based optimization algorithm has been used to solve the optimization problem of EDM process. A brief explanation of the activities performed by those honey bees has been given here to understand the algorithm. Several involved parameters in this algorithm have also been described here. Three types of honey bees exist in the hive, viz., *(i) Scout bee, (ii) Employed beeand (iii)Onlooker bee*. Scouts are those who search for new food source. In present algorithm only one scout is allowed. Once a scout finds a new food source becomes employed bee. These employed bees collect all information about food source, like, amount of food (nectar) quality of food, distance of food source from hive etc. With all information employed bees come to hive and perform waggle dance to share the information about food source to onlooker bees. Onlooker bees wait in the hive to acquire information of food source from employed bees. The total population is consists of these employed and onlooker bees. In this algorithm half of the population consists of employed bees and remaining half is onlooker bees. The selection of food source by the onlooker bees depends on several parameters of those food sources shared by employed bees. In this algorithm the food source means a set of input parameters. During initialization of this algorithm, a random set of food source (set of input parameters) is given to employed bees. The initial food source is produced randomly within the range of the boundaries of the parameters following the equation:

$$x_{ij} = x_j^{\min} + rand\ (0,1)\ \left(x_j^{\max} - x_j^{\min}\right) \tag{4}$$

where, $i = 1, 2, ...S_N$, and $j = 1, 2, ...D$. S_N is the number of food sources (equal to number of employed bees) and D is the number of optimization parameters (input parameters). The maximum value of *SN* is 3and *D* is 4 in this study. After having the idea of initial food source provided by employed bees, onlooker bees find the fitness of those food sources and the best food source is chosen. The fitness is calculated by the following equations:

For minimization problem,

$$\begin{aligned} fitness_i &= 1/\left(1 + R_i\right) && if, R_i \geq 0 \\ &= 1 + abs\left(R_i\right) && if, R_i < 0 \end{aligned} \tag{5}$$

For maximization problem,

$$fitness_i = 1 + R_i \tag{6}$$

R_i is the objective functions (responses), i.e., equations of *MRR*, *Ra* and *OC* shown in equation (1), (2) & (3).

After choosing the best food source based on fitness value, the onlooker bees become employed bees and go to the best food source site and search the neighbouring to that food source for better result. The neighbouring food sources are obtained by the following equation:

$$\nu_{ij} = x_{ij} + \xi_{ij}\left(x_{ij} - x_{kj}\right) \tag{7}$$

Within the neighbourhood of every food sources it represented by x_i, a food source ν_l is determined by changing one parameter of x_i. *j&k* are integers chosen randomly with the conditions:

$$j \in \left\{1,\ 2,\ \ldots\ D\right\}\ k \in \left\{1,\ 2,\ \ldots\ SN\right\} \text{and of course } k \neq i$$

ξ_{ij} is a uniformly distributed real random number in the range [-1,1].

The constrain in finding neighbouring food source is that, if the value of the new food source exceeds the boundary values, the boundary value is considered as new food source, i.e., if $x_i > x_i^{\max}$ then $x_i = x_i^{\max}$. If $x_i < x_i^{\min}$ then $x_i = x_i^{\min}$.

Here, the greedy selection method has been employed to select the best food source by calculating the probability values of fitness function. The best probable fit source is selected for the next cycle. The probability is calculated by the following equation:

$$p_i = \frac{fitness_i}{\sum_{i=1}^{SN} fitness_i} \tag{8}$$

Now, the onlooker bees, which have turned into employed bees, started consuming food from the best food source so far they have found. It is obvious, after some time this food source will become empty as a result the employed bee will become a scout and have to search for new food source. Thus the number of cycle after an employed bee will become a scout is determined by a parameter called 'Limit'. In this algorithm limit is chosen as 50. The scout then flies to find a new food source. As the scout finds a new food source, it becomes employed bee and the cycle goes on. The terminating criterion of this cycle is set by a parameter called MCN (Maximum Cycle Number). In this algorithm, the MCN has been varied from 200 to 500 depending on how fast the solution converges to optimum solution.

Whale Optimization Algorithm

Inspired by the intelligent foraging behaviour of humpback whales (*Megapteranovaeangliae*) during finding prey, Mirjalili and Lewis (2016), introduced the Whale Optimization Algorithm (WOA). It is a meta-heuristic algorithm, which follows the manoeuvre of humpback whales called bubble-netfeeding method. In this method practically the whale releases bubble and swims upward along a spiral path of decreasing radius to entrap the prey. This is mathematically modelled by Mirjalili and Lewis and has been used here to solve a practical manufacturing problem described as follows.

In searching mechanism, two approaches have been used (i) Shrinking encircling mechanism and (ii) Spiral updating position. In the first approach the search space linearly reduces and finally entrapped the target solution. In this method the new solution $\vec{v}_{new}$ and the distance between whale and prey $\vec{d}$ is obtained by the equation,

$$\vec{v}_{new}(i+1) = \vec{v}_{best}(i) - \vec{a} \cdot \vec{d} \tag{9}$$

$$\vec{d} = |\, \vec{c} \cdot \vec{v}_{best}(i) - \vec{v}(i) \,| \tag{10}$$

$$\vec{a} = 2\vec{\alpha} \cdot \vec{r} - \vec{\alpha} \tag{11}$$

$$\vec{c} = 2 \cdot \vec{r} \tag{12}$$

where, $\vec{v}_{best}$ is the best solution obtained at i^{th} iteration, $\vec{a}$ and $\vec{c}$ are coefficient vectors, $\vec{\alpha}$ is reduction factor which decreases linearly, $\vec{r}$ is a random vector between [0, 1] and '.' is element to element multiplication operator.

In the spiral updating position approach, it is considered the whale is moving upward along a logarithmic spiral and is modelled as:

$$\vec{v}_{new}(i+1) = \vec{d}_1.e^{bl}.\cos(2\pi l) + \vec{v}_{best}(i) \tag{13}$$

$$\vec{d}_1 = |\, \vec{v}_{best}(i) - \vec{v}(i) \,| \tag{14}$$

where, b is constant for defining the shape of the spiral and l is a random number between [-1, 1].

The probability of choosing these two approaches is equally distributed as 50%. The probability (p) value determines the choiceof either model to update the position of whales. It is modelled as:

$$\vec{v}_{new}(i+1) = \vec{v}_{best}(i) - \vec{a} \cdot \vec{d} \text{ if, } p < 0.5 \tag{15}$$

$$\vec{v}_{new}(i+1) = \vec{d}_1.e^{bl}.\cos(2\pi l) + \vec{v}_{best}(i) \text{ if, } p \geq 0.5 \tag{16}$$

This algorithm is used to find out the optimum value of the said objective functions imposing the boundary conditions of upper and lower limit of the input variables.

RESULTS AND DISCUSSION

Single-Objective Optimization

The single-objective optimization problem has been solved by artificial bee colony algorithm (ABC) and whale optimization algorithm (WOA). Both these methods fall under bio-inspired meta-heuristic, thus there is involvement of several parameters to control the optimum results. These control parameters of both algorithms are varied on trial basis and the best combination of control parameters have been taken for analysis. Both the optimizations have been performed using MATLAB.

The objective functions are:

- For material removal rate: Maximize *MRR* [*MRR* is shown in Equation (1)]
- For centre line average: Minimize R_a [R_a is shown in Equation (2)]
- For overcut: Minimize OC [*OC* is shown in Equation (3)]

These objective functions have been optimized by artificial bee colony and whale optimization algorithm separately imposing constrains. The optimum machining conditions with maximum/minimum response values obtained for each response from these analyses are shown in Table 3. The convergence plots ofboth analyses for each response are shown in Figure 2. Both the optimization methods gave exactly same solution for MRR and Ra, but a slight different solution in case of OC. However, ABC gave slight better result than WOA. Confirmation tests were carried out with the optimum machining parameter combinations obtained from the analyses. The experimental values of roughness obtained at optimum combinations are as: $R_a = 3.51\mu m$ $MRR = 2.62$ mg/s and $OC = 0.00$ for both ABC &WOA. As all instruments which have been used in this work, can measure up to two decimal places only, hence it was not possible to differentiate the results of experimental values very precisely. However, the experimental values of all the responses have good agreement with analytical values.Though the optimum response values obtained from the analyses are very much same, but it is noteworthy, that WOA converges more quickly than ABC. In figure 2(a) it can be seen that, WOA reached the minimum value of R_a after approximately 6 iterations and no further improvement took place with iterations. In case of ABC, in that same plot, it is observed, it took almost 50 iterations to reach the minimum value of R_a with no further modification. The same scenario can be observed in figure 2(b). In this case ABC took almost 100 iterations to reach the maximum value of MRR, while WOA took approximately 6 iterations to reach the result with no further modification. In figure 2(c) this difference is not much clear. Thus to save the time of analysis, WOA can be much better choice over ABC algorithm.

Multi-Objective Optimization

As it is clear from the results of single-objective optimizations, that optimum combination of machining parameters are different for different responses. Thus a certain combination of machining parameters cannot give best values of all the responsessimultaneously. Practically, when any component is machined with a certain combination of parameters, it is desired that all the responses must attain best possible values simultaneously. Hence, in practical situation a specific combination of machining parameter is needed, such that, all the responses attain a value as best as possible by compromising all of those responses equally. Multiple-objective optimization is practical solution to this problem.In this case also two optimization algorithms i.e., ABC and WOA have been employed and compared. The control parameters for both the algorithms are kept same as previously. The multi-response objective function for both artificial bee colony algorithm and whale optimization algorithm is generated by following the previous research work (Rao et al., 2008) and is shown in Equation (15).

$$R = \frac{w_1}{MRR_{\max}} MRR - \frac{w_2}{R_{a\min}} R_a - \frac{w_3}{OC_{\min}} OC \tag{15}$$

Table 3. Optimization results obtained from single-objective analysis

Parameters	ABC			WOA		
	Ra (μm)	*MRR (mg/s)*	*OC (mm²)*	*Ra (μm)*	*MRR (mg/s)*	*OC (mm²)*
POT (μs)	16.00000	16.00000	16.00000	16.00000	16.00000	16.3052
POF (μs)	12.00000	12.00000	21.00000	12.00000	12.00000	20.8857
GI (A)	10.00000	10.00000	7.94839	10.00000	10.00000	7.51212
SG (mm)	0.16000	0.19000	0.19000	0.16000	0.19000	0.186219
Optimum response value	3.51200	2.61519	5.38896E-05	3.512	2.615192	6.29122E-05

Figure 2. Convergence plots, (a) R_a, (b) MRR and (c) OC

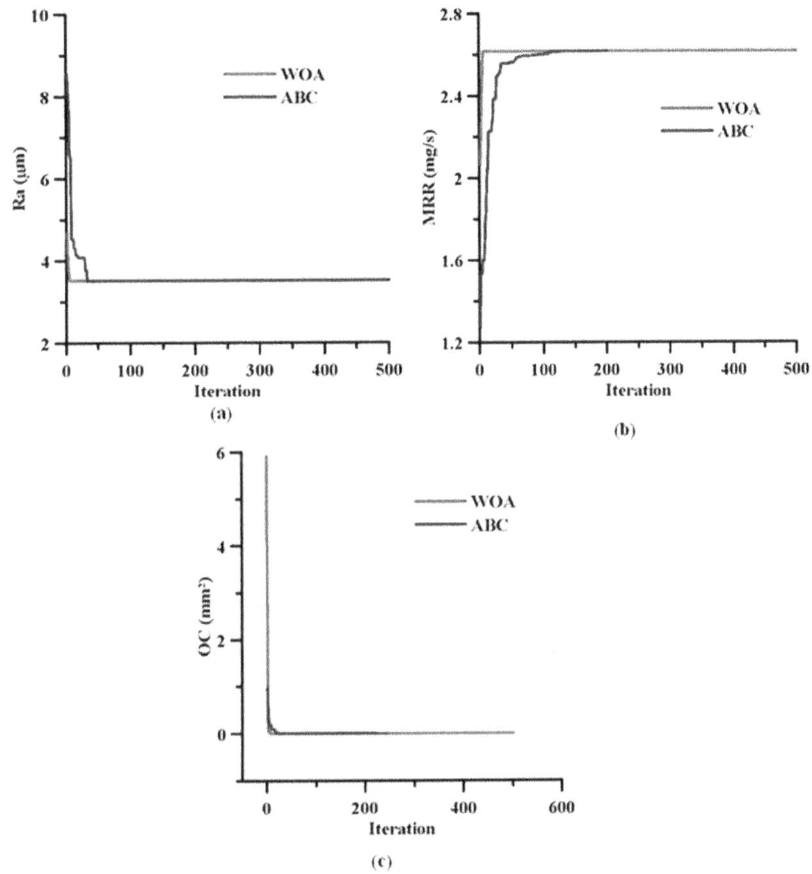

where, w_1, w_2 and w_3 are weights assigned to MRR, centre line average (Ra) and overcut (OC). Here MRR and R_a given equal importance by assigning equal weight value of 0.449 to both and OC is less significant than other two thus 0.102 is assigned to it. All the weights were calculated using Fuzzy set theory which has been described in another research paper (Bose & Pain, 2017). MRR_{max} represents the maximum material removal rate and $R_{a\,min}$ & OC_{min} represents the minimum values of centre line average and overcut respectively. MRR, R_a&OC are the regression equations as shown in Equations (1), (2) & (3) respectively. Constrains are same for this case also. The optimum machining conditions with best possible response values obtained from these analyses are shown in Table 4. Confirmation tests for MRR, centre line average and overcut gives value for MRR = 0.31 mg/s for ABC & 0.32 mg/s for WOA, R_a = 9.93 µm for ABC & 9.88µm for WOA and OC = 0.00 for ABC & WOA.

Significance of Machining Parameters

The significance of each machining parameter has been analyzed and expressed in terms of box and whisker plot shown in Figure 3 to 5. Figure 3 shows the variation of MRR with all the parameters. It is evident, at minimum and maximum value of pulse on time the height of box is very small, thus at these values of T_{ON}, MRR is mostly unresponsive to other parameters as the variation of MRR is very small.

Table 4. Optimization results obtained from multi-objective analysis

	Parameters	ABC	WOA
Machining	Pulse on time (μs)	16.80922	16.0532
	Pulse off time (μs)	20.40984	20.0938
	Gap current (A)	7.07531	7.02327
	Spark gap (mm)	0.19000	0.179731
Response	MRR (mg/s)	0.327144	0.307961
	R_a (μm)	9.930024	9.888188
	OC (mm^2)	9.02E-07	1.84E-05

But at other values of T_{ON}, MRR varies largely as heights of boxes are large. It is also notable, within the whole range of T_{ON}, the medians of all the boxes vary in a large range, and thus, T_{ON} has huge influence to MRR. For all other parameters like T_{OFF}, GI and SG, at all values the heights of boxes are varying largely, but the medians are quite close to each other, thus, these parameters are less significant to MRR. In case of centre line average shown in Figure 4, a little variation in median values is prominent for T_{OFF}, but for other parameters the median values are almost same. Thus for centreline average, T_{OFF} is most significant parameter compared to others. Figure 5 shows the variation of overcut with all machining parameters. In this case, for all the machining parameters, the median values vary within a range of 10 to 15 mm. Thus for overcut, no parameter is strongly significant than other.

Figure 3. Variation of MRR with machining parameters

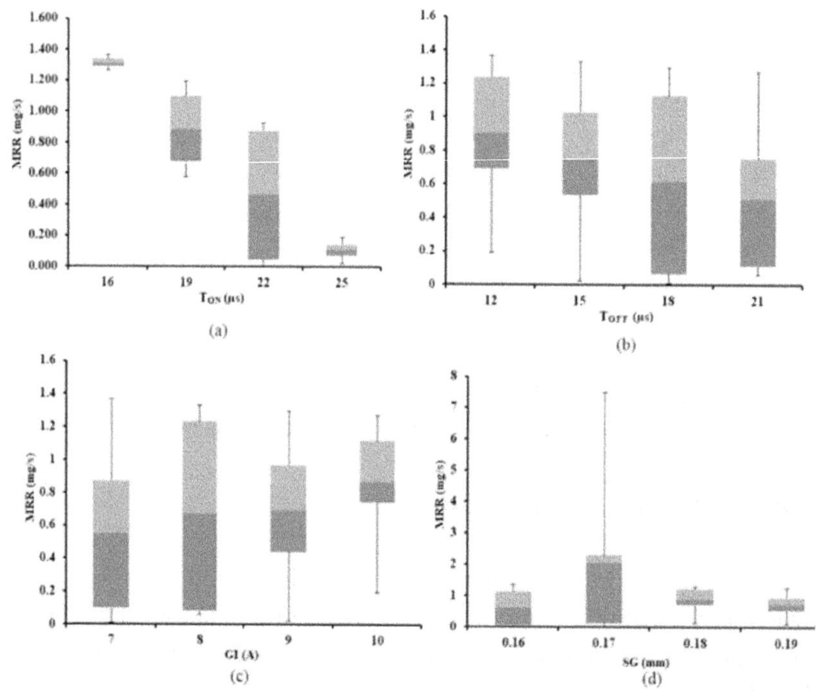

Figure 4. Variation of R_a with machining parameters

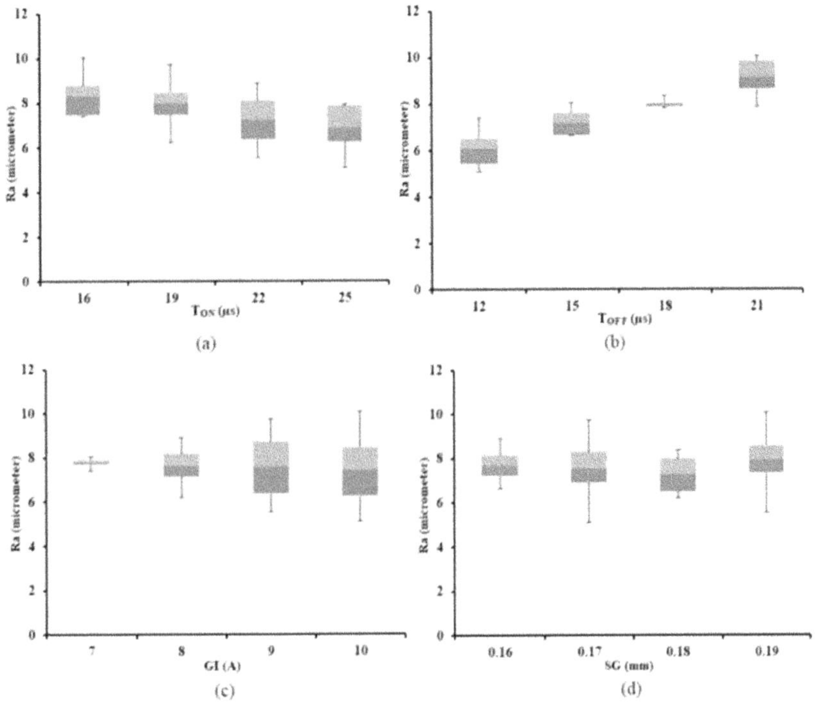

Figure 5. Variation of OC with machining parameters

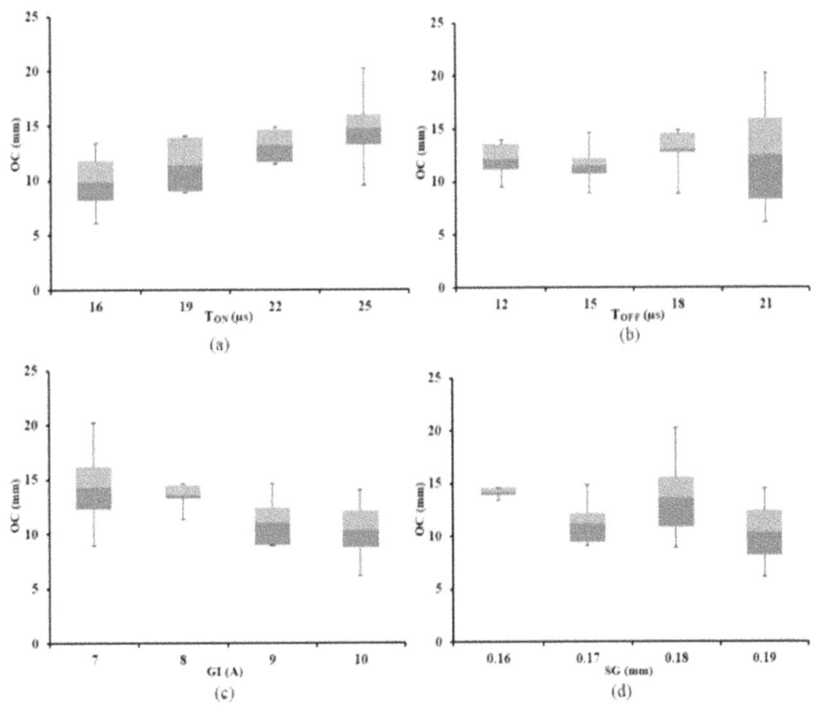

CONCLUSION

In this study, EDM has been performed on EN4 material by varying the pulse on time, pulse off time, gap current and spark gap. Material removal rate (*MRR*), centre line average (R_a) and overcut (*OC*) have been optimized and studied on the basis of machining parameters. The conclusions derived from this analysis are stated as follows:

- Artificial bee colony (ABC) and Whale optimization algorithm (WOA) have been successfully employed to find out the optimum combination of machining parameters for individual responses (single-objective) as well as taking all the responses simultaneously (multi-objective). It is observed that both these algorithms give similar results, however, in terms of high accuracy in some cases ABC gave a little better result than WOA. But WOA calculated the optimum value much faster than ABC.
- Significance of each machining parameters were investigated for all the responses. It is found, T_{ON} is more significant for MRR, T_{OFF} is more significant for R_a than other parameters. In case of overcut (OC), no machining parameters are significant.

REFERENCES

Bharaskar, S. S., Banwait, S. S., & Laroiya, S. C. (2013). Multiobjective Optimization of Electrical Discharge Machining Process Using a Hybrid Method. *Materials and Manufacturing Processes*, *28*(4), 348–354. doi:10.1080/10426914.2012.700152

Bose, G. K., & Pain, P. (2016). Parametric Analysis of Different Grades of Steel Materials Used in Plastic Industries through Die Sinking EDM Process. *International Journal of Materials Forming and Machining Processes*, *3*(1), 45–74. doi:10.4018/IJMFMP.2016010104

Bose, G. K., & Pain, P. (2017). Optimization Through Nature-Inspired Soft-Computing and Algorithm on ECG Process. In Handbook of Research on Soft Computing and Nature-Inspired Algorithms (pp. 489-519). IGI Global. doi:10.4018/978-1-5225-2128-0.ch017

Das, K. M., Kumar, K., Barman, T. K., & Sahoo, P. (2014). Application of Artificial Bee Colony Algorithm for Optimization of MRR and Surface Roughness in EDM of EN31 tool steel. *Procedia Materials Science, 6*, 741-751.

Gholipoor, A., Baseri, H., & Shabgard, M. Z. (2015). Investigation of near dry EDM compared with wet and dry EDM process. *Journal of Mechanical Science and Technology*, *29*(5), 2213–2218. doi:10.100712206-015-0441-2

Gostimirovic, M., Kovac, P., Sekulic, M., & Skoric, B. (2012). Influence of discharge energy on machining characteristics in EDM. *Journal of Mechanical Science and Technology*, *26*(1), 173–179. doi:10.100712206-011-0922-x

Kanyongo, G. Y., Certo, J., & Launcelot, B. I. (2006). Using regression analysis to established the relation between home environment and reading achievement: A case of Zimbabwe. *International Education Journal*, *7*(5), 632–641.

Karaboga, D. (2005). *An idea based on honey bee swarm for numerical optimization* (Vol. 200). Technical report-tr06, Erciyes University, Engineering Faculty, Computer Engineering Department.

Karaboga, D., & Basturk, B. (2007). A powerful and efficient algorithm for numerical function optimization: Artificial bee colony (ABC) algorithm. *Journal of Global Optimization*, *39*(3), 459–471. doi:10.100710898-007-9149-x

Karaboga, D., & Basturk, B. (2008). On the performance of artificial bee colony (ABC) algorithm. *Applied Soft Computing*, *8*(1), 687–697. doi:10.1016/j.asoc.2007.05.007

Kaur, G., & Arora, S. (2018). Chaotic Whale Optimization Algorithm. *Journal of Computation Design and Engineering.*, *5*(3), 275–284. doi:10.1016/j.jcde.2017.12.006

Kevah, A., & Ghazaan, M. I. (2017). Enhanced whale optimization algorithm for sizing optimization of skeletal structures. *Mechanics Based Design of Structures and Machines*, *45*(3), 345–362. doi:10.1080/15397734.2016.1213639

Kumar, V. S., & Kumar, P. M. (2014). Optimization of cryogenic cooled EDM process prameters using grey relation analysis. *Journal of Mechanical Science and Technology*, *28*(9), 3777–3784. doi:10.100712206-014-0840-9

Majumder, A. (2013). Process parameter optimization during EDM of AISI 316 LN stainless steel by using fuzzy based multi-objective PSO. *Journal of Mechanical Science and Technology*, *27*(2), 2143–2151. doi:10.100712206-013-0524-x

Mirjalili, S., & Lewis, A. (2016). The Whale Optimization Algorithm. *Advances in Engineering Software*, *96*, 51–67. doi:10.1016/j.advengsoft.2016.01.008

Mukherjee, R., & Chakraborty, S. (2012). Selection of EDM Process Parameters Using Biogeography-Based Optimization Algorithm. *Materials and Manufacturing Processes*, *27*(9), 954–962. doi:10.1080/10426914.2011.610089

Nasiri, J., & Khiyabani, F. M. (2018). *A Whale Optimization Algorithm (WOA) approach for Clustering. Cogent Mathematics & Statistics, 1483565.*

Rajesh, R., & Anand, M. D. (2012). The Optimization of the Electro-Discharge Machining Process using Response Surface Methodology and Genetic Algorithms. *Procedia Engineering*, *38*, 3941–3950. doi:10.1016/j.proeng.2012.06.451

Rao, R. V., Pawar, P. J., & Shankar, R. (2008). Multi-objective optimization of electrochemical machining process parameters using a particle swarm optimization algorithm. *Proceedings of the Institution of Mechanical Engineers. Part B, Journal of Engineering Manufacture*, *222*(8), 949–958. doi:10.1243/09544054JEM1158

Reddy, P. D. P., Reddy, V. C. V., & Manohar, T. G. (2017). Whale optimization algorithm for optimal sizing of renewable resources for loss reduction in distribution system. Renewables. *Wind, Water, and Solar*, *4*(3), 1–13.

Song, K. Y., & Chu, C. N. (2015). Effect of machining area on material removal rate in strip EDM. *International Journal of Precision Engineering and Manufacturing*, *16*(12), 2435–2440. doi:10.100712541-015-0313-9

Chapter 15
Hybrid Multi-Criteria Decision-Making Optimization Strategy for RP Material Selection:
A Case Study

Jagadish
National Institute of Technology Raipur, India

Sumit Bhowmik
https://orcid.org/0000-0002-7787-756X
National Institute of Technology Silchar, India

Suresh Gudala
National Institute of Technology Silchar, India

ABSTRACT

Rapid prototyping (RP) is an advanced manufacturing technique that uses a computer-aided design data for designing a prototype. To design a part, the RP uses many kinds of materials that best suit the application. Due to the existence of a large number of RP materials that make the selection process quite challenging and considered to be a multi-criteria decision-making (MCDM) optimization problem. The chapter proposed a hybrid MCDM optimization strategy for optimal material selection of the RP process. The hybrid strategy consists of modified DEMATEL with TOPSIS where the extraction of criteria weights using modified DEMATEL strategy while ranking of RP alternatives considering the effect of beneficial and non-beneficial using TOPSIS strategy. A real-life case study on RP material selection is demonstrated to validate the proposed strategy.

DOI: 10.4018/978-1-5225-8223-6.ch015

INTRODUCTION

In the advancement in manufacturing made the rapid prototyping (RP) process quite important and beneficial in the technical world. Basically, RP uses three-dimensional computer-aided design data (CAD) to replicate a prototype model of a planned designed part. Such a manufacturing process is carried by the successive layering of the prototype model using various kinds of materials. Due to the rapid growth of RP technology, the selection of the most appropriate material to meet the user requirements from among a number of RP materials has become increasingly important. Improper selection leads to failure of the RP product, dissatisfaction of customer, lower demands etc. Also, during the selection, designer should keep into account the influence of factors which affects the process of material selection for the RP process (Jaromin, 2013; Gibson et al. 2015; Gupta et al. 2017). Moreover, the selection of optimal RP material is very difficult because it depends on many attributes and constrains and also depends on different mechanical, physical and chemical properties. So RP material selection is a complex multi-attribute or criteria optimization problem that cannot be solved using conventional optimization techniques. Furthermore, RP material selection process includes contradictory criterias which made essential to implement the sophisticated MCDM methods (Chang, 2011; Lan, 2009). In which a decision maker has to choose the most optimal RP material from several sets of RP materials.

In the literature, various MCDM methods have been proposed for optimal material selection such as Jaromin (2013) a developed web material selector (WMS) application is designed for computer-aided materials selection The WMS allows to work with very large databases through the use of a professional tool to manage complex data structures. At the same time the application is designed to optimize the search process, what significantly reduces the number of materials whose properties a designer should carefully consider. Chang et al. (2011) pioneered in using the fuzzy DEMATEL method to find influential factors in selecting SCM suppliers. The DEMATEL method evaluates supplier performance to find key factor criteria to improve performance and provides a novel approach of decision making information in SCM supplier selection. Lan (2009) proposed a web-based computer aided material selection system for aircraft design, applying a material selection strategy combined screening and ranking methods. This combined strategy could make good use of selection experience and material testing data, thus making the selection results more reasonable and bringing more standardization to the material selection process. Dursun and Karsak (2013) proposed a fuzzy multi-criteria group decision making approach that makes use of the QFD concept is developed for supplier selection process. The proposed methodology initially identifies the features that the purchased product should possess in order to satisfy the company's needs, and then it seeks to establish the relevant supplier assessment criteria. Moreover, the proposed algorithm enables to consider the impacts of inner dependence among supplier assessment criteria. Attri and Grover (2015) presented the application of a novel MCDM method i.e. Preference selection index (PSI) method to solve various decision-making problems that are generally encountered in the design stage of production system life cycle. Tabriz et al. (2014) proposed a novel hybrid multi-criteria decision making (MCDM) methodology is to encompass the complex relationships and cope with the trade-offs among criteria. The ANP method used for determining the appropriate weightings to each sub-criterion was developed to overcome the problems of dependence and feedback among criteria. For identifying relationship network among criteria, DEMATEL method is employed as a supportive tool for ANP. Then TOPSIS is used to rank all competing alternatives according their performance. Shahroodi (2010) presented a methodology which deals with a brief review of the literature regarding AHP technique and its relevancy to its application in supplier selection process. Supplier selection is a complicated process.

This process needs evaluation of multiple criteria and various constraints associated with them. After analysis of the results we found that for manufacturing firms, supplier reliability, product quality and supplier experience are the top three supplier selection problems that needs to be taken up on priority for effective vendor selection. Pratihar (2015) presented a review on soft computing based expert systems developed to establish input-output relationships of various manufacturing processes. To determine these relationships, both fuzzy logic- and neural network based approaches were tried. Reasonably good results were obtained using the developed approaches.

Additionally, many investigators have also worked on optimal selection of RP material and the RP process such as Arenas et al. (2012) applied analytic hierarchy process (AHP) for optimal selection of five different families of adhesives (cyanoacrylate, polyurethane, epoxy, acrylic, and silicone) for FDM process. They found that, epoxy kind adhesives gave better results for the FDM process compared to other adhesives. Prasad and Chakraborty (2013) proposed a methodology to solve the material selection problems using a quality function deployment (QFD) based approach that can integrate the voice of the customers for a product with its technical requirements. To ease out the materials selection decision-making process, a user-friendly software prototype in VISUAL BASIC 6.0 is also developed. To demonstrate the model a real-life four case studies are considered. The results show that, proposed model provides better/ optimal materials for the selected case studies and dominance the existing models. Kumar and Singal (2015) presented multiple attribute decision making methods for solving the material selection problem for penstock in the FDM process. Work used three multi attribute decision making methods like analytic hierarchy process (AHP), a technique for order preference by similarity to ideal solution (TOPSIS) and Modified TOPSIS methods. To justify the models, a real life case study on material selection for FDM process is considered. They found that, methods dominance the existing methods and provide better and comparable results. Rao and Padmanabhan (2007) proposed a methodology for selection of a RP process that best suits the end use of a given product or part using graph theory and matrix approach. They found that, proposed method is a general method and can consider any number of quantitative and qualitative RP process selection attributes simultaneously and offers a more objective and simple RP process selection approach. Another research of Rao (2006) proposed a graph theory and matrix approach (GTMA) for RP material selection. Work considered seven materials as alternatives and seven properties as a criterias. They found that material 3(SS 301–FH 39.112294) found to be best among the other. Similarly, the research of Masood and Soo (2202) presented an expert system using fuzzy logic method for selection of best RP process. Work considered 39 RP systems commercially available from 21 RP manufacturers worldwide. They found that proposed method works well in all the constraints and provide better and comparable results. Shende and Kulkarni (2014) introduced a methodology for decision support system for RP process selection. This study demonstrated the potential benefits of using a structured approach to the selection of alternative rapid prototyping processes. Three different decision making methods like AHP, matrix approach and TOPSIS are employed. They found that, used methodologies can provide guidance not only for RP process selection but also for other areas of decision making in manufacturing areas. The above literature reveals the following research gaps,

- Most of the studies on RP process slecetion rather than RP materail slecetion.
- RP process / material selection are based on subjective criterias. Therefore, consideration of both subjective and objective criterias is essential in the selection process.
- The extraction of precise priority weights of the RP process / material criterias from the vagueness of information is absent in the traditional research.

- Failed to consider the interrelationship between the criterias and influence of both beneficial and non-beneficial criteria during the RP process / material selection process.
- Lack of correlation of criterias and consistency in the judgment.

Therefore, the need for hybrid decision-making approach consists of modified- decision-making technique and evaluation laboratory (DEMATEL) with technique for order preference by similarity to ideal solution (TOPSIS) has been presented to overcome the above research gaps and solve the multi-criteria decision making optimization in RP material selection. In this M-DEMATEL method is used for extraction of precise priority weights from the vagueness of information while modified TOPSIS is used for optimal selection of RP material considering the influence of both beneficial and non-beneficial citeria's into account. At last, the validation of the proposed method is done by considering the case study of material selection for rapid prototyping process.

PROPOSED STRATEGY

In this section, an integrated strategy consists of M-DEMATEL with TOPSIS methods is presented for optimal matreail slecetion of RP process. Here, M-DEMATEL strategy (Gabus and Fontela, 1972; Gabus and Fontela, 1973) is used for extraction of precise priority weights of the material criterias which influences the ranking of RP materials while TOPSIS (Hwang and Yoon, 1981) is used for optimal RP material selection with consideration of interrelationship aspects and influence of both beneficial and non-beneficial criterias of RP criteria and alternatives into account. The detailed steps of the proposed integrated MCDM strategy are discussed below:

Step 1: Identification of the material selection attributes.

RP material selection is starts with the material attributes and its selection criterias. First, identification of material selection criteria/sub-criteria/alternatives required for evaluation of optimal RP material is done. These criteria and sub-criteria consist of both qualitative (non-beneficial) and quantitative (beneficial) criterias which have a significant influence on the RP material selection.

Step 2: Development of a decision-making matrix and its normalization

After identification of selection attributes and criterias, the development of decision matrix is carriedout using Eq. (1). The decision matrix includes the performance values of each alternative with respect to the number of criteria/sub-criteria. The decision matrix is formulated using following expression.

$$N_{ij} = \begin{bmatrix} & X_1 & X_2 & \cdots & X_n \\ Y_1 & a_{11} & a_{12} & \cdots & a_{1n} \\ Y_2 & a_{21} & a_{22} & \cdots & a_{2n} \\ \cdots & \cdots & \cdots & \cdots & \cdots \\ Y_m & a_{m1} & a_{m2} & \cdots & a_{mn} \end{bmatrix} \tag{1}$$

where N_{ij} represents the decision matrix which includes performance values of *i*th alternatives on *j*th criterias, *m* represents the number of alternatives and *n* represents the number of criterias/sub criterias, $X_1, X_2, X_3, \ldots X_n$ represent the criterias, $Y_1, Y_2, Y_3 \ldots Y_m$ represent the alternatives and $a_{11}, a_{12, \ldots}\ a_{mn}$ represents the performance values of *i*th alternatives on *j*th criterias.

After that, normalization of the decision matrix is done using Eq. (2). The normalization process is done to transfer the relative performance values for of *i*th alternatives with a different data measurement unit into a compatible unit. The normalization is carried out using the following equation.

$$X_{ij} = \frac{X_{ij}}{\sqrt{\sum_{i=1}^{m} X_{ij}^2}} \tag{2}$$

where the value X_{ij} represents the normalized performance values of *i*th alternatives on *j*th criterias for $i = 1,2,3 \ldots\ldots n$ and $j = 1,2,3, \ldots\ldots m$.

Step 3: Determination of precise priority weights using DEMATEL.

Next, determination of precise priority weights for each of the criterias is done using DEMTAL method. In this first, pairwise comparison matrix between the criterias is formulated based on the expert (H) and a number of criterias (n). Each of the experts (H) provides the degree of importance between the *i*th numbers of criteria to the *j*th criterion. The DEMTAL scale ranging from 0-4, representing 'No influence (0),' 'Low influence (1),' 'Medium influence (2),' 'High influence (3),' and 'Very high influence (4),' are used for degree of importance values. Then, scores of each expert (H) formulate the (n x n) matrix and determine the average comparison decision matrix *a* for all experts (H) opinions using the following equations,

$$A_{ij} = \begin{bmatrix} a_{11} & a_{12} & \cdot\cdot & \cdot\cdot & a_{1n} \\ a_{21} & a_{22} & \cdot\cdot & \cdot\cdot & a_{2n} \\ \cdot\cdot & \cdot\cdot & \cdot\cdot & \cdot\cdot & \cdot\cdot \\ \cdot\cdot & \cdot\cdot & \cdot\cdot & \cdot\cdot & \cdot\cdot \\ a_{n1} & a_{n2} & \cdot\cdot & \cdot\cdot & a_{nn} \end{bmatrix} \tag{3}$$

$$a_{ij} = \frac{1}{H} \sum_{k=1}^{H} x_{ij}^k \tag{4}$$

where a_{ij} represents the average pairwise comparison matrix, *k* is the number of criterias, *H* denotes the number of expert and *x* is the degree of importance allotted by the each of the experts.

Further, computation of the total relation matrix is determined based on the pairwise comparison matrix using Eq. (5). Similarly, the determination of precise priority weights for each of the criteria by summation of column (i.e. vector *r*) and summation of rows (i.e. vector *c*) of matrix *K* using Eqs. (3.8) and (3.9) respectively.

$$K = a_{ij} + a_{ij}^2 + ... + a_{ij}^m = a_{ij}(I - a_{ij})^{-1} \tag{5}$$

$$. \ r = [r_i]_{n\times 1} = (\sum_{j=1}^{n} K_{ij})_{n\times 1} \tag{6}$$

$$c = [c_j]_{1\times n} = (\sum_{i=1}^{n} K_{ij})_{1\times n} \tag{7}$$

where, *K* is the total relation matrix, *I* is the identity matrix r_i and c_j denotes the sum of rows and columns for each total relation matrix (*K*), n is the total no criteria's.

After that, vector (r_i + c_i) and vector (r_i – c_i) values for each of the criterias are calculated to get the precise priority weights (w_j). Here, elements in vector (r_i + c_i) column values indicate the degree of importance of criterion (i) in the system, while vector (r_i – c_i) values represent the net affect which criterion /output parameters (i) make on the system. At last, consistency of the pairwise comparison matrix and precision of the priority weights (Prasad and Chakraborty, 2012) obtained from DEMATEL method are verified by consistency index value (CI) using the Eq. (8).

$$CI = \frac{1}{n^2}\sum_{i=1}^{n}\sum_{j=1}^{n}\frac{K_{ij}^{p} - K_{ij}^{p-1}}{K_{ij}^{p}} \times 100\% \tag{8}$$

where, *n* represents the number of criteria, K_{ij}^{p} denote the influential element of matrix *K* and K_{ij}^{p-1} represent all elements of *K*.

Step 4: Determine the weighted decision matrix.

In this step, weighted decision matrix is carried out by multiplying the weights of each criterias obtained from DEMATEL method in to normalized performance values of the decision matrix (X_{ij}). The calculation of weighted decision matrix using the following expression:

$$T = w_j \times N_{ij} \tag{9}$$

where, *T* represents the performance weights of *i*th alternatives on *j*th criterias and *n* represent the number of criterias. W_j precise weights of the criterias.

Step 5: Identification of positive and negative ideal solution

In this step, the positive ideal solution (PI^+) and negative ideal solution (NI^-) are identified from the weighted decision matrix using Eqs. (10) and (11). The positive ideal solution indicates the maximum value for beneficiary criteria and minimum value for non-beneficiary criteria whereas negative ideal solution indicates the minimum value for beneficiary criteria and maximum value for non-beneficiary criteria in the weighted decision matrix.

$$PI^+ = \{ C_1^+, C_2^+, ...C_n^+ \}, \text{ where } C_j^+ = \{(\text{maxi} (C_{ij} \text{ if j=J}) ;(\text{mini } C_{ij} \text{ if j = J}')\} \tag{10}$$

$$NI^- = \{ C_1^-, C_2^- ....C_n^- \}, \text{ where } C_j^- = \{(\text{mini} (C_{ij} \text{ if j=J}) ;(\text{maxi } C_{ij} \text{ if j=J}')\} \tag{11}$$

where, PI^+ and NI^- are positive and negative ideal solution matrices respectively. C_{ij} is the criteria values of the weighted decision matrix with J being associated with beneficial criteria and J^1 associated with non-beneficial criteria.

Step 6: Calculation of separation distance for each of the criterion

After calculation of PI^+ and NI^- values for each of the criteria, the evaluation of separation distance i.e. D^+ and D^- from PI^+ and NI^- is done using Eqs. (12) and (13) respectively.

$$D^+ = \sqrt{\sum_{j=1}^{n} (C_j^+ - C_{ij})^2} \tag{12}$$

$$D^- = \sqrt{\sum_{j=1}^{n} (C_j^- - C_{ij})^2} \tag{13}$$

where, $D^+$and D^- are the separation distance of the alternative from positive ideal solution(PI^+) and negative ideal solution (NI^-) respectively and j is the criterion index.

Step 9: Evaluation of relative closeness values of each alternative and Ranking

Next, closeness values (CV_i) are computed based on the separation distance of the alternative obtained in the previous step using Eq. (14). The closeness values (CV_i) measures the closeness of the each alternatives among other others.

$$CV_i = D_i^- / (D_i^+ + D_i^-), 0 \leq C_i \leq 1 \tag{14}$$

Based on the closeness value (C_i) ranking of alternatives is done. The alternatives with maximum closeness value (*Ci*) are the most optimal among other alternatives.

CASE STUDY: OPTIMAL MATERIAL SELECTION FOR RAPID PROTOTYPING PROCESS

Background

To show the applicability and the strength of the proposed method in selection of optimal material, the present research work selected a case study. The case study is on the selection of optimal material for a RP process (Rao, 2006). In this direction, present work considered seven criteria's namely, toughness index (TI), yield strength (YS), Young's modulus (YM), density (D), thermal expansion (TE), thermal conductivity (TC), and specific heat(SH) and seven number of alternatives such as Material 1: Al 2024–T6 17.289715, Material 2: Al 5052–O, Material 3: SS 301–FH 39.112294, Material 4: SS 310–3AH 30.631578, Material 5: Ti–6Al–4V 34.055403, Material 6: Inconel 718 29.037740, Material 7: 70Cu–30Zn 20.037740. The number of crietraia and altenatives for the RP process are shown in Table 1 & 2.

Modeling of RP Material Process Using Proposed Strategy

In this section modeling of material selection process for RP proces is done using proposed strategy. In this, first, identification of number of criterias and alternatives for RP process is done. The number of criterias and alternatives for the RP process is tabulated in Table 1 and 2 respectively. Second, development of decision matrix which includes the number of criteria/sub-criteria and its corresponding alternatives

Table 1. List of criterias (Rao, 2006)

Influential factors (criteria)	
Toughness index	TI
Yield strength	YS
Young's modulus	YM
Density	D
Thermal expansion	TE
Thermal conductivity	TC
Specific heat	SH

Table 2. List of alternative material (Rao, 2006)

Alternatives Material	
Material 1	Al 2024–T6 17.289715
Material 2	Al 5052–O
Material 3	SS 301–FH 39.112294
Material 4	SS 310–3AH 30.631578
Material 5	Ti–6Al–4V 34.055403
Material 6	Inconel 718 29.037740
Material 7	70Cu–30Zn 20.037740

using Eq. (1) is carriedout (Rao, 2006). Then, normalization of decision matrix is performed using Eq. (2) in order to convert the relative importance values for each of the alternatives with a different data measurement unit in the decision model into a compatible unit. The results of decision matrix and normalized decision matrix for the RP materials selection problem are tabulated in Table 3 and 4 respectively.

After the normalization, determination of priority weights for RP material selection criteria is carried out using Modified DEMATEL strategy. In this first, computation of *n x n* average matrix i.e. Matrix [A] by taking the opinions of all expert opinions and by averaging the H expert's score using Eq. (3) and resulting matrix is shown in Table 5. Then, the detramination of normalized initial direct-relation matrix [D] and matrix [K] using Eqs. (4 and 5) and their corresponding results are depicted in Tables 6-7 respectively.

Then after, priority weights of the criteria ($r_j + c_j$) are calculated using Eqs. (6) and (7) as shown in Table 8. The weights for each of the RP material criterias as obtained are 16.326, 16.013, 15.970, 15.210, 15.075, 14.055, and 12.759 for TI, YS, YM, D, TE, TC, and SH respectively. The obtained weights show that, the parameter TI i.e. toughness index is the most influential criteria compared to the other. In addition, consistency check is carriedout to show the accuracy and consistency of the decision matrix [A] using Eq. (8). The consistency index for the matrix [A] is obtained to 0.03% which is less than 5%, hence, the matrix [A] and weights are consistent and can be used for further analysis. The precise priority weights for each of the RP materials criterias are tabulated in Tables 8.

Table 3. Decision matrix for RP material selection (Rao, 2006)

RP materials	TI	YS	YM	D	TE	TC	SH
Material 1	75.5	420	74.2	2.8	21.4	0.37	0.16
Material 2	95	91	70	2.68	22.1	0.33	0.16
Material 3	770	1365	189	7.9	16.9	0.04	0.08
Material 4	187	1120	210	7.9	14.4	0.03	0.08
Material 5	179	875	112	4.43	9.4	0.016	0.09
Material 6	239	1190	217	8.51	11.5	0.31	0.07
Material 7	273	200	112	8.53	19.9	0.29	0.06

Table 4. Normalized data of the RP material selection attributes

RP materials	TI	YS	YM	D	TE	TC	SH
Material 1	0.0840	0.1786	0.1841	0.1604	0.4710	0.5650	0.5650
Material 2	0.1057	0.0387	0.1737	0.1535	0.4870	0.5040	0.5650
Material 3	0.8575	0.5807	0.4689	0.4520	0.3720	0.0610	0.2820
Material 4	0.2082	0.4765	0.5211	0.4520	0.3170	0.0450	0.2820
Material 5	0.1993	0.3722	0.2779	0.2530	0.2070	0.0240	0.3180
Material 6	0.2660	0.5063	0.5384	0.4870	0.2530	0.4740	0.2470
Material 7	0.3040	0.0850	0.2779	0.4880	0.4380	0.4430	0.2120

Table 5. Data of average matrix [A]

	TI	YS	YM	D	TE	TC	SH
TI	0.000	1.5263	2.1053	1.8947	2.2632	2.0000	1.3684
YS	1.589	0.0000	2.0526	2.3158	2.3684	2.1053	1.5789
YM	1.974	1.9474	0.0000	2.0526	2.4737	2.5263	1.9474
D	1.384	2.1579	2.0526	0.0000	2.2105	2.2105	1.5789
TE	1.847	2.0000	2.2105	1.7368	0.0000	2.2105	1.4211
TC	2.153	1.8947	2.1579	1.8947	2.2105	0.0000	1.6316
SH	1.153	1.7368	1.8421	1.5789	1.5789	1.7368	0.0000

Table 6. Normalized initial direct-relation matrix [D]

	TI	YS	YM	D	TE	TC	SH
TI	0.000	0.1164	0.1606	0.1445	0.1726	0.1526	0.1044
YS	0.104	0.0000	0.1566	0.1767	0.1807	0.1606	0.1204
YM	0.185	0.1485	0.0000	0.1566	0.1887	0.1927	0.1485
D	0.144	0.1646	0.1566	0.0000	0.1686	0.1686	0.1204
TE	0.145	0.1526	0.1686	0.1325	0.0000	0.1686	0.1084
TC	0.106	0.1445	0.1646	0.1445	0.1686	0.0000	0.1245
SH	0.043	0.1325	0.1405	0.1204	0.1204	0.1325	0.0000

Table 7. Data of matrix [K]

	TI	YS	YM	D	TE	TC	SH
TI	0.8417	1.0300	1.1441	1.0631	1.2037	1.1674	0.8903
YS	1.0027	0.9857	1.2058	1.1483	1.2775	1.2401	0.9544
YM	1.0789	1.1740	1.1353	1.1930	1.3509	1.3301	1.0272
D	0.942	1.0980	1.1740	0.9684	1.2353	1.2134	0.9296
TE	0.9900	1.0816	1.1763	1.0794	1.0842	1.2065	0.9147
TC	1.0259	1.1021	1.2022	1.1150	1.2586	1.0916	0.9497
SH	0.8113	0.9237	0.9989	0.9245	1.0274	1.0185	0.6940

After that, weighted decision matrix is calculated using Eq. (9). The weighted decision matrix is calculated by multiplying the precise priority weights obtained via modified DEMATEL strategy to the each of the performance value in the decision matrix. He resulting weighted normalized decision matrix is shown in Table 9.

Then, determination of positive ideal (PI^+) solutions and the negative ideal (NI^-) solutions and ideal (D^+) and non-ideal (D^-) solution for each of the RP material criterias are done using Eqs. (10-13) respectively. The values of D+ and D^- for each of the criterias are tells about the ho far the criterias from the reference position. The RP criteria with larger separation distance means the criteria have least influ-

Table 8. Priority weights of the criteria

Criteria's	$r_j + c_j$
TI	16.326
YS	16.013
YM	15.970
D	15.210
TE	15.075
TC	14.055
SH	12.759

Table 9. Weighted Decision Matrix

RP materials	TI	YS	YM	D	TE	TC	SH
Mat 1	1.3727	2.8614	2.9405	2.4400	7.1147	7.9421	7.1904
Mat. 2	1.7273	0.6199	2.7740	2.3354	7.3474	7.0835	7.1904
Mat. 3	14.000	9.2997	7.4899	6.8842	5.6186	0.8586	3.5952
Mat. 4	3.400	7.6305	8.3221	6.8842	4.7874	0.6439	3.5952
Mat. 5	3.2545	5.9614	4.4385	3.8604	3.1251	0.3434	4.0446
Mat. 6	4.3454	8.1074	8.5995	7.4158	3.8233	6.6542	3.1458
Mat. 7	4.9636	1.3626	4.4385	7.4332	6.616	6.2249	2.6964

ence on the output while smaller distance means, the parameter have greater influence on the output results. The result shows that, the separation distance of Mat. 6 from positive ideal solution ($D^{+)}$ is the least with 0.3434 and Mat. 2 from the negative ideal solution (D^{-}) is the least with 0.6199. The results of PI^{+} and NI^{-} solutions and D^{+} and D^{-}) for each of the alternatives are depicted in Tables 10-11.

At last, the closeness values (CV) values of each alternative are calculated using Eq. (14). This converts the multi-objective problem into a single objective problem (Brauers and Zavadskas, 2006; Prasad

Table 10. PI^{+} and NI^{-} solution values

RP materials	Positive ideal (PI^{+})	Negative ideal (NI^{-})
Mat. 1	14.000	1.3727
Mat. 2	9.2997	0.6199
Mat. 3	8.5995	2.7740
Mat. 4	2.3354	7.4332
Mat. 5	3.1251	7.3474
Mat. 6	0.3434	7.9421
Mat. 7	2.6964	7.1904

Table 11. Data of separation distance

RP materials	Ideal solution (D^{+})	Non- ideal solution (D^{-})
Mat. 1	121.16	197.01
Mat. 2	104.04	189.75
Mat. 3	390.52	170.88
Mat. 4	231.08	149.80
Mat. 5	168.98	103.94
Mat. 6	250.22	193.53
Mat. 7	167.77	192.66

and Chakraborty, 2012). The closeness values (CV) values for each of the RP material alternatives are tabulated in Table 12. The results shows that, Mat. 3 shows highest closeness values (CV) values compared to the other RP material which signifies that optimal RP materials among the other.

RESULT AND DISCUSSIONS

Optimization of RP Materials

The optimal selection of RP material mainly helps the designer in acquiring the most favorable RP material. The optimal RP materials selection is done based on the closeness values (CV) values. The result of optimization is tabulated in Table 13. The result shows that, the RP material (Mat.) 3: SS 301–FH 39.112294 yields optimal than the others. The optimal materials 3: provides higher toughness index (TI), yield strength (YS) and young's' Modulus (YM) compared to the other materials for the RP process. Also, the optimal RP material gives directly or indirectly improves the product quality, reduces cost and improves the performance of the RP process.

Table 12. Data of relative closeness

RP materials	Closeness values (CV)
Mat. 1	75.842
Mat. 2	85.716
Mat. 3	219.63
Mat. 4	81.285
Mat. 5	65.034
Mat. 6	56.689
Mat. 7	24.887

Table 13. The overall results of optimum material selection of RP

Materials		Closeness values (CV)	Ranks
1	Al 2024–T6 17.289715	75.842	4
2	Al 5052–O	85.716	2
3	**SS 301–FH 39.112294**	219.63	**1**
4	SS 310–3AH 30.631578	81.285	3
5	Ti–6Al–4V 34.055403	65.034	5
6	Inconel 718 29.037740	56.689	6
7	70Cu–30Zn 20.037740	24.887	7

Comparative Analysis

Additionally, comparative analysis is done via comparing the results of proposed strategy with that of the published results (Rao, 2006). The result of the comparative analysis is depicted in Figure 1. It is observed that, the proposed strategy i.e. M-DEMATEL-TOPSIS dominance the existing method and provide same results i.e. material (Mat.3). Therefore, it is concluded that the proposed strategy is favorable for selection of optimum material for RP process can be utilized for selection of optimal materials for other advanced manufacturing processes.

Figure 1. Comparative analysis with primary data
(Rao, 2006)

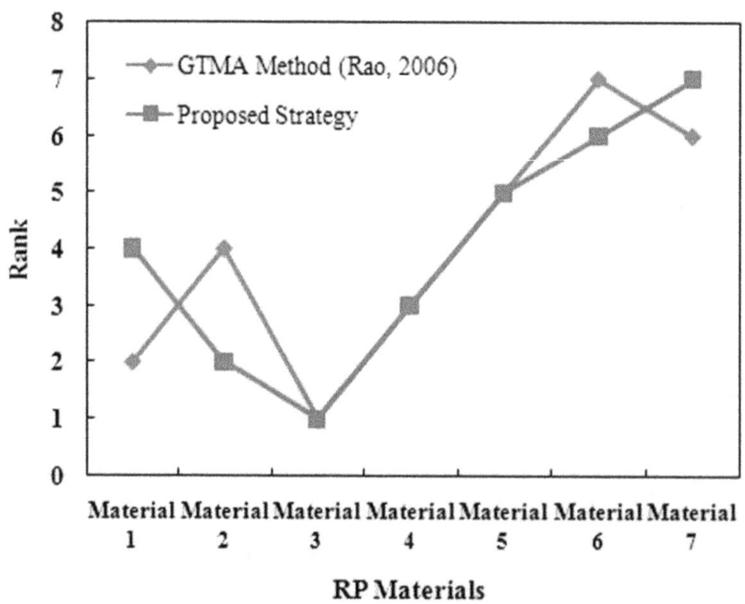

CONCLUSION

This chapter presented a novel MCDM strategy like modified TOPSIS-DEMATEL for the selection of optimum material for RP. Here, M-DEMATEL strategy is used for the extraction of precise priority weights from the vagueness of information while TOPSIS is used for optimal selection of RP material considering the influence of both beneficial and non-beneficial citeria's into account. To validate the proposed method, a case study on selection of optimum materials for RP is selected. The results show that, the Material 3 i.e. SS 301–FH 39.112294 is yields the optimal material for RP process. Further, results obtained from the proposed strategy is compared with published data (Rao, 2006), found comparable and acceptable. Finally, it is concluded that, the proposed strategy can be used as systematic frame work model for selection of any materials for other RP processes.

FUTURE SCOPE

Although, the present book chapter provided the application of M-DEMATEL with TOPSIS based MCDM method for selection of optimum material for RP process. However, development of other MCDM methods such as MOOSRA, PROMETHEE, VIKOR, OCRA, Entropy, SAW, MOORA etc., and their applications in material selection for RP process have been investigated yet. Apart from the MCDM methods, the research on the specific materials and its properties characterization for individual RP process are yet to be explored.

REFERENCES

Alam, A., & Tabriz, E. T. (2014). An Integrated Fuzzy DEMATEL-ANP-TOPSIS Methodology for Supplier Selection Problem. *Global J Manag. Stud Resear.*, *1*(2), 85–92.

Arenas, J. M., Alia, C., Blaya, F., & Sanz, A. (2012). Multi-criteria selection of structural adhesives to bond ABS parts obtained by rapid prototyping. *International Journal of Adhesion and Adhesives*, *33*, 67–74. doi:10.1016/j.ijadhadh.2011.11.005

Attri, R., & Grover, S. (2015). Application of preference selection index method for decision making over the design stage of production system life cycle. *J King Saud Univ. Engineering and Science*, *27*, 207–216.

Brauers, W. K. M., & Zavadskas, E. K. (2006). The MOORA method and its application to privatization in a transition economy. *Control and Cybernetics*, *35*(2), 445–469.

Chang, B., Chang, C. W., & Wu, C. H. (2011). Fuzzy DEMATEL method for developing supplier selection criteria. *Expert Systems with Applications*, *38*(3), 1850–1858. doi:10.1016/j.eswa.2010.07.114

Dursun, M. E., & Karsak, E. A. (2013). QFD-based fuzzy MCDM approach for supplier selection. *Applied Mathematical Modelling*, *37*(8), 5864–5875. doi:10.1016/j.apm.2012.11.014

Gabus, A., & Fontela, E. (1972). *World Problems an Invitation to Further Thought within the Framework of DEMATEL*. Geneva, Switzerland: Battelle Geneva Research Centre.

Gabus, A., & Fontela, E. (1973). *Perceptions of the World Problematique: Communication Procedure, Communicating with those Bearing Collective Responsibilities* (DEMATEL Report No. 1). Geneva: Battelle Geneva Research Centre.

Gibson, I., Rosen, D., & Stucker, B. (2015). *Additive Manufacturing Technologies: 3D Printing, Rapid Prototyping, and Direct digital manufacturing* (2nd ed.). New York: Springer Science and Business Media Pvt Ltd. doi:10.1007/978-1-4939-2113-3

Gupta, K., Jain, N. K., & Laubscher, R. F. (2017). Advances in Gear Manufacturing. In Advanced Gear Manufacturing and Finishing- Classical and Modern Processes. Academic Press Inc.

Hwang, C. L., & Yoon, K. (1981). *Multiple Attribute Decision Making: Methods and Applications*. New York: Springer-Verlag. doi:10.1007/978-3-642-48318-9

Jaromin, M. (2013). Computer aided material selection in design process. *Computer Science and Information Systems*, 3–7.

Kumar, N. R., & Singal, S. K. (2015). Penstock material selection in small hydropower plants using MADM methods. *Renewable & Sustainable Energy Reviews*, *52*, 240–255. doi:10.1016/j.rser.2015.07.018

Lan, H. (2009). Web-based rapid prototyping and manufacturing systems: A review. *Computers in Industry*, *60*(9), 643–656. doi:10.1016/j.compind.2009.05.003

Masood, S. H., & Soo, A. (2002). A rule-based expert system for rapid prototyping system selection. *Rob. Comput. Inte. Eng.*, *18*, 267–274.

Prasad, K., & Chakraborty, S. (2012). Application of multi-objective optimization on the basis of ratio analysis (MOORA) method for materials selection. *Materials & Design*, *37*, 317–324. doi:10.1016/j.matdes.2012.01.013

Prasad, K., & Chakraborty, S. (2013). A quality function deployment-based model for materials selection. *Materials & Design*, *49*, 525–535. doi:10.1016/j.matdes.2013.01.035

Rao, R. V. (2006). a material selection model using graph theory and matrix approach. *Materials Science and Engineering*, *431*(1-2), 248–255. doi:10.1016/j.msea.2006.06.006

Rao, R. V., & Padmanabhan, K. K. (2007). Rapid prototyping process selection using graph theory and matrix approach. *Journal of Materials Processing Technology*, *194*(1-3), 81–88. doi:10.1016/j.jmatprotec.2007.04.003

Shahroodi, K., Amin, K., Amini, S., & Najibzadeh, M. (2012). Application of Analytical Hierarchy Process (AHP) Technique to Evaluate and Selecting Suppliers in an Effective Supply Chain. Arabian J Busin and Manag. *RE:view*, *1*(6), 119–132.

Shende, V., & Kulkarni, P. (2014). Decision support system for rapid prototyping process selection. *Int. J. Sci. Res. Publ.*, *4*, 2250–3153.

Compilation of References

Aazam, M., Zeadally, S., & Harras, K. A. (2018). Deploying Fog Computing in Industrial Internet of Things and Industry 4.0. *IEEE Transactions on Industrial Informatics*, 1–1. doi:10.1109/TII.2018.2855198

Abbas, G., Gu, J., Farooq, U., Asad, M. U., & El-Hawary, M. (2017). Solution of an Economic Dispatch Problem through Particle Swarm Optimization: A Detailed Survey–Part I. *IEEE Access: Practical Innovations, Open Solutions*, *5*, 15105–15141. doi:10.1109/ACCESS.2017.2723862

Abdul Ghani, K., Jayabalan, V., & Sugumar, M. (2002). Impact of advanced manufacturing technology on organizational structure. *The Journal of High Technology Management Research*, *13*(2), 157–175. doi:10.1016/S1047-8310(02)00051-2

Abdullah, N. H., Shamsuddin, A., Wahab, E., & Hamid, N. A. (2012). Preliminary Qualitative Findings on Technology Adoption of Malaysian SMEs. *IEEE Conference Publications*, 15 - 20. 10.1109/CHUSER.2012.6504273

Abdul-Razaq, T. S., Potts, C. N., & Van Wassenhove, L. N. (1990). A survey of algorithms for the single machine total weighted tardiness scheduling problem. *Discrete Applied Mathematics*, *26*(2), 235–253. doi:10.1016/0166-218X(90)90103-J

Absil, P. A., Sluysmans, B., & Stevens, N. (2018, June). MIQP-Based Algorithm for the Global Solution of Economic Dispatch Problems with Valve-Point Effects. In 2018 Power Systems Computation Conference (PSCC) (pp. 1-7). IEEE. doi:10.23919/PSCC.2018.8450877

Accenture. (2014). *Industrial Internet Insights Report, For 2015, General Electric*. Accenture. Available at: https://www.accenture.com/ch-en/_acnmedia/Accenture/next-gen/reassembling-industry/pdf/Accenture-Industrial-Internet-Changing-Competitive-Landscape-Industries.pdf

Acur, N., Kandemir, D., & Boer, H. (2012). Strategic alignment and new product development: Drivers and performance effects. *Journal of Product Innovation Management*, *29*(2), 304–318. doi:10.1111/j.1540-5885.2011.00897.x

Adarsh, B. R., Raghunathan, T., Jayabarathi, T., & Yang, X. S. (2016). Economic dispatch using chaotic bat algorithm. *Energy*, *96*, 666–675. doi:10.1016/j.energy.2015.12.096

Agarwal, S. (2014). Economic order quantity model, a review. *VSRD International Journal of Mechanical, Civil, Automobile and Production Engineering*, 233.

Aguilar, P. A. (2012). Un modelo de clasificación de inventarios para incrementar el nivel de servicio al cliente y la rentabilidad de la empresa. *Pensamiento & Gestión*, 152-153.

Aguirre, S., & Franco, C. (2005). Diseño de un modelo de inventarios para la operación logística de una compañía farmacéutica. *Ingeniería y Universidad*, 30.

Ahmad, J., Tahir, M., & Mazumder, S. K. (2018). Dynamic Economic Dispatch and Transient Control of Distributed Generators in a Microgrid. *IEEE Systems Journal*, (99): 1–11.

Akgün, A. E., Byrne, J. C., Lynn, G. S., & Keskin, H. (2007). New product development in turbulent environments: Impact of improvisation and unlearning on new product performance. *Journal of Engineering and Technology Management, 24*(3), 203–230. doi:10.1016/j.jengtecman.2007.05.008

Akkiraju, R., Farrell, J., Miller, J., Nagarajan, M., Schmidt, M., Sheth, A., & Verma, K. (2005). *Web Service Semantics - WSDL-S, A joint UGA-IBM Technical Note, version 1.0*. Retrieved from http://lsdis.cs.uga.edu/projects/METEOR-S/WSDL-S

Alam, A., & Tabriz, E. T. (2014). An Integrated Fuzzy DEMATEL-ANP-TOPSIS Methodology for Supplier Selection Problem. *Global J Manag. Stud Resear., 1*(2), 85–92.

Alaoui, K. B., Naimi, Z., Benlarabi, A., & Outzourhit, A. (2015). Pilot line preindustrial reactor installation for applied research in vacuum deposition techniques for the preparation and characterization of photovoltaic cells. *IEEE Conference Publications*, 1 – 3.

Al-Betar, M. A., Awadallah, M. A., Khader, A. T., Bolaji, A. L. A., & Almomani, A. (2018). Economic load dispatch problems with valve-point loading using natural updated harmony search. *Neural Computing & Applications, 29*(10), 767–781. doi:10.100700521-016-2611-2

Albliwi, S., Antony, J., Abdul Halim Lim, S., & van der Wiele, T. (2014). Critical failure factors of Lean Six Sigma: A systematic literature review. *International Journal of Quality & Reliability Management Vol., 31*(9), 1012–1030. doi:10.1108/IJQRM-09-2013-0147

Alexy, O., West, J., Klapper, H., & Reitzig, M. (2018). Surrendering control to gain advantage: Reconciling openness and the resource-based view of the firm. *Strategic Management Journal, 39*(6), 1704–1727. doi:10.1002mj.2706

Alves, M., & Climaco, J. (2000). An Interactive Method for 0—1 Multiobjective Problems Using Simulated Annealing and Tabu Search. *Journal of Heuristics, 6*(3), 385–403. doi:10.1023/A:1009686616612

AMAC. (2013). *Maquiladora Association AC - Maquiladora Overview 2016*. Retrieved from www.indexjuarez.org

Amadi-Echendu, J. E., & Rasetlola, R. T. (2011). Technology commercialization factors, frameworks and models. *IEEE Conference Publications*, 144 – 148.

Anderson, J.C., Cleveland, G., & Schroeder, R.G. (1989). *Operations Strategy a Literature Review*. Academic Press.

Andersson, R., Eriksson, H., & Torstensson, H. (2006). Similarities and differences between TQM, six sigma and lean. *The TQM Magazine, 18*(3), 282–296. doi:10.1108/09544780610660004

Antoniou, G., & Harmelen, F. V. (2008). *A Semantic Web Primer*. Cambridge, MA: The MIT Press.

Antony, J. (2011). Six Sigma vs Lean: Some perspectives from leading academics and practitioners. *International Journal of Productivity and Performance Management, 60*(2), 185–190. doi:10.1108/17410401111101494

Antony, J., Gupta, S., Sunder, M. V., & Gijo, E. V. (2018). Ten commandments of Lean Six Sigma: A practitioners' perspective. *International Journal of Productivity and Performance Management, 67*(6), 1033–1044. doi:10.1108/IJPPM-07-2017-0170

Antony, J., Setijono, D., & Dahlgaard, J. J. (2016). Lean Six Sigma and Innovation–an exploratory study among UK organisations. *Total Quality Management & Business Excellence, 27*(1–2), 124–140. doi:10.1080/14783363.2014.959255

Antony, J., Snee, R., & Hoerl, R. (2017). Lean Six Sigma: Yesterday, today and tomorrow. *International Journal of Quality & Reliability Management, 34*(7), 1073–1093. doi:10.1108/IJQRM-03-2016-0035

Arenas, J. M., Alia, C., Blaya, F., & Sanz, A. (2012). Multi-criteria selection of structural adhesives to bond ABS parts obtained by rapid prototyping. *International Journal of Adhesion and Adhesives, 33*, 67–74. doi:10.1016/j.ijadhadh.2011.11.005

Asadzadeh, L. (2015). A local search genetic algorithm for the job shop scheduling problem with intelligent agents. *Computers & Industrial Engineering, 85*, 376–383. doi:10.1016/j.cie.2015.04.006

Asano, M., & Ohta, H. (2002). A heuristic for job shop scheduling to minimize total weighted tardiness. *Computers & Industrial Engineering, 42*(2–4), 137–147. doi:10.1016/S0360-8352(02)00019-0

Asefeso, A. (2014). Lean Six Sigma: Cost Reduction Strategies (2nd ed.). Academic Press.

Asghar, I., & Usman, M. (2013). Motivational and De-motivational Factors for Software Engineers: An Empirical Investigation. *IEEE Conference Publications*, 66 – 71. 10.1109/FIT.2013.20

Åstebro, T. (2004). Key Success Factors for Technological Entrepreneurs' R&D Projects. IEEE Transactions on Engineering Management, 51(3), 314 - 321. doi:10.1109/TEM.2004.830863

Attri, R., & Grover, S. (2015). Application of preference selection index method for decision making over the design stage of production system life cycle. *J King Saud Univ. Engineering and Science, 27*, 207–216.

Atzori, L., Iera, A., & Morabito, G. (2010). The internet of things: A survey. *Computer Networks, 54*(15), 2787–2805. doi:10.1016/j.comnet.2010.05.010

Avelar-Sosa, L., García-Alcaraz, J., Cedillo-Campos, M., & Adarme-Jaimes, W. (2014). Effects of regional infrastructure and offered services in the supply chains performance: Case Ciudad Juarez. *DYNA - Colombia, 81*(186), 208-217.

Awan, F., Michalska, H., & Joos, G. (2017, June). Economic dispatch in microgrids using compromise solution method. In PowerTech, 2017 IEEE Manchester (pp. 1-6). IEEE.doi:10.1109/PTC.2017.7981250

Ayag, Z. (2005). An integrated approach to evaluating conceptual design alternatives in a new product development environment. *International Journal of Production Research, 43*(4), 687-713.

Azaiez, S., Boc, M., Cudennec, L., Simoes, M. D. S., Haupert, J., Kchir, S., ... Tortech, T. (2016). Towards Flexibility in Future Industrial Manufacturing: A Global Framework for Self-organization of Production Cells. *Procedia Computer Science, 83*, 1268–1273. doi:10.1016/j.procs.2016.04.264

Azapagic, A. (2004). Developing a framework for sustainable development indicators for the mining and minerals industry. *Journal of Cleaner Production, 12*(6), 639–662. doi:10.1016/S0959-6526(03)00075-1

Azizipanah-Abarghooee, R., Dehghanian, P., & Terzija, V. (2016). Practical multi-area bi-objective environmental economic dispatch equipped with a hybrid gradient search method and improved Jaya algorithm. *IET Generation, Transmission & Distribution, 10*(14), 3580–3596. doi:10.1049/iet-gtd.2016.0333

Baader, F., & Nutt, W. (2003). Basic Description Logics. In The description logic handbook (pp. 43-95). Cambridge University Press.

Baldwin, C. Y., & Clark, K. B. (2000). *Design Rules.* Cambridge, MA: MIT Press.

Barney, J. (1991). Firm resources and sustained competitive advantage. *Journal of Management, 17*(1), 99–120. doi:10.1177/014920639101700108

Barve, A., & Muduli, K. (2011). Challenges to environmental management practices in Indian mining industries. *Proceedings of 2011 International Conference on Innovation, Management and Service*, 297-302.

Bei, H., & Li, M. (2006). Research on the Motivation Factors for the Development of Talent in the Industry Cluster. *IEEE Conference Publications*, 1237 – 1242. 10.1109/ICMSE.2006.314221

Belton, P. (2017). *Competitive Strategy: Creating and Sustaining Superior Performance*. Academic Press.

Belton, V., & Gear, T. (1985). The legitimacy of rank reversal da comment. *Omega*, *13*(3), 143–144. doi:10.1016/0305-0483(85)90052-0

Berger, C., Blauth, R., Boger, D., Bolster, C., Burchill, G., DuMouchel, W., ... Walden, D. (1993). Kano's method for understanding customer-defined quality. *Center for Quality of Management Journal*, *2*(4), 3–35.

Bergmann, R., & Schaaf, M. (2003). Structural Case-Based Reasoning and Ontology-Based Knowledge Management: A Perfect Match? *Journal of Universal Computer Science, UCS*, *9*(7), 608–626.

Berkel, R. V. (2007). Eco-efficiency in the Australian minerals processing sector. *Journal of Cleaner Production*, *15*(8-9), 772–781. doi:10.1016/j.jclepro.2006.06.017

Berners-Lee, T., Hendler, J., & Lassila, O. (2001). The Semantic Web. *Scientific American*, *284*(May), 34–43. doi:10.1038cientificamerican0501-34 PMID:11323639

Besrour, Bin Ab Rahim, & Dominic (2015). The Study of the Relation between Requirement Engineering Techniques and Challenges in Software Industry. *IEEE Conference Publications*, 49 – 53. 10.1109/ISMSC.2015.7594026

Beyea, S. C., & Nicoll, L. H. (1998). Writing an integrative review. *AORN Journal*, *67*(4), 877–880. doi:10.1016/S0001-2092(06)62653-7

Bharaskar, S. S., Banwait, S. S., & Laroiya, S. C. (2013). Multiobjective Optimization of Electrical Discharge Machining Process Using a Hybrid Method. *Materials and Manufacturing Processes*, *28*(4), 348–354. doi:10.1080/10426914.2012.700152

Bhasin, S., & Burcher, P. (2006). Lean viewed as a philosophy. *Journal of Manufacturing Technology Management*, *17*(1), 56–72. doi:10.1108/17410380610639506

Bhattacharjee, V., & Khan, I. (2018). A non-linear convex cost model for economic dispatch in microgrids. *Applied Energy*, *222*, 637–648. doi:10.1016/j.apenergy.2018.04.001

Bhatt, G. D., & Grover, V. (2005). Types of information technology capabilities and their role in competitive advantage: An empirical study. *Journal of Management Information Systems*, *22*(2), 253–277. doi:10.1080/07421222.2005.11045844

Bhongade, S., & Agarwal, S. (2016, January). An optimal solution for Combined Economic and Emission Dispatch problem using Artificial Bee Colony Algorithm. In *Power and Energy Systems: Towards Sustainable Energy (PESTSE), 2016 Biennial International Conference on* (pp. 1-7). IEEE.10.1109/PESTSE.2016.7516478

Bhui, P., & Senroy, N. (2016, March). A unified method for economic dispatch with valve point effects. In *Power Systems (ICPS), 2016 IEEE 6th International Conference on* (pp. 1-5). IEEE.10.1109/ICPES.2016.7584108

Bhuiyan, N. (2011). A framework for successful new product development. *Journal of Industrial Engineering and Management, 4*(4), 746-770.

Bianchini, D., Antonellis, V. A., Melchiori, M., & Salvi, D. (2006). Semantic Enriched Service Discovery. *International Conference in Data Engineering Workshop*, 38.

Bierwirth, C., & Kuhpfahl, J. (2017). Extended GRASP for the job shop scheduling problem with total weighted tardiness objective. *European Journal of Operational Research*, *261*(3), 835–848. doi:10.1016/j.ejor.2017.03.030

Bingi, K., Ibrahim, R., Karsiti, M. N., & Hassan, S. M. (2017). Fractional Order Set-point Weighted PID Controller for pH Neutralization Process Using Accelerated PSO Algorithm. *Arabian Journal for Science and Engineering*, 1–15.

Bingi, K., Ibrahim, R., Karsiti, M. N., & Hassan, S. M. (2017). Fractional-order filter design for set-point weighted PID controlled unstable systems. *International Journal of Mechanical and Mechatronics Engineering*, *17*(5), 173–179.

Birbil, S. I., Bülbül, K., & Frenk, J. (2009). *On The Economic Order Quantity Model With Transportation Costs*. Istanbul: H.M. Mulder.

Blackstone, J. H., Phillips, D. T., & Hogg, G. L. (1982). A state-of-the-art survey of dispatching rules for manufacturing job shop operations. *International Journal of Production Research*, *20*(1), 27–45. doi:10.1080/00207548208947745

Blindenbach-Driessen, F. (2015). The (In) Effectiveness of Cross-Functional Innovation Teams: The Moderating Role of Organizational Context. *IEEE Transactions on Engineering Management*, *62*(1), 29–38. doi:10.1109/TEM.2014.2361623

Blum, C. (2005). Ant colony optimization: Introduction and recent trends. *Physics of Life Reviews*, *2*(4), 353–373. doi:10.1016/j.plrev.2005.10.001

Blum, C., & Roli, A. (2003). Metaheuristics in combinatorial optimization: Overview and conceptual comparison. *ACM Computing Surveys*, *35*(3), 268–308. doi:10.1145/937503.937505

Boos, D., Guenter, H., Grote, G., & Kinder, K. (2013). Controllable accountabilities: The internet of things and its challenges for organisations. *Behaviour & Information Technology*, *32*(5), 449–467. doi:10.1080/0144929X.2012.674157

Booz, A. (1982). *New product management for the 1980's. Boo*, Allen & Hamilton, Inc.

Borgia, E. (2014). The Internet of Things vision: Key features, applications and open issues. *Computer Communications*, *54*, 1–31. doi:10.1016/j.comcom.2014.09.008

Bose, G. K., & Pain, P. (2017). Optimization Through Nature-Inspired Soft-Computing and Algorithm on ECG Process. In Handbook of Research on Soft Computing and Nature-Inspired Algorithms (pp. 489-519). IGI Global. doi:10.4018/978-1-5225-2128-0.ch017

Bose, G. K., & Pain, P. (2016). Parametric Analysis of Different Grades of Steel Materials Used in Plastic Industries through Die Sinking EDM Process. *International Journal of Materials Forming and Machining Processes*, *3*(1), 45–74. doi:10.4018/IJMFMP.2016010104

Bourke, J., & Roper, S. (2016). AMT adoption and innovation: An investigation of dynamic and complementary effects. *Technovation*, *55-56*, 42–55. doi:10.1016/j.technovation.2016.05.003

Brauers, W. K. M., & Zavadskas, E. K. (2006). The MOORA method and its application to privatization in a transition economy. *Control and Cybernetics*, *35*(2), 445–469.

Broadband. (2017). *Working Group on Education: Digital skills for life and work, Broadband Commission Working Group on Education, September 2017*. Available at: http://unesdoc.unesco.org/images/0025/002590/259013e.pdf

Brockhoff, D., & Zitzler, E. (2006). Are all objectives necessary? On dimensionality reduction in evolutionary multiobjective optimization. In PPSN IX, Reykjavik. LNCS 4193 (pp. 533–542). Springer. doi:10.1007/11844297_54

Brucker, P., Jurisch, B., & Sievers, B. (1994). A branch and bound algorithm for the job-shop scheduling problem. *Discrete Applied Mathematics*, *49*(1–3), 107–127. doi:10.1016/0166-218X(94)90204-6

Bruner, J. (2013). *Industrial Internet*. O'Reilly Media, Inc.

Bruno, Ferrari, & Lopes de Oliveira e Souza (2016). A Requirements Engineering and Management Process in Concept Phase of Complex Systems. *IEEE Conference Publications*, 1 - 6.

Bülbül, H., Ömürbek, N., Paksoy, T., & Bektaş, T. (2013). An empirical investigation of advanced manufacturing technology investment patterns: Evidence from a developing country. *Journal of Engineering and Technology Management*, *30*(2), 136–156. doi:10.1016/j.jengtecman.2013.01.002

Burke, L. A., & Hutchins, H. M. (2007). Training transfer: An integrative literature review. *Human Resource Development Review*, *6*(3), 263–296. doi:10.1177/1534484307303035

Byrne, B. M. (2010). *Structural Equation Modeling with AMOS: Basic Concepts, Applications, and Programming*. New York: Taylor and Francis Group LLC.

Byrne, G., Lubowe, D., & Blitz, A. (2007). Using a Lean Six Sigma approach to drive innovation. *Strategy and Leadership*, *35*(2), 5–10. doi:10.1108/10878570710734480

Cai, M. Zhang, W. Y., Zhang, K. & Li, S. T. (2010). SWMRD: A Semantic Web-based manufacturing resource discovery system for cross-enterprise collaboration. *International Journal of Production Research*, *48*(120), 3445-3460.

Cardoso, J. (2006). Discovering Semantic Web Services with and without a Common Ontology Commitment. *IEEE Service Computing Workshop*, 183-190.

Cardoso, J., & Sheth, A. (2003). Semantic e-Workflow Composition. *Journal of Intelligent Information Systems*, *21*(3), 191–225. doi:10.1023/A:1025542915514

Carnevalli, J. A., & Miguel, P. C. (2008). Review, analysis and classification of the literature on QFD-Types of research, difficulties and benefits. *International Journal of Production Economics*, *114*(2), 737–754. doi:10.1016/j.ijpe.2008.03.006

Cassar, G., Barnaghi, P., & Moessner, K. (2014). Probabilistic matchmaking methods for automated service discovery. *IEEE Transactions on Services Computing*, *7*(4), 654–666. doi:10.1109/TSC.2013.28

Causado, E. (2015). Modelo de inventarios para control económico de pedidos en empresa comercializadora de alimentos. *Revista Ingenierías Universidad de Medellín*, 165.

Cejas, M., & Garrido, I. Y. (2017). La gestión del inventario como factor estrátegico en la administración de las empresas. *Revista Científica Electrónica de Ciencias Gerenciales*, 112-113.

Cepeda Valero, Ó. M., & Jiménez Sánchez, L. F. (2016). Modelo de control óptimo para el sistema Producción-Inventarios. *Ingeniería Industrial. Actualidad y Nuevas Tendencias*, 35-36.

Chamba, M., & Ano, O. (2013). Economic dispatch of energy and reserve in competitive markets using meta-heuristic algorithms. *IEEE Latin America Transactions*, *11*(1), 473–478. doi:10.1109/TLA.2013.6502848

Chang, B., Chang, C. W., & Wu, C. H. (2011). Fuzzy DEMATEL method for developing supplier selection criteria. *Expert Systems with Applications*, *38*(3), 1850–1858. doi:10.1016/j.eswa.2010.07.114

Chan, L. K., & Wu, M. L. (2002). Quality function deployment: A comprehensive review of its concepts and methods. *Quality Engineering*, *15*(1), 23–35. doi:10.1081/QEN-120006708

Chansareewittaya, S. (2017, March). Hybrid BA/TS for economic dispatch considering the generator constraint. In *Digital Arts, Media and Technology (ICDAMT), International Conference on* (pp. 115-119). IEEE.

Chaudha, A., Jain, R., Singh, A. R., & Mishra, P. K. (2011). Integration of Kano's model into quality function deployment (QFD). *International Journal of Advanced Manufacturing Technology*, *53*(5), 689–698. doi:10.100700170-010-2867-0

Chaudhuri, A., & Boer, H. (2016). The impact of product-process complexity and new product development order winners on new product development performance: The mediating role of collaborative competence. *Journal of Engineering and Technology Management*, *42*, 65–80. doi:10.1016/j.jengtecman.2016.10.002

Chaulya, S. K. (2004). Improving EHS management in India's small-scale mining sector. *Environmental Quality Management*, *13*(3), 65–80. doi:10.1002/tqem.20005

Chen, Y., & Atherton, D. P. (2007). *Linear feedback control: analysis and design with MATLAB* (Vol. 14). Siam.

Chen, Y., Petras, I., & Xue, D. (2009). *Fractional order control-a tutorial.* Paper presented at the American Control Conference, 2009. ACC'09.

Chen, D., Qiu, W., Min, Y., & Mei-yung, L. (2007). Activity Flow Optimization and Risk Evaluation of Complex Project. *IEEE Conference Publications*, 5196 - 5199. 10.1109/WICOM.2007.1272

Chen, G., & Yang, Q. (2018). An ADMM-based distributed algorithm for economic dispatch in islanded microgrids. *IEEE Transactions on Industrial Informatics*, *14*(9), 3892–3903. doi:10.1109/TII.2017.2785366

Cheng, R., Gen, M., & Tsujimura, Y. (1996). A tutorial survey of job-shop scheduling problems using genetic algorithms—I. representation. *Computers & Industrial Engineering*, *30*(4), 983–997. doi:10.1016/0360-8352(96)00047-2

Chen, M., & Lyu, J. (2009). A Lean Six-Sigma approach to touch panel quality improvement. *Production Planning and Control*, *20*(5), 445–454. doi:10.1080/09537280902946343

Cherniwchan, J. (2017). Trade liberalization and the environment: Evidence from NAFTA and U.S. manufacturing. *Journal of International Economics*, *105*, 130–149. doi:10.1016/j.jinteco.2017.01.005

Cherres, S. L. (2010). Un caso de aplicación del sistema ABC en una empresa peruana: Frenosa. *Contabilidad y Negocios*, 30.

Cherukuri, A., & Cortés, J. (2017). Distributed coordination of DERs with storage for dynamic economic dispatch. *IEEE Transactions on Automatic Control.*

Ching, T. W., & Xu, T. (2014). Performance Study of Electric Vehicles in Macau. *IEEE Conference Publications*, 1 – 5. 10.1109/CCECE.2014.6900957

Chiou, J. P. (2007). Variable scaling hybrid differential evolution for large-scale economic dispatch problems. *Electric Power Systems Research*, *77*(3), 212–218. doi:10.1016/j.epsr.2006.02.013

Chokri, M., Aymen, F., & Lassaad, S. (2017). Prototype design of a Compact Plug-in Solar Electric Vehicle. *IEEE Conference Publications*, 1 – 4.

Chuang & Wu. (2007). User-Based Evaluation of Search Engines: Hygiene Factors and Motivation Factors. *IEEE Conference Publications*, 1 – 10.

Chuang, Y., & Wu, L. (2007). User-Based Evaluations of Search Engines: Hygiene Factors and Motivation Factors. *IEEE Conference Publications*, 1 – 10. 10.1109/HICSS.2007.590

Chung, C. A. (1996). Human issues influencing the successful implementation of advanced manufacturing technology. *Journal of Engineering and Technology Management*, *13*(3–4), 283–299. doi:10.1016/S0923-4748(96)01010-7

Ciornei, I., & Kyriakides, E. (2013). Recent methodologies and approaches for the economic dispatch of generation in power systems. *International Transactions on Electrical Energy Systems*, *23*(7), 1002–1027. doi:10.1002/etep.1635

Coello, C. (2000). An Updated Survey of GA-Based Multiobjective Optimization Techniques. *ACM Computing Surveys*, *32*(2), 109–143. doi:10.1145/358923.358929

Co, H. C., Patuwo, B. E., & Hu, M. Y. (1998). The human factor in advanced manufacturing technology adoption: An empirical analysis. *International Journal of Operations & Production Management*, *18*(1), 87–106. doi:10.1108/01443579810192925

Cook, J. S., & Cook, L. L. (1994). Achieving Competitive Advantages of Advanced Manufacturing Technology. *Benchmarking for Quality Management & Technology*, *1*(2), 42–63. doi:10.1108/14635779410063329

Cooper, R. G., & Kleinschmidt, E. J. (1995). Benchmarking the firm's critical success factors in new product development. *Journal of Product Innovation Management*, *12*(5), 374–391. doi:10.1016/0737-6782(95)00059-3

Copacino, W., & Anderson, D. (2003). Connecting with the Bottom Line: A Global Study of Supply Chain Leadership and its Contribution to the High Performance Business. *Accenture*, 1.

Corbett, J. M. (1988). Ergonomics in the development of human-centred AMT. *Applied Ergonomics*, *19*(1), 35–39. doi:10.1016/0003-6870(88)90196-2 PMID:15676645

Cortez, P. (2014). *Modern Optimization with R*. London: Springer. doi:10.1007/978-3-319-08263-9

Damousis, I. G., Bakirtzis, A. G., & Dokopoulos, P. S. (2003). Network-constrained economic dispatch using real-coded genetic algorithm. *IEEE Transactions on Power Systems*, *18*(1), 198–205. doi:10.1109/TPWRS.2002.807115

Das, K. M., Kumar, K., Barman, T. K., & Sahoo, P. (2014). Application of Artificial Bee Colony Algorithm for Optimization of MRR and Surface Roughness in EDM of EN31 tool steel. *Procedia Materials Science, 6*, 741-751.

Daugherty, P., & Berthon, B. (2015). *Winning with the Industrial Internet of Things: How to Accelerate the Journey to Productivity and Growth.* Dublín: Accenture.

de Freitas, J. G., Costa, H. G., & Ferraz, F. T. (2017). Impacts of Lean Six Sigma over organizational sustainability: A survey study. *Journal of Cleaner Production*, *156*, 262–275. doi:10.1016/j.jclepro.2017.04.054

de Oliveira & Mata. (2005). *Ninteger v. 2.3 Fractional control toolbox for MATLAB.* Lisboa, Universidade Technical.

de Sousa Mendes, G. H., & Miller Devós Ganga, G. (2013). Predicting success in product development: The application of principal component analysis to categorical data and binomial logistic regression. *Journal of Technology Management & Innovation*, *8*(3), 83–97.

de Souza Andrade, Soto Urbina, & de olivera N. Follador (2016). Process Proposal for the Intellectual Property Protection Management in a Technology Licensing Office from a Brazilian Scientific and Technological Institution. *IEEE Conference Publications*, 1672 – 1680.

Deb, K. (2014) Multi-objective Optimization. In Search Methodologies (pp. 403-449). New York, NY: Springer Science+Business Media. doi:10.1007/978-1-4614-6940-7_15

Deb, K., & Saxena, D. K. (2005). On finding pareto-optimal solutions through dimensionality reduction for certain large-dimensional multi-objective optimization problems. Kangal report no. 2005011, Kanpur Genetic Algorithms Laboratory (KanGAL).

Deb, K. (2001). *Multi-Objective Optimization Using Evolutionary Algorithms*. Chichester, UK: Wiley.

Delahaye, D., Chaimatanan, S., & Mongeau, M. (2019). Simulated Annealing: From Basics to Applications. In M. Gendreau & J.-Y. Potvin (Eds.), *Handbook of Metaheuristics* (pp. 1–35). Cham, Switzerland: Springer International Publishing AG. doi:10.1007/978-3-319-91086-4_1

Deloitte. (2015). *Industry 4.0. Challenges and solutions for the digital transformation and use of exponential technologies*. Deloitte AG. Available at: https://www2.deloitte.com/content/dam/Deloitte/ch/Documents/manufacturing/ch-en-manufacturing-industry-4-0-24102014.pdf

Deniz, F. N., Alagoz, B. B., Tan, N., & Atherton, D. P. (2016). An integer order approximation method based on stability boundary locus for fractional order derivative/integrator operators. *ISA Transactions*, *62*, 154–163. doi:10.1016/j.isatra.2016.01.020 PMID:26876378

Deshmukh, S. V., & Chavan, A. (2012). Six Sigma and SMEs: A critical review of literature. *International Journal of Lean Six Sigma*, *3*(2), 157–167. doi:10.1108/20401461211243720

Diabat, A., & Govindan, K. (2011). An analysis of the drivers affecting the implementation of green supply chain management. *Resources, Conservation and Recycling*, *55*(6), 659–667. doi:10.1016/j.resconrec.2010.12.002

Dierickx, I., & Cool, K. (1989). Asset stock accumulation and sustainability of competitive advantage. *Management Science*, *35*(12), 1504–1511. doi:10.1287/mnsc.35.12.1504

Divya, C.M., Divya, S., Ratheesh, K. & Volga, R. (2012). *Environmental issues in stone crushers.* Academic Press.

Dixit, G. P., Dubey, H. M., Pandit, M., & Panigrahi, B. K. (2011). Economic load dispatch using artificial bee colony optimization. *International Journal of Advances in Electronics Engineering*, *1*(1), 119–124.

Djouambi, A., Charef, A., & Besançon, A. (2007). Optimal approximation, simulation and analog realization of the fundamental fractional order transfer function. *International Journal of Applied Mathematics and Computer Science*, *17*(4), 455–462. doi:10.2478/v10006-007-0037-9

Doerner, K., Gutjahr, W., Hartl, R., Strauss, C., & Stummer, C. (2004). Pareto Ant Colony Optimization: A Metaheuristic Approach to Multiobjective Portfolio Selection. *Annals of Operations Research*, *131*(1-4), 79–99. doi:10.1023/B:ANOR.0000039513.99038.c6

Domingue, J., Cabral, L., Galizia, S., Tanasescu, V., Gugliotta, A., Norton, B., & Pedrinaci, C. (2008). IRS-III: A Broker-based Approach to Semantic Web Service. *Journal of Web Semantics*, *6*(2), 109–132. doi:10.1016/j.websem.2008.01.001

Dominici, G., & Palumbo, F. (2013). The drivers of customer satisfaction in the hospitality industry: Applying the Kano model to Sicilian hotels. *International Journal of Leisure and Tourism Marketing*, *3*(3), 215–236. doi:10.1504/IJLTM.2013.052623

Dong-dong, G. A. O., & Ping, L. V. (2009). A Sorting-Based Management Model to Support Early Supplier Involvement in New Product Development. *IEEE Conference Publications*, 496 - 501.

Dorigo, M., & Gambardella, L. (1997). Ant Colony System: A Cooperative Learning Approach to the Travelling Salesman Problem. *IEEE Transactions on Evolutionary Computation*, *1*(1), 53–66. doi:10.1109/4235.585892

Dormido, S., Pisoni, E., & Visioli, A. (2010). An interactive tool for loop-shaping design of fractional-order PID controllers. *Proceedings of the 4th IFAC Workshop on Fractional Differentiation and Its Applications (FDA'10).*

Dormido, S., Pisoni, E., & Visioli, A. (2012). Interactive tools for designing fractional-order PID controllers. *International Journal of Innovative Computing, Information, & Control*, *8*(7), 4579–4590.

Du, B., Wei, Y., Liang, S., & Wang, Y. (2017). Rational approximation of fractional order systems by vector fitting method. *International Journal of Control, Automation, and Systems*, *15*(1), 186–195. doi:10.100712555-015-0351-1

Durga Prasad, K. G., Venkata Subbaiah, K., & Narayana Rao, K. (2014). Supply chain design through QFD-based optimization. *Journal of Manufacturing Technology Management*, *25*(5), 712–733. doi:10.1108/JMTM-03-2012-0030

Dursun, M. E., & Karsak, E. A. (2013). QFD-based fuzzy MCDM approach for supplier selection. *Applied Mathematical Modelling, 37*(8), 5864–5875. doi:10.1016/j.apm.2012.11.014

Dyer, J. H., & Singh, H. (1998). The relational view: Cooperative strategy and sources of interorganizational competitive advantage. *Academy of Management Review, 23*(4), 660–679. doi:10.5465/amr.1998.1255632

Dzielinski, A., & Sierociuk, D. (2008). Simulation and experimental tools for fractional order control education. *Proc. IFAC.* 10.3182/20080706-5-KR-1001.01975

Edmondson, M., & Ward, A. (2017). *Tackling the disconnect between universities, Small businesses and graduates in cities and regions*. Gradcore. Available at: https://www.eurashe.eu/library/mission-phe/EURASHE_AC_LeHavre_170330-31_pres_EDMONDSON-WARD.pdf

Efstathiades, A., Tassou, S., & Antoniou, A. (2002). Strategic planning, transfer and implementation of Advanced Manufacturing Technologies (AMT). Development of an integrated process plan. *Technovation, 22*(4), 201–212. doi:10.1016/S0166-4972(01)00024-4

Eiselt, H., & Sandblom, C.-L. (2007). Computational Complexity. In *Linear Programming and its Applications* (pp. 31–44). Berlin: Springer.

Elmuti, D., & Kathawala, Y. (1997). An overview of benchmarking process: A tool for continuous improvement and competitive advantage. *Benchmarking for Quality Management & Technology, 4*(4), 229–243. doi:10.1108/14635779710195087

Elsaiah, S., Benidris, M., Mitra, J., & Cai, N. (2014, July). Optimal economic power dispatch in the presence of intermittent renewable energy sources. In *PES General Meeting| Conference & Exposition, 2014 IEEE* (pp. 1-5). IEEE.10.1109/PESGM.2014.6939903

Enriquez, F. T., Osuna, A. J., & Bosch, V. G. (2004). Prioritizing customer needs at spectator events: Obtaining accuracy at a difficult QFD arena. *International Journal of Quality & Reliability Management, 21*(9), 984–990. doi:10.1108/02656710410561790

Ernst, H. (2002). Success factors of new product development: A review of the empirical literature. *International Journal of Management Reviews, 4*(1), 1–40. doi:10.1111/1468-2370.00075

Esa, M. M., Rahman, N. A. A., & Jamaludin, M. (2015). Reducing High Setup Time in Assembly Line: A Case Study of Automotive Manufacturing Company in Malaysia. *Procedia: Social and Behavioral Sciences, 211*, 215–220. doi:10.1016/j.sbspro.2015.11.086

Esmaeilian, B., Behdad, S., & Wang, B. (2016). The evolution and future of manufacturing: A review. *Journal of Manufacturing Systems, 39*, 79–100. doi:10.1016/j.jmsy.2016.03.001

Essafi, I., Mati, Y., & Dauzère-Pérès, S. (2008). A genetic local search algorithm for minimizing total weighted tardiness in the job-shop scheduling problem. *Computers & Operations Research, 35*(8), 2599–2616. doi:10.1016/j.cor.2006.12.019

Ettlie, J. E. (1995). Early Manufacturing Involvement in New product Development. *IEEE Conference Publications*, 104 – 109. 10.1109/IEMC.1995.523917

Evans, D., & Pearson, A. (2001). Systematic reviews: Gatekeepers of nursing knowledge. *Journal of Clinical Nursing, 10*(5), 593–599. doi:10.1046/j.1365-2702.2001.00517.x

Evermann, J., & Tate, M. (2016). Assessing the predictive performance of structural equation model estimators. *Journal of Business Research, 69*(10), 4565–4582. doi:10.1016/j.jbusres.2016.03.050

Farrag, T. A., Saleh, A. I., & Ali, H. A. (2013). Toward SWSS discovery: Mapping from WSDL to OWL-S based on ontology search and standardization engine. *IEEE Transactions on Knowledge and Data Engineering*, *25*(3), 1135–1147. doi:10.1109/TKDE.2012.25

Fensel, D., & Bussler, C. (2002). The Web Service Modeling Framework WSMF. *Electronic Commerce Research and Applications*, *1*(2), 113–137. doi:10.1016/S1567-4223(02)00015-7

Ferraz, A., Machado, J., & Carvalho, V. (2015). Prototype for determination of pre-transfusion tests based on image processing techniques. *IEEE Conference Publications*, 1 – 6. 10.1109/ENBENG.2015.7088849

Florén, H., Frishammar, J., Parida, V., & Wincent, J. (2018). Critical success factors in early new product development: A review and a conceptual model. *The International Entrepreneurship and Management Journal*, *14*(2), 411–427. doi:10.100711365-017-0458-3

Floudas, C. A., & Lin, X. (2005). Mixed Integer Linear Programming in Process Scheduling: Modeling, Algorithms, and Applications. *Annals of Operations Research*, *139*(1), 131–162. doi:10.100710479-005-3446-x

Fonseca, C., & Fleming, P. (1993). *Genetic Algorithms for Multiobjective Optimization: Formulation, Discussion and Generalization.* Retrieved from http://citeseerx.ist.psu.edu/viewdoc/download?doi=10.1.1.48.9077&rep=rep1&type=pdf

Fortnow, L. (2009). The Status of the P Versus NP Problem. *Communications of the ACM*, *52*(9), 78–86. doi:10.1145/1562164.1562186

FS. (2015). *Frost &Sullivan. Industry 4.0 Business Ecosystem to Change Dynamics in the Global Industrial Landscape.* FS.

Fu, C., Zheng, J., Zhao, J., & Xu, W. (2001). Application of grey relational analysis for corrosion failure of oil tubes. *Corrosion Science*, *43*(5), 881–889. doi:10.1016/S0010-938X(00)00089-5

Fugate, B., Sahin, F., & Mentzer, J. T. (2006). Supply Chain Management Coordination Mechanisms. *Journal of Business Logistics*, *27*(2), 129–161. doi:10.1002/j.2158-1592.2006.tb00220.x

Fung, C. P. (2003). Manufacturing process optimization for wear property of fiber-reinforced polybutylene terephthalate composites with grey relational analysis. *Wear*, *254*(3-4), 298–306. doi:10.1016/S0043-1648(03)00013-9

Gabus, A., & Fontela, E. (1973). *Perceptions of the World Problematique: Communication Procedure, Communicating with those Bearing Collective Responsibilities* (DEMATEL Report No. 1). Geneva: Battelle Geneva Research Centre.

Gabus, A., & Fontela, E. (1972). *World Problems an Invitation to Further Thought within the Framework of DEMATEL.* Geneva, Switzerland: Battelle Geneva Research Centre.

Ganeshan, R., & Harrison, T. P. (1995). *An introduction to supply chain management, in Supply Chain Management, Version 1.* Available from http://silmaril.smeal.psu.edu/misc/supply_chain_ intro.html

Gao, K. Z., Suganthan, P. N., Chua, T. J., Chong, C. S., Cai, T. X., & Pan, Q. K. (2015). A two-stage artificial bee colony algorithm scheduling flexible job-shop scheduling problem with new job insertion. *Expert Systems with Applications*, *42*(21), 7652–7663. doi:10.1016/j.eswa.2015.06.004

Garcia-Alcaraz, J. L., Iniesta, A., & Juárez, M. C. (2012). Benefits of advanced manufacturing technologies. *African Journal of Business Management*, *6*(16), 5524–5532. doi:10.5897/AJBM11.2777

García-Alcaraz, J. L., Maldonado, A. A., Iniesta, A. A., Robles, G. C., & Hernández, G. A. (2014). A systematic review/survey for JIT implementation: Mexican maquiladoras as case study. *Computers in Industry*, *65*(4), 761–773. doi:10.1016/j.compind.2014.02.013

García-Alcaraz, J. L., Prieto-Luevano, D. J., Maldonado-Macías, A. A., Blanco-Fernández, J., Jiménez-Macías, E., & Moreno-Jiménez, J. M. (2015). Structural equation modeling to identify the human resource value in the JIT implementation: Case maquiladora sector. *International Journal of Advanced Manufacturing Technology, 77*(5), 1483–1497. doi:10.100700170-014-6561-5

Gatignon, H., Tushman, M. L., Smith, W., & Anderson, P. (2002). A structural approach to assessing innovation: Construct development of innovation locus, type, and characteristics. *Management Science, 48*(9), 1103–1122. doi:10.1287/mnsc.48.9.1103.174

Gee, J. (2018). *The New Work Order*. Academic Press.

Genchev, S., & Willis, G. (2014). A note on manufacturing flexibility as a firm-specific dynamic capability. *Manufacturing Letters, 2*(4), 100–103. doi:10.1016/j.mfglet.2014.07.002

George, M.L., & George, M. (2003). *Lean Six Sigma for Service*. Academic Press.

Gholipoor, A., Baseri, H., & Shabgard, M. Z. (2015). Investigation of near dry EDM compared with wet and dry EDM process. *Journal of Mechanical Science and Technology, 29*(5), 2213–2218. doi:10.100712206-015-0441-2

Ghorbani, M., Mohammad Arabzad, S., & Shahin, A. (2013). A novel approach for supplier selection based on the Kano model and fuzzy MCDM. *International Journal of Production Research, 51*(18), 5469–5484. doi:10.1080/00207543.2013.784403

Ghose, M. K. (2003a). Promoting cleaner production in the Indian small-scale mining industry. *Journal of Cleaner Production, 11*(2), 167–174. doi:10.1016/S0959-6526(02)00036-7

Ghose, M. K. (2003b). Indian small-scale mining with special emphasis on environmental management. *Journal of Cleaner Production, 11*(2), 159–165. doi:10.1016/S0959-6526(02)00035-5

Ghose, M. K. (2003c). A perspective on small-scale mining in India. In G. M. Hilson (Ed.), *The Socio-economic Impacts of Artisanal and Small-scale Mining in Developing Countries* (pp. 449–457). Balkema Publishers. doi:10.1201/9780203971284.ch26

Gibson, I., Rosen, D., & Stucker, B. (2015). *Additive Manufacturing Technologies: 3D Printing, Rapid Prototyping, and Direct digital manufacturing* (2nd ed.). New York: Springer Science and Business Media Pvt Ltd. doi:10.1007/978-1-4939-2113-3

Giffi, C., McNelly, J., Dollar, B., Carrick, G., Drew, M., & Gangula, B. (2015). *The skills gap in US manufacturing: 2015 and beyond*. Washington, DC: Deloitte and Manufacturing Institute. Available at https://www2.deloitte.com/content/dam/Deloitte/us/Documents/manufacturing/us-pip-the-manufacturing-institute-and-deloitte-skills-gap-in-manufacturing-study.pdf

Gilbert, S. (2000). *Greening supply chain: Enhancing competitiveness through the top forum on enhancing competitiveness green productivity.* Report of through green productivity held in The Republic of China.

Giunchiglia, F., Shvaiko, P., & Yatskevich, M. (2004). S-Match: an algorithm and an implementation of semantic matching. *Proceedings of 1st European Semantic Web Symposium (ESWS)*, 3053, 61-75. 10.1007/978-3-540-25956-5_5

Gliesch, A., Ritt, M., & Moreira, M. C. O. (2018). A Multistart Alternating Tabu Search for Commercial Districting. *Lecture Notes in Computer Science, 10782*, 158-173. 10.1007/978-3-319-77449-7_glieas11

Gnanaraj, S. M., Devadasan, S. R., Murugesh, R., & Sreenivasa, C. G. (2012). Sensitisation of SMEs towards the implementation of Lean Six Sigma–an initialisation in a cylinder frames manufacturing Indian SME. *Production Planning and Control, 23*(8), 599–608. doi:10.1080/09537287.2011.572091

Gordon, G., & Bush, J. B. Jr. (1997). Management of Technical Innovation. *IEEE Conference Publications*, 199 – 122.

Gostimirovic, M., Kovac, P., Sekulic, M., & Skoric, B. (2012). Influence of discharge energy on machining characteristics in EDM. *Journal of Mechanical Science and Technology*, *26*(1), 173–179. doi:10.100712206-011-0922-x

Goudarzi, A., Swanson, A. G., Tooryan, F., & Ahmadi, A. (2017, February). Non-convex optimization of combined enviromental economic dispatch through the third version of the cultural algorithm (CA3). In *Power and Energy Conference (TPEC), IEEE Texas* (pp. 1-6). IEEE.10.1109/TPEC.2017.7868281

Graner, M. (2016). Are methods the key to product development success? An empirical analysis of method application in new product development. In *Impact of Design Research on Industrial Practice* (pp. 23–43). Springer International Publishing. doi:10.1007/978-3-319-19449-3_2

Grant, R. M. (1999). *The resource-based theory of competitive advantage: implications for strategy formulation*. Knowledge and Strategy.

Grimm, S., Monk, B., & Preist, C. (2006). Matching Semantic Service Descriptions with Local Closed-World Reasoning. *European Semantic Web Conference*, 575-589. 10.1007/11762256_42

Gruber, T. R. (1993). *A Translation Approach to Portable Ontology Specifications*. Stanford University, Computer Science Department, Knowledge Systems Laboratory, Technical Report KSL 92-71.

Guarino, N., & Giaretta, P. (1995). *Ontologies and Knowledge Base: Towards a Terminological Classification Toward Very Large Knowledge Base: Knowledge Building and Knowledge Sharing*. Amsterdam: IOS Press.

Guo, Y., Tong, L., Wu, W., Zhang, B., & Sun, H. (2017). Coordinated multi-area economic dispatch via critical region projection. *IEEE Transactions on Power Systems*, *32*(5), 3736–3746. doi:10.1109/TPWRS.2017.2655442

Gupta, K., Jain, N. K., & Laubscher, R. F. (2017). Advances in Gear Manufacturing. In Advanced Gear Manufacturing and Finishing- Classical and Modern Processes. Academic Press Inc.

Guzmán, L., Steinbach, S., Diebold, P., Zehler, T., Schneider, K., & Habbe, M. (2016). Evaluating the Benefits of Systematic Project Management in Large Public Sector Projects. *IEEE Conference Publications*, 32 – 35. 10.1145/2896839.2896841

Györgyi, P., & Kis, T. (2018). Minimizing the maximum lateness on a single machine with raw material constraints by branch-and-cut. *Computers & Industrial Engineering*, *115*, 220–225. doi:10.1016/j.cie.2017.11.016

Hadjimarcou, J., Brouthers, L. E., McNicol, J. P., & Michie, D. E. (2013). Maquiladoras in the 21st century: Six strategies for success. *Business Horizons*, *56*(2), 207–217. doi:10.1016/j.bushor.2012.11.005

Hair, J. F., Jr., Anderson, R. E., Tatham, R. L., & Black, W. C. (2005). Analisis Multivariante (5th ed.). Pearson, Prentice-Hall.

Hair, J. F., Anderson, R. E., Tatham, R. L., & Black, W. C. (1995). *Multivariate Data Analysis with Readings*. Englewood Cliffs, NJ: Prentice Hall.

Hair, J. F., Black, W. C., Babin, B. J., & Anderson, R. E. (2009). *Multivariate data analysis*. Upper Saddle River, NJ: Prentice Hall.

Halkin, A. (2017). Estimation of Economic Order Quantity with Variable Parameters (Ukraine Case Study). *Science Research*, 1-4.

Hampson, I. (1999). Lean Production and the Toyota Production System Or, the Case of the Forgoften Production Concepts. *Economic and Industrial Democracy*, *20*(3), 369–391. doi:10.1177/0143831X99203003

Hasan, Z., & El-Hawary, M. E. (2016, October). Economic dispatch at peak load using load reduction for smart grid network. In *Electrical Power and Energy Conference (EPEC), 2016 IEEE* (pp. 1-5). IEEE.10.1109/EPEC.2016.7771744

Hashim & Mimos. (2002). Development of a Silicon Sensor Wafer Fab for Local Industry Requirement. *IEEE Conference Publications*, 145 – 149.

Hatcher, S., & Samuels, M. (2000). Early Involvement of Operational Test: Value Added for the CH-60S/SH-60R1. *IEEE Conference Publications*, 35 – 42. 10.1109/AERO.2000.878211

Hauser, J. R., & Clausing, D. (1988). The House of Quality. *Harvard Business Review*, *66*(3), 63–73.

Haverila, M. J. (2012). Product–firm compatibility in new product development in technology companies. *The Journal of High Technology Management Research*, *23*(2), 130–141. doi:10.1016/j.hitech.2012.06.008

Hayes, A. F., & Preacher, K. J. (2010). Quantifying and Testing Indirect Effects in Simple Mediation Models When the Constituent Paths Are Nonlinear. *Multivariate Behavioral Research*, *45*(4), 627–660. doi:10.1080/00273171.2010.498290 PMID:26735713

Hayes, R. H., & Upton, D. M. (1998). Operations-based strategy. *California Management Review*, *40*(4), 8–25. doi:10.2307/41165962

He, Y., Keung Lai, K., Sun, H., & Chen, Y. (2014). The impact of supplier integration on customer integration and new product performance: The mediating role of manufacturing flexibility under trust theory. *International Journal of Production Economics, 147*(Part B), 260-270. doi:10.1016/j.ijpe.2013.04.044

Heirati, N., & O'Cass, A. (2016). Supporting new product commercialization through managerial social ties and market knowledge development in an emerging economy. *Asia Pacific Journal of Management*, *33*(2), 411–433. doi:10.100710490-015-9437-9

Helfat, C. E. (2017). *Stylized facts regarding the evolution of organizational resources and capabilities*. The SMS Blackwell Handbook of Organizational Capabilities.

Herry. (2016). Project Management: Model Research in Success Rate of A Digital Start-Up Project. *IEEE Conference Publications*, 1 - 6.

Hillier, F. S., & Lieberman, G. J. (2010). *Introduction to operations research*. Dubuque, IA: McGraw-Hill.

Hillier, F. S., & Lieberman, G. J. (2014). *Introduction to Operations Research* (10th ed.). New York: McGraw-Hill.

Hill, J., Thomas, A. J., Mason-Jones, R. K., & El-Kateb, S. (2018). The implementation of a Lean Six Sigma framework to enhance operational performance in an MRO facility. *Production & Manufacturing Research*, *6*(1), 26–48. doi:10.1080/21693277.2017.1417179

Hilson, G. (2000). Barriers to implementing cleaner technologies and cleaner production (CP) practices in the mining industry a case study of the Americas. *Minerals Engineering*, *13*(7), 699–717. doi:10.1016/S0892-6875(00)00055-8

Hilson, G., & Nayee, V. (2002). Environmental management system implementation in the mining industry: A key to achieving cleaner production. *International Journal of Mineral Processing*, *64*(1), 19–41. doi:10.1016/S0301-7516(01)00071-0

Hilton, R. J., & Sohal, A. (2012). A conceptual model for the successful deployment of Lean Six Sigma. *International Journal of Quality & Reliability Management*, *29*(1), 54–70. doi:10.1108/02656711211190873

Hirose, Y. (2012). Knowledge Transfer from Researches to Society: How to Offset the Cultural Gap? *IEEE Conference Publications*, 1107-1116.

Hitt, M. A., Carnes, C. M., & Xu, K. (2016). A current view of resource based theory in operations management: A response to Bromiley and Rau. *Journal of Operations Management*, *41*(10), 107–109. doi:10.1016/j.jom.2015.11.004

Hitt, M. A., Hoskisson, R. E., & Kim, H. (1997). International diversification: Effects on innovation and firm performance in product-diversified firms. *Academy of Management Journal*, *40*(4), 767–798.

Hitt, M. A., Xu, K., & Carnes, C. M. (2016). Resource based theory in operations management research. *Journal of Operations Management*, *41*, 77–94. doi:10.1016/j.jom.2015.11.002

Ho, S. C., & Chuah, K. B. (2017). *Determinants of ITF R&D Technology Commercialization in Logistics and Supply Chain Industries: R&D Technologist Perspective*. Retrieved from http://www.wikicfp.com/cfp/servlet/event.showcfp?eventid=62787

Ho, S. C., & Chuah, K. B. (2018). *A Handbook for ITF R&D Project Management*. Retrieved from https://www.igi-global.com/publish/call-for-papers/call-details/2827

Ho, S. C., & Chuah, K. B. (2018). *Application of MSTAM Methodology in Project Management – A Case Study of ITF Robotic Automation R&D Project*. Retrieved from https://www.igi-global.com/publish/call-for-papers/call-details/3000

Ho, S. C., & Chuah, K. B. (2018). *Critical Success Factors for Strategic Management of ITF R&D Projects Commercialization: An Industry Expert Perspective*. Retrieved from https://www.igi-global.com/publish/call-for-papers/call-details/3018

Hoke, A., Brissette, A., Chandler, S., Pratt, A., & Maksimović, D. (2013, August). Look-ahead economic dispatch of microgrids with energy storage, using linear programming. In *Technologies for Sustainability (SusTech), 2013 1st IEEE Conference on* (pp. 154-161). IEEE.10.1109/SusTech.2013.6617313

Holland, J. H. (1992). *Adaptation in Natural and Artificial Systems: An Introductory Analysis with Applications to Biology, Control and Artificial Intelligence*. Cambridge, MA: MIT Press.

Holmes, P., & Smith, P. (2001). *Introduction to structural equation modelling using LISREAL*. Perth: ACSPRI-Winter training Program.

Holtta, Eisto, & Mahlamaki. (2009). Benefits for cast product development through early supplier involvement. *IEEE Conference Publications*, 1 – 7.

Hota, P. K., & Sahu, N. C. (2015). Non-Convex Economic Dispatch with Prohibited Operating Zones through Gravitational Search Algorithm. *Iranian Journal of Electrical and Computer Engineering*, *5*(6).

Hsu, Y. H., & Fang, W. (2009). Intellectual capital and new product development performance: The mediating role of organizational learning capability. *Technological Forecasting and Social Change*, *76*(5), 664–677. doi:10.1016/j.techfore.2008.03.012

Huang, X., Soutar, G. N., & Brown, A. (2004). Measuring new product success: An empirical investigation of Australian SMEs. *Industrial Marketing Management*, *33*(2), 117–123. doi:10.1016/S0019-8501(03)00034-8

Huettemann, G., Gaffry, C., & Schmitt, R. H. (2016). Adaptation of Reconfigurable Manufacturing Systems for Industrial Assembly – Review of Flexibility Paradigms, Concepts, and Outlook. *Procedia CIRP*, *52*, 112–117. doi:10.1016/j.procir.2016.07.021

Hund, L. B., Campbell, D. L., & Newcomer, J. T. (2017). Statistical Guidance for Setting Product Specification Limits. *IEEE Conference Publications*, 1 – 6. 10.1109/RAM.2017.7889664

Huszak, A. & Imre, S. (2010). *Eliminating Rank Reversal Phenomenon in GRA e based Network Selection Method*. doi:10.1109/ICC.2010.5502475

Hwang, C. L., & Yoon, K. (1981). *Multiple Attribute Decision Making: Methods and Applications*. New York: Springer-Verlag. doi:10.1007/978-3-642-48318-9

Hynek, J., & Janecek, V. (2009). *Problems of advanced manufacturing technology benefits evaluation*. Paper presented at the 2009 International Conference on Intelligent Engineering Systems. 10.1109/INES.2009.4924746

IEC. (2017). *IEC International Standards. International Electrotechnical Commission*. Available at: https://webstore.iec.ch/home

IIC. (2017). *The Industrial Internet Reference Architecture Technical Paper*. Industrial Internet Consortium (IIC). Retrieved July 25, 2018, from http://www.iiconsortium.org/IIRA.htm

Imen, L., Mouhamed, B., & Djamel, L. (2013, November). Economic dispatch using classical methods and neural networks. In *Electrical and Electronics Engineering (ELECO), 2013 8th International Conference on* (pp. 172-176). IEEE.10.1109/ELECO.2013.6713826

Internetofthingsagenda. (2018). *Industrial Internet of Things (IIoT) Definition*. IoT Agenda, Tech Target. Available at: http://internetofthingsagenda.techtarget.com/definition/Industrial-Internet-of-Things-IIoT

Ishikawa, S., Kubota, R., & Horio, K. (2015). Effective hierarchical optimization by a hierarchical multi-space competitive genetic algorithm for the flexible job-shop scheduling problem. *Expert Systems with Applications*, *42*(24), 9434–9440. doi:10.1016/j.eswa.2015.08.003

Iskin, I. (2011). Literature review on adoption of energy efficient technologies from a demand side management perspective: Taxonomy of adoption drivers, barriers and policy tools. *IEEE Conference Publications*, 1 – 16.

Islam, R., Ahmed, M., & Alias, M. H. (2007). Application of Quality Function Deployment in redesigning website: A case study on TV3. *International Journal of Business Information Systems*, *2*(2), 195–216. doi:10.1504/IJBIS.2007.011619

ITF IP. (2017). *Guide on Intellectual Property Arrangements for Research and Development Projects Funded Under the Innovation and Technology Support Programme and the Midstream Research Programme for Universities of the Innovation and Technology Fund*. Retrieved from https://www.itf.gov.hk/l-eng/Forms/IP_Guideline_201710.pdf

Jabr, R. A., Coonick, A. H., & Cory, B. J. (2000). A homogeneous linear programming algorithm for the security constrained economic dispatch problem. *IEEE Transactions on Power Systems*, *15*(3), 930–936. doi:10.1109/59.871715

Jacobs, M. A., Yu, W., & Chavez, R. (2016). The effect of internal communication and employee satisfaction on supply chain integration. *International Journal of Production Economics*, *171*(Part 1), 60–70. doi:10.1016/j.ijpe.2015.10.015

Jaggi, C. K., & Mittal, M. (2011). Economic Order Quantity Model for Deteriorating Items with Imperfect Quality. Delhi: Revista Investigación Operacional.

Jaromin, M. (2013). Computer aided material selection in design process. *Computer Science and Information Systems*, 3–7.

Jasperneite. (2012). Jasperneite, J., 2012: Alter Wein in neuen Schläuchen. *Computer & Automation*, *12*, 24–28. Available at: http://www.ciit-owl.de/uploads/media/410- 10%20gh%20Jasperneite%20CA%202012-12_lowres1.pdf

Jayaraman, V., Klassen, R., & Linton, J. D. (2007). Supply chain management in a sustainable environment. *Journal of Operations Management*, *25*(6), 1071–1074. doi:10.1016/j.jom.2007.01.016

Jeyaraman, K., & Kee Teo, L. (2010). A conceptual framework for critical success factors of lean Six Sigma: Implementation on the performance of electronic manufacturing service industry. *International Journal of Lean Six Sigma*, *1*(3), 191–215. doi:10.1108/20401461011075008

Jiang, J. J., & Conrath, D. W. (1997). Semantic Similarity Based on Corpus Statistics and Lexical Taxonomy. *Proceedings of International Conference Research on Computational Lingustics.*

Johnson, B. D. (2017). *How to invent the future, Trend, Analysis of the Facts, Numbers, and Trends Shaping the World.* Retrieved from http://trend.pewtrusts.org/-/media/post-launch-images/trend-magazine/summer-2017/trend_summer_2017.pdf

Joreskog, K. G. (1973). A general method for estimating a linear structural equation system. In A. S. Goldberger & O. D. Duncan (Eds.), *Structural Equation Models in the Social Sciences* (pp. 85–112). New York: Academic Press.

Jumie Yuventi & Weiss. (2013). Value sensitivity of Quality Function Deployment approaches in systems engineering-driven construction projects. *IEEE Conference Publications*, 847 - 852.

Kaa. (2018). Available at: https://www.kaaproject.org

Kalakota, R., & Whiston, A. (1997). *Electronic commerce: a manager's guide*. Reading, MA: Addison Wesley.

Kanyongo, G. Y., Certo, J., & Launcelot, B. I. (2006). Using regression analysis to established the relation between home environment and reading achievement: A case of Zimbabwe. *International Education Journal*, *7*(5), 632–641.

Karaboga, D. (2005). *An idea based on honey bee swarm for numerical optimization* (Vol. 200). Technical report-tr06, Erciyes University, Engineering Faculty, Computer Engineering Department.

Karaboga, D., & Basturk, B. (2007). A powerful and efficient algorithm for numerical function optimization: Artificial bee colony (ABC) algorithm. *Journal of Global Optimization*, *39*(3), 459–471. doi:10.100710898-007-9149-x

Karaboga, D., & Basturk, B. (2008). On the performance of artificial bee colony (ABC) algorithm. *Applied Soft Computing*, *8*(1), 687–697. doi:10.1016/j.asoc.2007.05.007

Kar, B., & Roy, P. (2018). A Comparative Study Between Cascaded FOPI–FOPD and IOPI–IOPD Controllers Applied to a Level Control Problem in a Coupled Tank System. *Journal of Control, Automation and Electrical Systems*, *29*(3), 340–349. doi:10.100740313-018-0373-z

Karunakaran, S. (2016). Innovative application of LSS in aircraft maintenance environment. *International Journal of Lean Six Sigma*, *7*(1), 85–108. doi:10.1108/IJLSS-01-2015-0001

Kasemsap, K. (2018). *Applying Lean Production and Six Sigma in global operations. In Operations and Service Management: Concepts* (pp. 582–612). Methodologies, Tools, and Applications.

Kaufer, F., & Klusch, M. (2006). WSMO-MX: A Logic Programming Based Hybrid Service Matchmaker. *European Conference on Web Services*, 161-170. 10.1109/ECOWS.2006.39

Kaur, G., & Arora, S. (2018). Chaotic Whale Optimization Algorithm. *Journal of Computation Design and Engineering.*, *5*(3), 275–284. doi:10.1016/j.jcde.2017.12.006

Kaushik, P., & Khanduja, D. (2009). Application of Six Sigma DMAIC methodology in thermal power plants: A case study. *Total Quality Management*, *20*(2), 197–207. doi:10.1080/14783360802622995

Kavin, L., & Narasimhan, R. (2018). An Investigation of Contextual Influences on Innovation in Complex Projects. *Innovation and Supply Chain Management: Relationship, Collaboration and Strategies*, 51–77.

Kelley James, E. (1960). The cutting-plane method for solving convex programs. *Journal of the Society for Industrial and Applied Mathematics*, *8*(4), 703–712. doi:10.1137/0108053

Keneni, B., Austin, B., Elkin, C., & Devabhaktuni, V. (2017). Educational Prototype Demonstrating Frequency Spectrum Sharing Through Channel Borrowing and Priority Assignment. *IEEE Conference Publications*, 540 – 544. 10.1109/EIT.2017.8053422

Ketkar, S., Kock, N., Parente, R., & Verville, J. (2012). The impact of individualism on buyer–supplier relationship norms, trust and market performance: An analysis of data from Brazil and the U.S.A. *International Business Review, 21*(5), 782–793. doi:10.1016/j.ibusrev.2011.09.003

Kevah, A., & Ghazaan, M. I. (2017). Enhanced whale optimization algorithm for sizing optimization of skeletal structures. *Mechanics Based Design of Structures and Machines, 45*(3), 345–362. doi:10.1080/15397734.2016.1213639

Khanchanapong, T., Prajogo, D., Sohal, A. S., Cooper, B. K., Yeung, A. C. L., & Cheng, T. C. E. (2014). The unique and complementary effects of manufacturing technologies and lean practices on manufacturing operational performance. *International Journal of Production Economics, 153*, 191–203. doi:10.1016/j.ijpe.2014.02.021

Khan, N. A., Sidhu, G. A. S., & Gao, F. (2016). Optimizing combined emission economic dispatch for solar integrated power systems. *IEEE Access: Practical Innovations, Open Solutions, 4*, 3340–3348.

Kiefer, C., & Bernstein, A. (2008). The Creation and Evaluation of iSPARQL Strategies for Matchmaking. *European Semantic Web Conference*, 463-477. 10.1007/978-3-540-68234-9_35

Kim, B., & Kim, J. (2009). Structural factors of NPD (new product development) team for manufacturability. *International Journal of Project Management, 27*(7), 690–702. doi:10.1016/j.ijproman.2008.11.003

Kim, S.-K., Lee, B.-G., & Oh, K.-S. (2009). The Effect of R&D and Technology Commercialization Capabilities on the Innovation Performance of Korean IT SMEs: The Case of Direct and Indirect Recipients of Public R&D Funding. *IEEE Conference Publication*, 1531 - 1541. 10.1109/PICMET.2009.5261985

Kitson, A. (2006). The relevance of scholarship for nursing research and practice. *Journal of Advanced Nursing, 55*(5), 541–543. doi:10.1111/j.1365-2648.2006.04004_1.x

Klusch, M., Fries, B., & Sycara, K. (2008). OWLS-MX: A Hybrid Semantic Web Service Matchmaker for OWL-S Services. *Journal of Web Semantics.*

Kock, N. (2011). Using WarpPLS in e-collaboration studies: Mediating effects, control and second order variables, and algorithm choices. *International Journal of e-Collaboration, 7*(3), 1–13. doi:10.4018/jec.2011070101

Kock, N. (2018). *WarpPLS 6.0 User Manual.* Laredo, TX: ScriptWarp Systems.

Kock, N., & Lynn, G. S. (2012). Lateral collinearity and misleading results in variance-based SEM: An illustration and recommendations. *Journal of the Association for Information Systems, 13*(7), 546–580. doi:10.17705/1jais.00302

Koc, T., & Bozdag, E. (2009). The impact of AMT practices on firm performance in manufacturing SMEs. *Robotics and Computer-integrated Manufacturing, 25*(2), 303–313. doi:10.1016/j.rcim.2007.12.004

Kondili, E., Pantelides, C. C., & Sargent, R. W. H. (1993). A general algorithm for short-term scheduling of batch operations—I. MILP formulation. *Computers & Chemical Engineering, 17*(2), 211–227. doi:10.1016/0098-1354(93)80015-F

Kong, F., Wei, W. & Gong, J.H. (2016). Rank reversal and rank preservation in ANP method. *Journal of Discrete Mathematics Science Crypt to gr., 19*, 821-836.

Kong, D., Feng, Q., Zhou, Y., & Xue, L. (2016). Local implementation for green-manufacturing technology diffusion policy in China: From the user firms' perspectives. *Journal of Cleaner Production, 129*, 113–124. doi:10.1016/j.jclepro.2016.04.112

Kong, Y., Zhu, Y., & Wang, Y. (2019). A center-based modeling approach to solve the districting problem. *International Journal of Geographical Information Science, 33*(2), 368–384. doi:10.1080/13658816.2018.1474472

Kori, Pedaste, Altin, Tõnisson, & Palts. (2016). Factors That Influence Students' Motivation to Start and to Continue Studying Information Technology in Estonia. *IEEE Transactions on Education, 59*(4), 255–262.

Kotter, J.P. (2012). *Leading Change.* Academic Press.

Krishna, B. T. (2011). Studies on fractional order differentiators and integrators: A survey. *Signal Processing, 91*(3), 386–426. doi:10.1016/j.sigpro.2010.06.022

Kumari, M. S., & Sydulu, M. (2009). A fast computational genetic algorithm for economic load dispatch. *International Journal of Recent Trends in Engineering, 1*(1).

Kumar, M., Antony, J., Singh, R. K., Tiwari, M. K., & Perry, D. (2006). Implementing the Lean Sigma framework in an Indian SME: A case study. *Production Planning and Control, 17*(4), 407–423. doi:10.1080/09537280500483350

Kumar, N. R., & Singal, S. K. (2015). Penstock material selection in small hydropower plants using MADM methods. *Renewable & Sustainable Energy Reviews, 52*, 240–255. doi:10.1016/j.rser.2015.07.018

Kumar, V. S., & Kumar, P. M. (2014). Optimization of cryogenic cooled EDM process prameters using grey relation analysis. *Journal of Mechanical Science and Technology, 28*(9), 3777–3784. doi:10.100712206-014-0840-9

Kuo, M. T., Lu, S. D., & Tsou, M. C. (2017). Considering Carbon Emissions in Economic Dispatch Planning for Isolated Power Systems. *Natural Gas, 701*, 233–237.

Lachhab, N., Svaricek, F., Wobbe, F., & Rabba, H. (2013). *Fractional order PID controller (FOPID)-toolbox.* Paper presented at the Control Conference (ECC), 2013 European.

Lafou, M., Mathieu, L., Pois, S., & Alochet, M. (2015). Manufacturing System Configuration: Flexibility Analysis For automotive Mixed-Model Assembly Lines. *IFAC-PapersOnLine, 48*(3), 94–99. doi:10.1016/j.ifacol.2015.06.064

Lagrosen, S. (2005). Customer involvement in new product development: A relationship marketing perspective. *European Journal of Innovation Management, 8*(4), 424–436. doi:10.1108/14601060510627803

Lajmi, S., Ghedira, C., & Ghedira, K. (2006a). How to apply CBR method in web service composition. In *Second International Conference on Signal-Image Technology and Internet Based Systems (SITI'2006).* Springer Verlag.

Lajmi, S., Ghedira, C., Ghedira, K., & Benslimane, D. (2006b). Web_CBR: How to compose web service via case based reasoning. *IEEE International Symposium on Service-Oriented Applications, Integration and Collaboration held with the IEEE International Conference on eBusiness Engineering (ICEBE 2006).*

Laliwala, Z., Khosla, R., Majumdar, P., & Chaudhary, S. (2006). Semantic and Rule Based Event-Driven Dynamic Web Service Composition for Automation of Business Processes. *Proceedings of the IEEE Service Computing Workshop (SCW06).*

Lambert, D. M., & Cooper, M. C. (2000). Issues in Supply Chain Management. *Industrial Marketing Management, 29*(1), 65–83. doi:10.1016/S0019-8501(99)00113-3

Langer, W. H., Drew, L. J., & Sachs, J. S. (2004). Aggregates and the environment. American Geological Institute.

Lan, H. (2009). Web-based rapid prototyping and manufacturing systems: A review. *Computers in Industry, 60*(9), 643–656. doi:10.1016/j.compind.2009.05.003

Lanusse, P., Malti, R., & Melchior, P. (2013). CRONE control system design toolbox for the control engineering community: tutorial and case study. *Phil. Trans. R. Soc. A, 371*(1990), 20120149.

Lau, A. K. (2011). Critical success factors in managing modular production design: Six company case studies in Hong Kong, China, and Singapore. *Journal of Engineering and Technology Management, 28*(3), 168–183. doi:10.1016/j.jengtecman.2011.03.004

Laureani, A., & Antony, J. (2011). Standards for lean six sigma certification. *International Journal of Productivity and Performance Management, 61*(1), 110–120. doi:10.1108/17410401211188560

Laureani, A., & Antony, J. (2017). Leadership and Lean Six Sigma: A systematic literature review. *Total Quality Management & Business Excellence*, 1–29. doi:10.1080/14783363.2017.1288565

Lawler, E. L., & Wood, D. E. (1966). Branch-and-bound methods: A survey. *Operations Research, 14*(4), 699–719. doi:10.1287/opre.14.4.699

Lazzerini, B., & Pistolesi, F. (2015, November). A linear programming-driven MCDM approach for multi-objective economic dispatch in smart grids. In *SAI Intelligent Systems Conference (IntelliSys)*, 2015 (pp. 475-484). IEEE.10.1109/IntelliSys.2015.7361183

Lee, K., & EI-Sharkawee, M. (2008). *Modern Heuristic Optimization Techniques*. IEEE Press.

Lei, H., Laporte, G., Liu, Y., & Zhang, T. (2014). Dynamic design of sales territories. *Computers & Operations Research, 56*, 84–92. doi:10.1016/j.cor.2014.11.008

Lenina, S. V. B. (2016). Intellectual Property Protection: A Step to realize Make in India. *IEEE Conference Publications*, 2089 – 2092. 10.1109/SCOPES.2016.7955816

Lester, D. H. (1998). Critical success factors for new product development. *Research Technology Management, 41*(1), 36–43. doi:10.1080/08956308.1998.11671182

Li, S. H. (2002). *An Integrated Model for Supply Chain Management Practice, Performance and Competitive Advantage* (PhD Dissertation). University of Toledo, Toledo, OH.

Liang, H., Liu, Y., Shen, Y., Li, F., & Man, Y. (2018). A Hybrid Bat Algorithm for Economic Dispatch with Random Wind Power. *IEEE Transactions on Power Systems, 33*(5), 5052–5061. doi:10.1109/TPWRS.2018.2812711

Liang, T., & Konsynski, B. R. (1993). Modeling by analogy: Use of analogical reasoning in model management systems. *Decision Support Systems, 9*(1), 113–125. doi:10.1016/0167-9236(93)90026-Y

Li, J.-Q., Pan, Q.-K., & Tasgetiren, M. F. (2014). A discrete artificial bee colony algorithm for the multi-objective flexible job-shop scheduling problem with maintenance activities. *Applied Mathematical Modelling, 38*(3), 1111–1132. doi:10.1016/j.apm.2013.07.038

Li, J., Sikora, R., Shaw, M. J., & Woo Tan, G. A. (2006). Strategic analysis of inter-organizational information sharing. *Decision Support Systems, 42*(1), 251–266. doi:10.1016/j.dss.2004.12.003

Li, L., & Horrocks, I. (2003). A Software Framework for Matchmaking Based on Semantic Web Technology. *International Conference in World Wide Web*, 331-339. 10.1145/775152.775199

Li, M., Yu, B., Rana, O. F., & Wang, Z. (2008). Grid Service Discovery with Rough Sets. *IEEE Transactions on Knowledge and Data Engineering, 20*(6), 851–862. doi:10.1109/TKDE.2007.190744

Lin, J., Chen, C. L., Tsai, S. F., & Yuan, C. Q. (2015). New intelligent particle swarm optimization algorithm for solving economic dispatch with valve-point effects. *Journal of Marine Science and Technology, 23*(1), 44–53.

Lin, J., & Lin, Q. (2017). Analysis of the Key Factors of Intellectual Property Management at Art Institutions. *IEEE Conference Publications*, 206 – 208. 10.1109/ICASI.2017.7988609

Lin, M. C., Wang, C. C., & Chen, T. C. (2006). A strategy for managing customer-oriented product design. *Concurrent Engineering*, *14*(3), 231–244. doi:10.1177/1063293X06068390

Lin, M., Chin, K. S., Ma, L., & Tsui, K. L. (2018). (in press). A comprehensive multi-objective mixed integer non-linear programming model for an integrated elderly care service districting problem. *Annals of Operations Research*. doi:10.100710479-018-3078-6

Linton, J. D., Klassen, R., & Jayaraman, V. (2007). Sustainable supply chains: An introduction. *Journal of Operations Management*, *25*(6), 1075–1082. doi:10.1016/j.jom.2007.01.012

Liu, C., Ramirez-Serrano, A., & Yin, G. (2011). Customer-driven product design and evaluation method for collaborative design environments. *Journal of Intelligent Manufacturing*, *22*(5), 751–764. doi:10.100710845-009-0334-2

Liu, J., Cheng, Q., & Wang, J. (2015). Identification of geochemical factors in regression to mineralization endogenous variables using structural equation modeling. *Journal of Geochemical Exploration*, *150*, 125–136. doi:10.1016/j.gexplo.2014.12.021

Liu, W., & Atuahene-Gima, K. (2018). Enhancing product innovation performance in a dysfunctional competitive environment: The roles of competitive strategies and market-based assets. *Industrial Marketing Management*, *73*, 7–20. doi:10.1016/j.indmarman.2018.01.006

Li, Z., Liu, L., Dehghan, S., Chen, Y. Q., & Xue, D. (2017). A review and evaluation of numerical tools for fractional calculus and fractional order controls. *International Journal of Control*, *90*(6), 1165–1181. doi:10.1080/00207179.2015.1124290

Loeh, H. (2006). Project and Product Management in Collaborative Product Innovation. *IEEE Conference Publications*, 1 – 10. 10.1109/ICE.2006.7477079

Lou, Y., Fu, X., & Huang, L. (2010). Evaluation on the Commercialization Potential of Emerging Technologies Based on Structural Equation Model. *IEEE Conference Publications*, 329 – 333. 10.1109/UKSIM.2010.68

Lujan-Moreno, G. A., Howard, P. R., Rojas, O. G., & Montgomery, D. C. (2018). Design of Experiments and Response Surface Methodology to Tune Machine Learning Hyperparameters, with a Random Forest Case-Study. *Expert Systems with Applications*, *109*, 195–205. doi:10.1016/j.eswa.2018.05.024

Lynn, G. S., Abel, K. D., Valentine, W. S., & Wright, R. C. (1999). Key factors in increasing speed to market and improving new product success rates. *Industrial Marketing Management*, *28*(4), 319–326. doi:10.1016/S0019-8501(98)00008-X

Machado, J., Kiryakova, V., & Mainardi, F. (2011). Recent history of fractional calculus. *Communications in Nonlinear Science and Numerical Simulation*, *16*(3), 1140–1153. doi:10.1016/j.cnsns.2010.05.027

Macharis, C., Springael, J., De Brucker, K., & Verbeke, A. (2004). PROMETHEE and AHP: the design of operational synergies in multi-criteria analysis: strengthening PROMETHEE with ideas of AHP. *European Journal of Operational Research*, *153*, 307–317. doi:10.1016/S0377-2217(03)00153-X

Madhuri, Ch. B., & Chandulal, J. A. (2010). Evaluating websites using COPRAS-GRA combined with grey clustering. *International Journal of Engineering Science and Technology*, *2*(10), 5280–5294.

Majchrzak, A. (1988). The human infrastructure impact statement (HIIS) A tool for managing the effective implementation of advanced manufacturing technology. *Jisuanji Jicheng Zhizao Xitong*, *1*(2), 95–102. doi:10.1016/0951-5240(88)90093-6

Majumder, A. (2013). Process parameter optimization during EDM of AISI 316 LN stainless steel by using fuzzy based multi-objective PSO. *Journal of Mechanical Science and Technology*, *27*(2), 2143–2151. doi:10.100712206-013-0524-x

Malekpoor, H., Chalvatzis, K., Mishra, N., Mehlawat, M. K., Zafirakis, D., & Song, M. (2017). Integrated grey relational analysis and multi objective grey linear programming for sustainable electricity generation planning. *Annals of Operations Research*, 1–29.

Mali, A. V., Morey, N. N., & Khtri, A. P. (2016). Improvement in the efficiency of the Stone Crusher. *International Journal of Science. Engineering and Technology Research.*, *5*(6), 2265–2271.

Malti, R., & Victor, S. (2015). *CRONE Toolbox for system identification using fractional differentiation models.* Paper presented at the 17th IFAC Symposium on System Identification (SYSID 2015). 10.1016/j.ifacol.2015.12.223

Malti, R., Melchior, P., Lanusse, P., & Oustaloup, A. (2011). *Towards an object oriented CRONE toolbox for fractional differential systems.* Paper presented at the 18th IFAC world congress. 10.3182/20110828-6-IT-1002.02443

Malti, R., Melchior, P., Lanusse, P., & Oustaloup, A. (2012). Object-oriented crone toolbox for fractional differential signal processing. *Signal, Image and Video Processing*, *6*(3), 393–400. doi:10.100711760-012-0323-3

Mani, V., Gunasekaran, A., Papadopoulos, T., Hazen, B., & Dubey, R. (2016). Supply chain social sustainability for developing nations: Evidence from India. *Resources, Conservation and Recycling*, *111*, 42–52. doi:10.1016/j.resconrec.2016.04.003

Manser, M. H., Musirin, I., & Othman, M. M. (2017). Immune Log-Normal Evolutionary Programming (ILNEP) for solving economic dispatch problem with prohibited operating zones. In *Industrial Engineering and Applications (ICIEA), 2017 4th International Conference on.* IEEE.10.1109/IEA.2017.7939199

Manyika, J., Chui, M., Bisson, P., Woetzel, J., Dobbs, R., Bughin, J., & Aharon, D. (2015). Unlocking the Potential of the Internet of Things. *McKinsey Global Institute.* Available at: https://www.mckinsey.com/business-functions/digital-mckinsey/our-insights/the-internet-of-things-the-value-of-digitizing-the-physical-world

Manyika, J., Chui, M., & Bisson, P. (2015). *The Internet of Things: Mapping the value beyond the hype.* San Francisco: McKinsey & Company. Available at https://www.mckinsey.de/sites/mck_files/files/unlocking_the_potential_of_the_internet_of_things_full_report.pdf

Marín, C., Ramírez, G. S., & Muñoz, J. A. (2012). Sistema de costeo ABC para empresas del sector eléctrico que actúen como operadores de red. *Sciences et Techniques (Paris)*, 78–79.

Marinov, T. M., Ramirez, N., & Santamaria, F. (2013). Fractional integration toolbox. *Fractional Calculus & Applied Analysis*, *16*(3), 670–681. doi:10.247813540-013-0042-7 PMID:24812536

Martin, D., Burstein, M., Mcdermott, D., Mcilraith, S., Paolucci, M., Sycara, K., ... Srinivasan, N. (2007). Bringing Semantics to Web Services with OWL-S. *World Wide Web (Bussum)*, *10*(3), 243–277. doi:10.100711280-007-0033-x

Martin, D., Paolucci, M., McIlraith, S., Burstein, M., McDermott, D., McGunness, D., ... Sycara, K. (2004). Bringing Semantics to Web Services: The OWL-S Approach. *Proceeding of First International Workshop Semantic Web Services and Web Process Composition.*

Mason, S. J., Fowler, J. W., & Matthew Carlyle, W. (2002). A modified shifting bottleneck heuristic for minimizing total weighted tardiness in complex job shops. *Journal of Scheduling*, *5*(3), 247–262. doi:10.1002/jos.102

Masood, S. H., & Soo, A. (2002). A rule-based expert system for rapid prototyping system selection. *Rob. Comput. Inte. Eng.*, *18*, 267–274.

Mathiyazhagan, K., Govindan, K., Noorul, H. A., & Geng, Y. (2013b). An ISM approach for the barrier analysis in implementing green supply chain management. *Journal of Cleaner Production, 47*, 283–297. doi:10.1016/j.jclepro.2012.10.042

Mats & Jeffrey. (2000). Specification based Prototyping of Control Systems. *IEEE Conference Publications*, 1D3/1 - 1D3/8.

Matskin, M., Maigre, R., & Tyugu, E. (2007). Computational logical semantics for business process language. *Proceedings of second international conference on Internet and Web applications and services (ICIW 2007)*, 526-531.

Matušů, R. (2011). Application of fractional order calculus to control theory. *International Journal of Mathematical Models and Methods in Applied Sciences, 5*(7), 1162-1169.

Matzler, K., & Hinterhuber, H. H. (1998). How to make product development projects more successful by integrating Kano's model of customer satisfaction into quality function deployment. *Technovation, 18*(1), 25–38. doi:10.1016/S0166-4972(97)00072-2

Matzler, K., Hinterhuber, H. H., Bailom, F., & Sauerwein, E. (1996). How to delight your customers. *Journal of Product and Brand Management, 5*(2), 6–18. doi:10.1108/10610429610119469

McGee, J. (2015). *Resource-Based View*. Wiley Encyclopedia of Management.

McIlraith, S., & Martin, D. (2003). Bringing Semantics to Web Services. *IEEE Intelligent Systems, 18*(1), 90–93. doi:10.1109/MIS.2003.1179199

McLean, R. S., Antony, J., & Dahlgaard, J. J. (2017). Failure of Continuous Improvement initiatives in manufacturing environments: A systematic review of the evidence. *Total Quality Management & Business Excellence, 28*(3–4), 219–237. doi:10.1080/14783363.2015.1063414

Meng, W., & Wang, X. (2017). Distributed Energy Management in Smart Grid With Wind Power and Temporally Coupled Constraints. *IEEE Transactions on Industrial Electronics, 64*(8), 6052–6062. doi:10.1109/TIE.2017.2682001

Micro, Medium & Small Scale Industry. (2018, August 28). Retrieved from https://www.cii.in/Sectors.aspx

Miles, R. E., Snow, C. C., Meyer, A. D., & Coleman, H. J. Jr. (1978). Organizational strategy, structure, and process. *Academy of Management Review, 3*(3), 546–562. doi:10.5465/amr.1978.4305755

Miner, A. S., Bassof, P., & Moorman, C. (2001). Organizational improvisation and learning: A field study. *Administrative Science Quarterly, 46*(2), 304–337. doi:10.2307/2667089

Mirjalili, S., & Lewis, A. (2016). The Whale Optimization Algorithm. *Advances in Engineering Software, 96*, 51–67. doi:10.1016/j.advengsoft.2016.01.008

Mohamed, F., Abdel-Nasser, M., Mahmoud, K., & Kamel, S. (2018, February). Economic dispatch using stochastic whale optimization algorithm. In *Innovative Trends in Computer Engineering (ITCE), 2018 International Conference on* (pp. 19-24). IEEE.10.1109/ITCE.2018.8316594

Mokotoff, E., & Chrétienne, P. (2002). A cutting plane algorithm for the unrelated parallel machine scheduling problem. *European Journal of Operational Research, 141*(3), 515–525. doi:10.1016/S0377-2217(01)00270-3

Momoh, J. A., & Reddy, S. S. (2014, July). Combined economic and emission dispatch using radial basis function. In *PES General Meeting Conference & Exposition, 2014 IEEE* (pp. 1-5). IEEE.10.1109/PESGM.2014.6939506

Mönch, L., Schabacker, R., Pabst, D., & Fowler, J. W. (2007). Genetic algorithm-based subproblem solution procedures for a modified shifting bottleneck heuristic for complex job shops. *European Journal of Operational Research, 177*(3), 2100–2118. doi:10.1016/j.ejor.2005.12.020

Monje, C. A., Chen, Y., Vinagre, B. M., Xue, D., & Feliu-Batlle, V. (2010). Fractional-order systems and controls: fundamentals and applications. Springer Science & Business Media.

Moran, J., Granada, E., Míguez, J. L., & Porteiro, J. (2006). Use of grey relational analysis to assess and optimize small biomass boilers. *Fuel Processing Technology*, *87*(2), 123–127. doi:10.1016/j.fuproc.2005.08.008

Moreno, M. A., Rojas, O., Olivares-Benitez, E., Nucamendi-Guillén, S., & Garcia de Alba Valenzuela, H. R. (2018). Production Planning for a Company in the Industry of Compact Discs Mass Replications. In *New Perspectives on Applied Industrial Tools and Techniques* (pp. 497–516). Cham: Springer. doi:10.1007/978-3-319-56871-3_24

MRFR. (2018). Industrial Internet of Things Market Analysis 2018 To 2022. *Market Research Future*. Available to: https://www.marketresearchfuture.com/reports/industrial-iot-platform-market-2186

Mufazzal, S., & Muzakkir, S. M. (2018). A new multi-criterion decision making (MCDM) method based on proximity indexed value for minimizing rank reversals. *Computers & Industrial Engineering*, *119*, 427–438. doi:10.1016/j.cie.2018.03.045

Mukherjee, R., & Chakraborty, S. (2012). Selection of EDM Process Parameters Using Biogeography-Based Optimization Algorithm. *Materials and Manufacturing Processes*, *27*(9), 954–962. doi:10.1080/10426914.2011.610089

Mutingi, M., & Matope, S. (2013). Dynamics of Information Technology Adoption in a Complex Environment. *IEEE Conference Publications*, 1466 - 1471. 10.1109/ICIT.2013.6505888

Nain Sukhia, K., Ashraf Khan, A., & Bano, M. (2014). Introducing Economic Order Quantity Model for inventory control in web based point of sale applications and comparative analysis of techniques for demand forecasting in inventory management. *International Journal of Computers and Applications*, *107*(19), 1–8. doi:10.5120/18856-7385

Nasiri, J., & Khiyabani, F. M. (2018). *A Whale Optimization Algorithm (WOA) approach for Clustering. Cogent Mathematics & Statistics, 1483565.*

Naslund, D. (2008). Lean, six sigma and lean sigma: Fads or real process improvement methods? *Business Process Management Journal*, *14*(3), 269–287. doi:10.1108/14637150810876634

Nason, R. S., & Wiklund, J. (2018). An assessment of resource-based theorizing on firm growth and suggestions for the future. *Journal of Management*, *44*(1), 32–60. doi:10.1177/0149206315610635

Naveen, P., Chandel, A. K., Vedik, B., & Topwal, K. (2016, April). Economic dispatch with valve point effect using symbiotic oragnisms search algorithm. In *Energy Efficient Technologies for Sustainability (ICEETS), 2016 International Conference on* (pp. 430-435). IEEE.10.1109/ICEETS.2016.7583793

Nemhauser, G. L., & Savelsbergh, M. W. P. (1992). A cutting plane algorithm for the single machine scheduling problem with release times. In *Combinatorial optimization* (pp. 63–83). Springer. doi:10.1007/978-3-642-77489-8_4

NewGOV.HK. (2017). *LCQ7: Promoting development of innovation and technology*. Retrieved from: http://www.info.gov.hk/gia/general/201205/16/P201205160267.htm

Nicholas, J., Ledwith, A., Aloini, D., Martini, A., & Nosella, A. (2015). Searching for radical new product ideas: Exploratory and confirmatory factor analysis for construct validation. *International Journal of Technology Management*, *68*(1-2), 70–98. doi:10.1504/IJTM.2015.068779

Nikolaou, I. E., & Evangelinos, K. I. (2010). A SWOT analysis of environmental management practices in Greek mining and mineral industry. *Resources Policy*, *35*(3), 226–234. doi:10.1016/j.resourpol.2010.02.002

Nikou, S. A., & Economides, A. A. (2014). Acceptance of Mobile-Based Assessment from the Perspective of Self-Determination Theory of Motivation. *IEEE Conference Publications*, 454 – 458. 10.1109/ICALT.2014.136

Nischal, M. M., & Mehta, S. (2015). Optimal load dispatch using ant lion optimization. *Int J Eng Res Appl*, *5*(8), 10–19.

Nishad, I. (2018). Inventory Management "ABC Analysis and Economic Order Quantity Approach": A case study. *Jasc: Journal of Applied Science and Computations*, 132-137.

Nitzl, C. (2016). The use of partial least squares structural equation modelling (PLS-SEM) in management accounting research: Directions for future theory development. *Journal of Accounting Literature*, *37*, 19–35. doi:10.1016/j.acclit.2016.09.003

Nivedha, R. R., Singh, J. G., & Ongsakul, W. (2018, January). PSO based economic dispatch of a hybrid microgrid system. In *2018 International Conference on Power, Signals, Control and Computation (EPSCICON)* (pp. 1-5). IEEE.10.1109/EPSCICON.2018.8379595

O'Halloran, D., & Kvochko, E. (2015). Industrial internet of things: unleashing the potential of connected products and services. In *World Economic Forum* (p. 40). Available at: http://www3.weforum.org/docs/WEFUSA_IndustrialInternet_Report2015.pdf

OASIS. (2004). *Introduction to UDDI: Important Features and Functional Concepts*. Organization for the Advancement of Structured Information Standards.

Oke, S. A. (2013). Manufacturing quality function deployment: Literature review and future trends. *Engineering Journal (New York)*, *17*(3), 79–103.

Oktaviani, A., Subawanto, H., & Hardi, H. (2017). The implementation of the ABC Classification and (Q,R) with Economic Order Quantity (EOQ) model on the travel agency. *ResearchGate*, 45.

Olivos, S., & Penagos, J. W. (2013). Modelo de Gestión de Inventarios: Conteo Cíclico por Análisis ABC. *INGENIARE, Universidad Libre-Barranquilla*, 108-109.

OMG. (2009). *Business Process Model and Notation*. Retrieved from http://www.omg.org/spec/BPMN/1.2/(2009)

Ong, C. S., Lai, J. Y., & Wang, Y. S. (2004). Factors affecting engineers' acceptance of asynchronous e-learning systems in high-tech companies. *Information & Management*, *41*(6), 795–804. doi:10.1016/j.im.2003.08.012

Ordoobadi, S. M., & Mulvaney, N. J. (2001). Development of a justification tool for advanced manufacturing technologies: System-wide benefits value analysis. *Journal of Engineering and Technology Management*, *18*(2), 157–184. doi:10.1016/S0923-4748(01)00033-9

Osman, T., Thakker, D., & Al-Dabass, D. (2006a). Semantic-Driven Matching of Web services using Case-Based Reasoning. *Fourth IEEE International Conference on Web Services (ICWS 2006)*, 29-36.

Osman, T., Thakker, D., & Al-Dabass, D. (2006b). S-CBR: Semantic Case Based Reasoner for Web services discovery and matchmaking. *20th European Conference on Modeling and Simulation (ECMS2006)*, 723-729.

Oustaloup, A., Melchior, P., Lanusse, P., Cois, O., & Dancla, F. (2000). *The CRONE toolbox for Matlab*. Paper presented at the Computer-Aided Control System Design, 2000. CACSD 2000. IEEE International Symposium on. 10.1109/CACSD.2000.900210

Öztürk, H., Eroglu, A., & Lee, G. M. (2015). An economic order quantity model for lots containing defective items with rework option. *International Journal of Industrial Engineering*.

Pal, K. (2018). A Big Data Framework for Decision Making in Supply Chain Management. In Emerging Applications in Supply Chains for Sustainable Business Development. IGI Global.

Paliwal, A. V., Shafiq, B., Vaidya, J., Xiong, H., & Adam, N. (2012). Semantics-based automated service discovery. *IEEE Transactions on Services Computing*, *5*(2), 260–275. doi:10.1109/TSC.2011.19

Pal, K., & Karakostas, B. (2014). A Multi Agent-based Service Framework for supply Chain Management, The Fifth International Conference on Ambient System, Networks and Technologies (ANT-2014), Hasselt, Belgium. *Procedia Computer Science*, *32*, 53–60. doi:10.1016/j.procs.2014.05.397

Pal, K., & Palmer, O. (2000). A decision-support systems for business acquisition. *Decision Support Systems*, *27*(4), 411–429. doi:10.1016/S0167-9236(99)00083-4

Pan, S., Jian, J., & Yang, L. (2018). A hybrid MILP and IPM approach for dynamic economic dispatch with valve-point effects. *International Journal of Electrical Power & Energy Systems*, *97*, 290–298. doi:10.1016/j.ijepes.2017.11.004

Papadimitriou, C. H., & Steiglitz, K. (1998). *Combinatorial optimization: algorithms and complexity*. Courier Corporation.

Papazoglou, M. (2012). *Web Services and SOA: Principles and Technology*. Pearson.

Park, T., & Kim, K. J. (1998). Determination of an optimal set of design requirements using house of quality. *Journal of Operations Management*, *16*(5), 569–581. doi:10.1016/S0272-6963(97)00029-6

Pathak, J., Koul, N., Caragea, D., & Honavar, V. G. (2005). A Framework for Semantic Web Services Discovery. *ACM International Workshop on Web Information and Data Management*, 45-50. 10.1145/1097047.1097057

Patil, A., Oundhaka, S., Sheth, A., & Verma, K. (2004). METEOR-S Web service Annotation Framework. *The Proceedings of the Thirteenth International World Wide Web Conference*, 553-562.

Pearson. (2017). *Pearson correlation coefficient*. Retrieved from https://en.wikipedia.org/wiki/Pearson_correlation_coefficient

Percival, J. C., & Cozzarin, B. P. (2010). Complementarities in the implementation of advanced manufacturing technologies. *The Journal of High Technology Management Research*, *21*(2), 122–135. doi:10.1016/j.hitech.2010.05.002

Pérez, F., Irisarri, E., Orive, D., Marcos, M., & Estevez, E. (2015). A CPPS Architecture approach for Industry 4.0. *Emerging Technologies & Factory Automation (ETFA), 2015 IEEE 20th Conference on*, 1–4. 10.1109/ETFA.2015.7301606

Peteraf, M. A. (1993). The cornerstones of competitive advantage: A resource-based view. *Strategic Management Journal*, *14*(3), 179–191. doi:10.1002mj.4250140303

Petrás, I. (2011). *Fractional derivatives, fractional integrals, and fractional differential equations in Matlab. In Engineering education and research using MATLAB*. InTech.

Petráš, I. (2011). *Fractional-order nonlinear systems: modeling, analysis and simulation*. Springer Science & Business Media. doi:10.1007/978-3-642-18101-6

Pezzella, F., Morganti, G., & Ciaschetti, G. (2008). A genetic algorithm for the flexible job-shop scheduling problem. *Computers & Operations Research*, *35*(10), 3202–3212. doi:10.1016/j.cor.2007.02.014

Phaal, R., O'Sullivan, E., Routley, M., Ford, S., & Probert, D. (2011). A framework for mapping industrial emergence. *Technological Forecasting and Social Change*, *78*(2), 217–230. doi:10.1016/j.techfore.2010.06.018

Pieńkowski, M. (2014). Waste measurement techniques for Lean companies. *International Journal of Lean Thinking*, *5*(1), 9–24.

Pilkington, A., & Meredith, J. (2009). The evolution of the intellectual structure of operations management—1980–2006: A citation/co-citation analysis. *Journal of Operations Management*, *27*(3), 185–202. doi:10.1016/j.jom.2008.08.001

Pinedo, M. L. (2008). *Scheduling. Theory, Algorithms and Systems (3rd ed.).* New York: Springer.

Pinedo, M., & Singer, M. (1999). A shifting bottleneck heuristic for minimizing the total weighted tardiness in a job shop. *Naval Research Logistics*, *46*(1), 1–17. doi:10.1002/(SICI)1520-6750(199902)46:1<1::AID-NAV1>3.0.CO;2-#

Pine, J. B. II. (1997). *Mass Customization: The New Frontier in Business Competition.* Cambridge, MA: Harvard Business School Press.

Piplani, R., Pujawan, N., & Ray, S. (2008). Sustainable supply chain management. *International Journal of Production Economics*, *111*(2), 193–194. doi:10.1016/j.ijpe.2007.05.001

Pirkkalainen, H., Jokinen, J. P. P., & Pawlowski, J. M. (2014). Understanding Social OER Environments—A Quantitative Study on Factors Influencing the Motivation to Share and Collaborate. *IEEE Conference Publications*, 388 – 400. 10.1109/TLT.2014.2323970

Pisoni, E., Visioli, A., & Dormido, S. (2009). *An interactive tool for fractional order PID controllers.* Paper presented at the Industrial Electronics, 2009. IECON'09. 35th Annual Conference of IEEE. 10.1109/IECON.2009.5414720

Pollard, E. (2015). *Understanding employers' graduate recruitment and selection practices. BIS Research Paper 231.* London, UK: Department for Business, Innovation and Skills. Available at https://derby.openrepository.com/derby/handle/10545/609154

Porter, M. E. (1996). *What is strategy.* Academic Press.

Poyhonen, M., & Hamalainen, R. P. (2001). On the convergence of multiattribute weighting methods. *European Journal of Operational Research*, *129*(3), 569–585. doi:10.1016/S0377-2217(99)00467-1

Prabhaker, P. R., Goldhar, J. D., & Lei, D. (1995). Marketing implications of newer manufacturing technologies. *Journal of Business and Industrial Marketing*, *10*(2), 48–58. doi:10.1108/08858629510087373

Prasad, B. (1998). Review of QFD and related deployment techniques. *Journal of Manufacturing Systems*, *17*(3), 221–234. doi:10.1016/S0278-6125(98)80063-0

Prasad, K. D., Subbaiah, K. V., & Rao, K. N. (2014). Multi-objective optimization approach for cost management during product design at the conceptual phase. *Journal of Industrial Engineering International*, *10*(1), 48. doi:10.100740092-014-0048-8

Prasad, K. G. D., Subbaiah, K. V., Rao, K. N., & Sastry, C. V. R. (2010). Prioritization of Customer Needs in House of Quality Using Conjoint Analysis. *International Journal of Qualitative Research*, *4*(2), 145–154.

Prasad, K., & Chakraborty, S. (2012). Application of multi-objective optimization on the basis of ratio analysis (MOORA) method for materials selection. *Materials & Design*, *37*, 317–324. doi:10.1016/j.matdes.2012.01.013

Prasad, K., & Chakraborty, S. (2013). A quality function deployment-based model for materials selection. *Materials & Design*, *49*, 525–535. doi:10.1016/j.matdes.2013.01.035

Prasad, K., Subbaiah, K., & Prasad, M. (2017). Supplier evaluation and selection through DEA-AHP-GRA Integrated approach-A case study. *Uncertain Supply Chain Management*, *5*(4), 369–382. doi:10.5267/j.uscm.2017.4.001

Project Network. (2015). Retrieved from https://en.wikipedia.org/wiki/Project_network

Przybysz, P. M., Duckwitz, S., Mütze-Niewöhner, S., & Schlick, C. M. (2013). Investigation of Team Composition and Task-related Conflict as Determinants of Engineering Service Productivity. *IEEE Conference Publications*, 250 – 254. 10.1109/IEEM.2013.6962412

Pyoun, Y. S., & Choi, B. K. (1994). Quantifying the flexibility value in automated manufacturing systems. *Journal of Manufacturing Systems*, *13*(2), 108–118. doi:10.1016/0278-6125(94)90026-4

Qing, T., Hui-yu, L., & Ming-jian, Z. (2011). Research on the Transaction Memory System in the Team of Early Supplier Involvement. *IEEE Conference Publications*, 1503 - 1507.

Rainie, L. (2017). *Technological innovation? That's the easy part.* Trend Summer. Available at http://trend.pewtrusts.org/-/media/post-launch-images/trend-magazine/summer-2017/trend_summer_2017.pdf

Rajesh, R., & Anand, M. D. (2012). The Optimization of the Electro-Discharge Machining Process using Response Surface Methodology and Genetic Algorithms. *Procedia Engineering*, *38*, 3941–3950. doi:10.1016/j.proeng.2012.06.451

Ramadan, B. M., Logenthiran, T., Naayagi, R. T., & Su, C. (2016, November). Hybridization of genetic algorithm and priority list to solve economic dispatch problems. In Region 10 Conference (TENCON), 2016 IEEE (pp. 1467-1470). IEEE.

Ramberg, J. S. (2000). Six sigma: Fad or fundamental. *Quality Digest*, *6*(5), 30–31.

Ranjan, S., Jha, V. K., & Pal, P. (2016). Application of emerging technologies in ERP implementation in Indian manufacturing enterprises: An exploratory analysis of strategic benefits. *International Journal of Advanced Manufacturing Technology*, 1–12. doi:10.100700170-016-8770-6

Rao, R. V. (2006). a material selection model using graph theory and matrix approach. *Materials Science and Engineering*, *431*(1-2), 248–255. doi:10.1016/j.msea.2006.06.006

Rao, R. V., & Padmanabhan, K. K. (2007). Rapid prototyping process selection using graph theory and matrix approach. *Journal of Materials Processing Technology*, *194*(1-3), 81–88. doi:10.1016/j.jmatprotec.2007.04.003

Rao, R. V., Pawar, P. J., & Shankar, R. (2008). Multi-objective optimization of electrochemical machining process parameters using a particle swarm optimization algorithm. *Proceedings of the Institution of Mechanical Engineers. Part B, Journal of Engineering Manufacture*, *222*(8), 949–958. doi:10.1243/09544054JEM1158

Rao, S. S. (2009). *Engineering Optimization. Theory and Practice (4th ed.).* New York: Wiley. doi:10.1002/9780470549124

Reategui, E. B., & Campbell, J. A. (1995). A Classification System for Credit Card Transactions. In J. P. Haton, M. Keane, & M. Mango (Eds.), *Advances in Case-Based Reasoning: Second European Workshop (EWCBR-94)*. Springer. 10.1007/3-540-60364-6_43

Reddy, P. D. P., Reddy, V. C. V., & Manohar, T. G. (2017). Whale optimization algorithm for optimal sizing of renewable resources for loss reduction in distribution system. Renewables. *Wind, Water, and Solar*, *4*(3), 1–13.

Regan, G., Flood, D., & McCaffery, F. (2016). Research Findings from an Industrial Trial of a Traceability Assessment and Implementation Framework. *IEEE Conference Publications*, 91 – 95. 10.1145/2904354.2904365

Resende, M., & Werneckz, R. (2004). A hybrid heuristic for the p-median problem. *Journal of Heuristics*, *10*(1), 59–88. doi:10.1023/B:HEUR.0000019986.96257.50

Rigdon, E. E. (1998). Structural Equation Modelling. In G. A. Marcoulides (Ed.), *Mahwah, Modern methods for business research* (pp. 251–294). Lawrence Erlbaum Associates Publishers.

Rios Mercado, R., & Fernández, E. (2009). A reactive GRASP for a commercial territory design problem with multiple balancing requirements. *Computers & Operations Research*, *36*(3), 755–776. doi:10.1016/j.cor.2007.10.024

Robertson, D., & Ulrich, K. (1998). Planning for product platforms. *Sloan Management Review*, *39*(4), 19.

Rockart, J. F. (1979). Chief executives define their own data needs. *Harvard Business Review*, *57*(2), 81–93. PMID:10297607

Roisko, V., Kämppi, P., & Luojus, S. (2013). Touch Screen Based TETRA Vehicle Radio: Preliminary Results of Multi-methodology Usability Testing Prototype. *IEEE Conference Publications*, 951 – 952. 10.1109/ICCVE.2013.6799938

Roman, D., Keller, U., Lausen, H., de Bruijn, J., Lara, R., Stollberg, M., ... Fensel, D. (2005). Web service modeling ontology. *Applied Ontology*, *1*(1), 77–106.

Roman, D., Kopecky, J., Vitvar, T., Domingue, J., & Fensel, D. (2015). WSMO-Lite and hRESTS: Lightweight semantic annotations for Web services and RESTful APIs. *Journal of Web Semantics*, *31*, 39–58. doi:10.1016/j.websem.2014.11.006

Rosing, K., & Revelle, C. (1997). Heuristic concentration: Two stage solution construction. *European Journal of Operational Research*, *97*(1), 75–86. doi:10.1016/S0377-2217(96)00100-2

Ross, M., Abbey, C., Bouffard, F., & Joos, G. (2018). Microgrid economic dispatch with energy storage systems. *IEEE Transactions on Smart Grid*, *9*(4), 3039–3047. doi:10.1109/TSG.2016.2624756

Roy, M. J., & Thérin, F. (2008). Knowledge acquisition and environmental commitment in SMEs. *Corporate Social Responsibility and Environmental Management*, *15*(5), 249–259. doi:10.1002/csr.145

Roy, S., Dan, P. K., & Modak, N. (2018). Cascading effects of management actions on NPD in the manufacturing sector: The Indian context. *Journal of Manufacturing Technology Management*, *29*(7), 1115–1137. doi:10.1108/JMTM-11-2017-0231

Roy, S., Dan, P., & Modak, N. (2018). Effect of teamwork culture on NPD team's capability in Indian engineering manufacturing sector. *Management Science Letters*, *8*(7), 767–784. doi:10.5267/j.msl.2018.5.009

Saaty, T. L. (2008). Decision making with analytic hierarchy process. *International Journal of Services Sciences*, *1*(1), 83–98. doi:10.1504/IJSSCI.2008.017590

Saaty, T.-L. (1980). *The analytic hierarchy process*. McGraw-Hill.

Sako, M., & Kato, K. (2017). Research on the Harmonization of Intellectual Property Systems from the Point of View of Japanese Inventors. *IEEE Conference Publications*, 245 – 251. 10.1109/TEMSCON.2017.7998384

Samvedi, A., Jain, V., & Chan, F. T. (2012). An integrated approach for machine tool selection using fuzzy analytical hierarchy process and grey relational analysis. *International Journal of Production Research*, *50*(12), 3211–3221. doi:10.1080/00207543.2011.560906

Sankar, S. (2012). An Economic Order Quantity Model for Nonconforming Quality Products. *Service Science*, 331–332.

Sanny, L., & Felicia, M. (2014). Strategy of optimization inventory: Case of Study in private manufacturing in construction field company in Indonesia. *Journal of Applied Sciences (Faisalabad)*, *14*(24), 3538–3546. doi:10.3923/jas.2014.3538.3546

Sarason, Y., & Tegarden, L. F. (2003). The erosion of the competitive advantage of strategic planning: A configuration theory and resource based view. *Journal of Business and Management*, *9*(1), 1.

Sarkis, J. (2012). A boundaries and flows perspective of green supply chain management. *Supply Chain Management*, *17*(2), 202–216. doi:10.1108/13598541211212924

Sayah, S., & Zehar, K. (2008). Using evolutionary computation to solve the economic load dispatch problem. *Leonardo Journal of Sciences*, *12*(12), 67–78.

Schoute, M. (2011). The relationship between product diversity, usage of advanced manufacturing technologies and activity-based costing adoption. *The British Accounting Review*, *43*(2), 120–134. doi:10.1016/j.bar.2011.02.002

Schubring, S., Lorscheid, I., Meyer, M., & Ringle, C. M. (2016). The PLS agent: Predictive modeling with PLS-SEM and agent-based simulation. *Journal of Business Research*, *69*(10), 4604–4612. doi:10.1016/j.jbusres.2016.03.052

Segura-Ramiro, A., Ríos-Mercado, R., Alvarez, A., & De Alba Romenus, K. (2007). *A location-allocation heuristic for a territory design problem in a beverage distribution firm. International Conference of Industrial Engineering*, Cancun, México.

Seuring, S. (2004). Industrial ecology, life cycles, supply chains: Differences and interrelations. *Business Strategy and the Environment*, *13*(5), 306–319. doi:10.1002/bse.418

Seuring, S., Sarkis, J., Muller, M., & Rao, P. (2008). Sustainability and supply chain management e an introduction to the special issue. *Journal of Cleaner Production*, *16*(15), 1545–1551. doi:10.1016/j.jclepro.2008.02.002

Shah, P., & Agashe, S. (2016). Review of fractional PID controller. *Mechatronics*, *38*, 29–41. doi:10.1016/j.mechatronics.2016.06.005

Shahroodi, K., Amin, K., Amini, S., & Najibzadeh, M. (2012). Application of Analytical Hierarchy Process (AHP) Technique to Evaluate and Selecting Suppliers in an Effective Supply Chain. Arabian J Busin and Manag. *RE:view*, *1*(6), 119–132.

Shahzadi, R., Sadeghi, M., & Aghaz, A. (2014). Managing Conflict in Distributed Projects. *IEEE Conference Publications*, 1061 – 1065.

Sharma Asha-Maria. (2018). *INDUSTRIE 4.0. Smart manufacturing for the future*. Germany Trade & Invest (GTAI). Retrieved May 3, 2018, from https://www.gtai.de/GTAI/Navigation/EN/Invest/Industries/Industrie-4-0/Industrie-4-0/industrie-4-0-what-is-it.html

Sharma, J. R., Rawani, A. M., & Barahate, M. (2008). Quality function deployment: A comprehensive literature review. *International Journal of Data Analysis Techniques and Strategies*, *1*(1), 78–103. doi:10.1504/IJDATS.2008.020024

Shay, J. P., & Rothaermel, F. T. (1999). Dynamic competitive strategy: Towards a multi-perspective conceptual framework. *Long Range Planning*, *32*(6), 559–572. doi:10.1016/S0024-6301(99)00073-4

Shende, V., & Kulkarni, P. (2014). Decision support system for rapid prototyping process selection. *Int. J. Sci. Res. Publ.*, *4*, 2250–3153.

Shi, L. B., Wang, R., & Yao, L. Z. (2016). Modelling and solutions of coordinated economic dispatch with wind–hydro–thermal complex power source structure. *IET Renewable Power Generation*, *11*(3), 262–270. doi:10.1049/iet-rpg.2016.0429

Shin, B. C. (2010). Preliminary test results of prototype urban maglev train. *IEEE Conference Publications*, 32 – 35.

Shingo, S. (1988). *Non-Stock Production*. Cambridge, UK: Productivity Press.

Shinno, H., & Hashizume, H. (2002). Structured method for identifying success factors in new product development of machine tools. *CIRP Annals-Manufacturing Technology*, *51*(1), 281–284. doi:10.1016/S0007-8506(07)61517-0

Sholeh, M., Ghasemi, A., & Shahbazi, M. (2018). A new systematic approach in new product development through an integration of general morphological analysis and IPA. *Decision Science Letters*, *7*(2), 181–196. doi:10.5267/j.dsl.2017.5.004

Shtangey, S., & Tereshchenko, A. (2015). Laboratory Information Management System - Information Technology for Production Quality. *IEEE Conference Publications*, 112 – 114.

Sierociuk, D. (2005). *Fractional order discrete state-space system simulink toolkit user guide.* Retrieved from http://www.ee.pw.edu.pl/~dsieroci/fsst/fsst.htm

Singh Sunhal, A., & Mangal, D. (2017). Analysis of inventory management in a supply chain by using economic order quantity (EOQ) model. *International Journal of Engineering sciences & Research Technology*, 303-308.

Singhal, P. K., Naresh, R., Sharma, V., & Kumar, G. (2014, May). Enhanced lambda iteration algorithm for the solution of large scale economic dispatch problem. In Recent Advances and Innovations in Engineering (ICRAIE), 2014 (pp. 1-6). IEEE.doi:10.1109/ICRAIE.2014.6909294

Singhry, H. B., Abd Rahman, A., & Imm, N. S. (2016). Effect of advanced manufacturing technology, concurrent engineering of product design, and supply chain performance of manufacturing companies. *International Journal of Advanced Manufacturing Technology*, *86*(1), 663–669. doi:10.100700170-015-8219-3

Sirmon, D. G., Hitt, M. A., Ireland, R. D., & Gilbert, B. A. (2011). Resource orchestration to create competitive advantage: Breadth, depth, and life cycle effects. *Journal of Management*, *37*(5), 1390–1412. doi:10.1177/0149206310385695

Sisinni, E., Saifullah, A., Han, S., Jennehag, U., & Gidlund, M. (2018). Industrial Internet of Things: Challenges, Opportunities, and Directions. *IEEE Transactions on Industrial Informatics*, 1–1. doi:10.1109/TII.2018.2852491

Siswanto, J., & Maulida, A. (2017). Validated ERP Modules Requirement for Micro, Small and Medium Enterprise Fashion Industry. *IEEE Conference Publications*, 1 – 6.

Skoutas, D., Simitsis, A., & Sellis, T. (2007). A Ranking Mechanism for Semantic Web Service Discovery. *IEEE Congress on Services*, 41-48.

Slaper, T. F., & Hall, T. J. (2011). The triple bottom line: What is it and how does it work. *Indiana Business Review*, *86*(1), 4–8.

Slatter, R. R., Husband, T. M., Besant, C. B., & Ristic, M. R. (1989). A Human-Centred Approach to the Design of Advanced Manufacturing Systems. *CIRP Annals - Manufacturing Technology, 38*(1), 461-464. doi:10.1016/S0007-8506(07)62746-2

Small, M. H., & Chen, I. J. (1997). Economic and strategic justification of AMT inferences from industrial practices. *International Journal of Production Economics*, *49*(1), 65–75. doi:10.1016/S0925-5273(96)00120-X

Smparounis, K., Mavrikios, D., Pappas, M., Xanthakis, V., Viganò, G. P., & Pentenrieder, K. (2008). A virtual and augmented reality approach to collaborative product design and demonstration. *IEEE Conference Publications*, 1 – 8.

Snee, R. D. (2010). Lean Six Sigma–getting better all the time. *International Journal of Lean Six Sigma*, *1*(1), 9–29. doi:10.1108/20401461011033130

Snell, S. A., & Dean, J. W. (1992). Integrated Manufacturing and Human Resource Management: A Human Capital Perspective. *Academy of Management Journal*, *35*(3), 467–504. doi:10.2307/256484

Soh, P.-H., & Roberts, E. B. (2005, November). Technology Alliances and Networks: An External Link to Research Capability. *IEEE Transactions on Engineering Management*, *52*(4), 419–428. doi:10.1109/TEM.2005.850727

Song, K. Y., & Chu, C. N. (2015). Effect of machining area on material removal rate in strip EDM. *International Journal of Precision Engineering and Manufacturing*, *16*(12), 2435–2440. doi:10.100712541-015-0313-9

Song, X. M., & Montoya-Weiss, M. M. (1998). Critical development activities for really new versus incremental products. *Journal of Product Innovation Management*, *15*(2), 124–135. doi:10.1016/S0737-6782(97)00077-5

Sony, M. (2018). Industry 4.0 and lean management: A proposed integration model and research propositions. *Production & Manufacturing Research*, *6*(1), 416–432. doi:10.1080/21693277.2018.1540949

Sony, M., & Naik, S. (2011). Successful implementation of Six Sigma in services: An exploratory research in India Inc. *International Journal of Business Excellence*, *4*(4), 399–419. doi:10.1504/IJBEX.2011.041059

Sony, M., & Naik, S. (2012). Six Sigma, organizational learning and innovation: An integration and empirical examination. *International Journal of Quality & Reliability Management*, *29*(8), 915–936. doi:10.1108/02656711211258535

Sotskov, Y. N., & Shakhlevich, N. V. (1995). NP-hardness of shop-scheduling problems with three jobs. *Discrete Applied Mathematics*, *59*(3), 237–266. doi:10.1016/0166-218X(95)80004-N

Stahre, J. (1995). Evaluating human/machine interaction problems in advanced manufacturing. *Jisuanji Jicheng Zhizao Xitong*, *8*(2), 143–150. doi:10.1016/0951-5240(95)00008-H

Studer, R., Benjamins, V. R., & Fensel, D. (1998). Knowledge engineering: Principles and methods. *Data & Knowledge Engineering*, *25*(1-2), 161–197. doi:10.1016/S0169-023X(97)00056-6

Subbaiah, K. V., Durga Prasad, K. G., & Narayana Rao, K. (2011). Customer-driven product planning using conjoint analysis and QFD-ANP methodology. *International Journal of Productivity and Quality Management*, *7*(3), 374–394. doi:10.1504/IJPQM.2011.039353

Suebsin, C., & Gerdsri, N. (2009). Key Factors Driving the Success of Technology Adoption: Case Examples of ERP Adoption. *IEEE Conference Publications*, 2638 - 2643. 10.1109/PICMET.2009.5261818

Suleiman, K. (2002). Factors Differentiating the Commercialization of Disruptive and Sustaining Technologies. *IEEE Transactions on Engineering Management*, *49*(4), 375–387.

Sultan, F., & Chan, L. (2000). The adoption of new technology: The case of object-oriented computing in software companies. *IEEE Transactions on Engineering Management*, *47*(1), 106–126. doi:10.1109/17.820730

Sun, Wu, Huang, & Lin. (2013). A Computing Resources Market Model and Supply-Demand Matching Mechanism. *IEEE Conference Publications*, 217 – 224.

Sun, X., & Li, Y. (2010). An intelligent multi-criteria decision support system for systems design. 10th AIAA Aviation Technology, Integration, and Operations (ATIO) Conference, 1–11. doi:10.2514/6.2010-9222

Sun, H., & Wing, W. C. (2005). Critical success factors for new product development in the Hong Kong toy industry. *Technovation*, *25*(3), 293–303. doi:10.1016/S0166-4972(03)00097-X

Su, Y., & Yang, C. (2010). A structural equation model for analyzing the impact of ERP on SCM. *Expert Systems with Applications*, *37*(1), 456–469. doi:10.1016/j.eswa.2009.05.061

Swink, M., & Nair, A. (2007). Capturing the competitive advantages of AMT: Design–manufacturing integration as a complementary asset. *Journal of Operations Management*, *25*(3), 736–754. doi:10.1016/j.jom.2006.07.001

Sycara, K. P., Paolucci, M., Ankolekar, A., & Srinivasan, N. (2003). Automated Discovery, Interaction and Computation of Semanticmweb Services. *Journal of Web Semantics*, *1*(1), 27–46. doi:10.1016/j.websem.2003.07.002

Sycara, K., Widoff, S., Klusch, M., & Lu, J. (2002). LARKS: Dynamic Matching Among Heterogeneous Software Agents in Cyberspace. *Autonomous Agents and Multi-Agent Systems*, *5*(2), 173–203. doi:10.1023/A:1014897210525

Tabassum, Siddik, Shoyaib, & Khaled. (2014). Determining Interdependency Among Non-functional Requirements to Reduce Conflict. *IEEE Conference Publications*, 1 – 6.

Takada, N. (2016). Untangling the Boundaries in Technology Collaborations: The Deviation Effects of "Project Autonomy" on Innovations through Collaborations. *IEEE Conference Publications*, 399 – 409. 10.1109/PICMET.2016.7806633

Tang, Z., Hill, D. J., & Liu, T. (2018). A Novel Consensus-Based Economic Dispatch for Microgrids. *IEEE Transactions on Smart Grid*, *9*(4), 3920–3922. doi:10.1109/TSG.2018.2835657

Taoudi, A., Bounabat, B., & Elmir, B. (2013). Quality based Project Control using Interoperability degree as a Quality Factor. *IEEE Conference Publications*, 1 - 6. 10.1109/ISKO-Maghreb.2013.6728117

Tavares-Pereira, F., Figueira, J., Mousseau, V., & Roy, B. (2007). Multiple criteria districting problems: The public transportation network pricing system of the Paris region. *Annals of Operations Research*, *154*(1), 69–92. doi:10.100710479-007-0181-5

Teixeira, A. A., Jabbour, C. J. C., de Sousa Jabbour, A. B. L., Latan, H., & de Oliveira, J. H. C. (2016). Green training and green supply chain management: Evidence from Brazilian firms. *Journal of Cleaner Production*, *116*, 170–176. doi:10.1016/j.jclepro.2015.12.061

Tepljakov, A., Petlenkov, E., Belikov, J., & Finajev, J. (2013). *Fractional-order controller design and digital implementation using FOMCON toolbox for MATLAB*. Paper presented at the Computer Aided Control System Design (CACSD), 2013 IEEE Conference on. 10.1109/CACSD.2013.6663486

Tepljakov, A. (2017a). *FOMCON: Fractional-Order Modeling and Control Toolbox. In Fractional-order Modeling and Control of Dynamic Systems* (pp. 107–129). Springer. doi:10.1007/978-3-319-52950-9_6

Tepljakov, A. (2017b). *Fractional-order modeling and control of dynamic systems*. Springer. doi:10.1007/978-3-319-52950-9

Tepljakov, A., Petlenkov, E., & Belikov, J. (2011). FOMCON: A MATLAB toolbox for fractional-order system identification and control. *International Journal of Microelectronics and Computer Science*, *2*(2), 51–62.

Teunter, R. H., Babai, M. Z., & Syntetos, A. A. (2010). *ABC Classification: Service Levels and Inventory Costs*. Production and Operations Management Society.

Thames, L., & Schaefer, D. (2017). Industry 4.0: an overview of key benefits, technologies, and challenges. In Cybersecurity for Industry 4.0 (pp. 1–33). Springer.

Thomas, A. J., Barton, R., & John, E. G. (2008). Advanced manufacturing technology implementation: A review of benefits and a model for change. *International Journal of Productivity and Performance Management*, *57*(2), 156–176. doi:10.1108/17410400810847410

Thomas, A. J., Mason-Jones, R., Davies, A., & John, E. G. (2015). Reducing turn-round variability through the application of Six Sigma in aerospace MRO facilities. *Journal of Manufacturing Technology Management*, *26*(3), 314–332. doi:10.1108/JMTM-05-2013-0052

Tieng, K., Jeenanunta, C., Rittippant, N., Chongpisal, P., & Hamada, R. (2016). The influences of knowledge management on innovation within formal and non-formal R&D manufacturing firms, Thailand. *IEEE Conference Publications*, 208 – 213. 10.1109/ICMIT.2016.7605035

Timans, W., Ahaus, K., van Solingen, R., Kumar, M., & Antony, J. (2016). Implementation of continuous improvement based on Lean Six Sigma in small-and medium-sized enterprises. *Total Quality Management & Business Excellence*, *27*(3–4), 309–324. doi:10.1080/14783363.2014.980140

Toro, L. A., & Bastidas, V. E. (2011). Metodología para el control y la gestión de inventarios en una empresa minorista de electrodomésticos. *Sciences et Techniques (Paris)*, 85.

Tosun, N. (2006). Determination of optimum parameters for multi-performance characteristics in drilling by using grey relational analysis. *International Journal of Advanced Manufacturing Technology, 28*(5-6), 450–455. doi:10.100700170-004-2386-y

Tripathy, S., Kumar Ray, P., & Sahu, S. (2012). Factors governing R&D practices in Indian manufacturing firms: Structural equation modelling. *International Journal of Modelling in Operations Management, 2*(1), 45–68. doi:10.1504/IJMOM.2012.043960

Troffaes, M., & Sahlin, U. (2017). *Imprecise swing weighting for multi-attribute utility elicitation based on partial preferences.* PMLR.

Tsekouras, G. J., Kanellos, F. D., Mastorakis, N. E., & Mladenov, V. (2013, September). Optimal Operation of Electric Power Production System without Transmission Losses Using Artificial Neural Networks Based on Augmented Lagrange Multiplier Method. In *International Conference on Artificial Neural Networks* (pp. 586-594). Springer. 10.1007/978-3-642-40728-4_73

Tung, N. S., & Chakravorty, S. (2015). Grey Wolf Optimization for Active Power Dispatch Planning Problem Considering Generator Constraints and Valve Point Effect. *International Journal of Hybrid Information Technology, 8*(12), 117–134. doi:10.14257/ijhit.2015.8.12.07

Tzokas, N., Hultink, E. J., & Hart, S. (2004). Navigating the new product development process. *Industrial Marketing Management, 33*(7), 619–626. doi:10.1016/j.indmarman.2003.09.004

Ulrich, K. T., & Eppinger, S. D. (2000). *Product Design and Development* (2nd ed.). Boston: Irwin/McGraw-Hill.

Valencia, J., Lambán, M. P., & Royo, J. (2014). Modelo analítico para determinar lotes óptimos de producción considerando diversos factores productivos y logísticos. Red de Revistas Científicas de América Latina, el Caribe, España y Portugal, 62.

Valério, D., & Da Costa, J. S. (2004). Ninteger: a non-integer control toolbox for MatLab. *Proceedings of the Fractional Differentiation and its Applications.*

Valério, D., Trujillo, J. J., Rivero, M., Machado, J. A. T., & Baleanu, D. (2013). Fractional calculus: A survey of useful formulas. *The European Physical Journal. Special Topics, 222*(8), 1827–1846. doi:10.1140/epjst/e2013-01967-y

Varelas, G., Voutsakis, E., Raftopoulou, P., Petrakis, E. G. M., & Milios, E. (2005). Semantic Similarity methods in WordNet and their application to information retrieval on the Web. *Proceedings of the 7th annual ACM international workshop on web information and data management.* 10.1145/1097047.1097051

Vélez, M., & Castro, C. (2002). Modelo de Revisión Periódica para el Control del Inventario en Artículos con Demanda Estacional una Aproximación desde la Simulación. Red de Revistas Científicas de América Latina, el Caribe, España y Portugal, 23-24.

Verma, K., Sivashanmugam, K., Sheth, A., Patil, A., Oundhakar, S., & Miller, J. (2005). METEOR-S WSDI: A Scalable P2P Infrastructure of Registries for Semantic Publication and Discovery of Web Services. *Information Technology Management, 6*(1), 17–39. doi:10.100710799-004-7773-4

Verworn, B. (2009). A structural equation model of the impact of the "fuzzy front end" on the success of new product development. *Research Policy, 38*(10), 1571–1581. doi:10.1016/j.respol.2009.09.006

Vijayaraj, S., & Santhi, R. K. (2016, March). Multi-area economic dispatch using flower pollination algorithm. In *Electrical, Electronics, and Optimization Techniques (ICEEOT), International Conference on* (pp. 4355-4360). IEEE. 10.1109/ICEEOT.2016.7755541

Vila, M., & Pereira, J. (2014). A branch-and-bound algorithm for assembly line worker assignment and balancing problems. *Computers & Operations Research*, *44*, 105–114. doi:10.1016/j.cor.2013.10.016

Vinagre, B. M., Podlubny, I., Hernandez, A., & Feliu, V. (2000). Some approximations of fractional order operators used in control theory and applications. *Fractional Calculus & Applied Analysis*, *3*(3), 231–248.

Vinodh, S., Gautham, S. G., & Ramiya, R. A. (2011). Implementing lean sigma framework in an Indian automotive valves manufacturing organisation: A case study. *Production Planning and Control*, *22*(7), 708–722. doi:10.1080/09537287.2010.546980

Vinodh, S., Kumar, S. V., & Vimal, K. E. K. (2014). Implementing lean sigma in an Indian rotary switches manufacturing organisation. *Production Planning and Control*, *25*(4), 288–302. doi:10.1080/09537287.2012.684726

Vrijhoef, R., & Koskela, L. (1999). Role of supply chain management in construction. *Proceedings of the Seventh Annual Conference of the International Group for Lean Construction*, 133-146.

Waldeck, N. E., & Leffakis, Z. M. (2007). HR perceptions and the provision of workforce training in an AMT environment: An empirical study. *Omega*, *35*(2), 161–172. doi:10.1016/j.omega.2005.05.001

Walsh, Kirchhoff, & Newbert. (2002). Differentiating market strategies for disruptive technologies. IEEE Transactions on Engineering Management, 49(4), 106 – 126.

Wang, C. H. (2014). Integrating correspondence analysis with Grey relational model to implement a user-driven STP product strategy for smart glasses. *Journal of Intelligent Manufacturing*, *27*(5), 1007–1016. doi:10.100710845-014-0931-6

Wang, C.-H., Huang, S.-Z., Chang, C.-H., Lin, P.-J., & Chiew, Y.-Y. (2015). Co-innovation network driven entrepreneurship in high-tech technology-evidences from China. *IEEE Conference Publications*, 1002 – 1015. 10.1109/PICMET.2015.7273122

Wang, J., Liang, Z., & Xue, L. (2014). Multinational R&D in China: Differentiation and integration of global R&D networks. *International Journal of Technology Management*, *65*(1-4), 96–124. doi:10.1504/IJTM.2014.060959

Wang, L., & Cao, J. (2007). Web Services Semantic Searching enhanced by Case Based Reasoning. *18th International Workshop on Database and Expert Systems Applications.*

Wang, L., & Shen, B. (2016). A Study of Design Collaboration between the Designer and Supplier in the Fashion Supply Chain. *IEEE Conference Publications*, 1 – 3.

Wang, P., Jin, Z., Liu, L., & Cai, G. (2008). Building Toward Capability Specifications of Web Services Based on an Environment Ontology. *IEEE Transactions on Knowledge and Data Engineering*, *20*(4), 547–561. doi:10.1109/TKDE.2007.190719

Wang, T., & Ji, P. (2010). Understanding customer needs through quantitative analysis of Kano's model. *International Journal of Quality & Reliability Management*, *27*(2), 173–184. doi:10.1108/02656711011014294

Wang, X., & Chan, H. K. (2013). A hierarchical fuzzy TOPSIS approach to assess improvement areas when implementing green supply chain initiatives. *International Journal of Production Research*, *51*(10), 37–41. doi:10.1080/00207543.2012.754553

Wang, X., & Triantaphyllou, E. (2008). Ranking irregularities when evaluating alternatives by using some ELECTRE methods. *Omega*, *36*(1), 45–63. doi:10.1016/j.omega.2005.12.003

Wang, Y. M., & Luo, Y. (2009). On rank reversal in decision analysis. *Mathematical and Computer Modelling*, *49*(5-6), 1221–1229. doi:10.1016/j.mcm.2008.06.019

Wan, W. K., & Monsef, S. (2012). New Product Development Through Open Innovation: Role of organization structure and contextual factors. *IEEE Conference Publications*, 446 – 449.

Watson, I. (1997). *Applying Case-Based Reasoning: Techniques for Enterprise Systems*. Morgan Kaufman.

Weber, M., & Borcherding, K. (1993). Behavioral influences on weight judgments in multiattribute decision making. *European Journal of Operational Research*, *67*(1), 1–12. doi:10.1016/0377-2217(93)90318-H

WEF. (2016). The future of jobs: Employment, skills and workforce strategy for the fourth industrial revolution. *World Economic Forum*.

Wei, C., & Jun-fu, C. (2011). Theoretical discussion of applying grey system theory in neuropsychological studies. *Grey Systems. Theory and Application*, *1*(3), 268–273.

Wen, K. L. (2004). The grey system analysis and its application in gas breakdown and var compensator finding. *International Journal of Computational Cognition*, *2*(1), 21–44.

Werner, F. (2013). A survey of genetic algorithms for shop scheduling problems. In Heuristics: Theory and Applications. Nova Science Publishers.

Wernerfelt, B. (1984). A resource-based view of the firm. *Strategic Management Journal*, *5*(2), 171–180. doi:10.1002mj.4250050207

Whitley, D. (1994). A genetic algorithm tutorial. *Statistics and Computing*, *4*(2), 65–85. doi:10.1007/BF00175354

Whittemore, R., & Knafl, K. (2005). The integrative review: Updated methodology. *Journal of Advanced Nursing*, *52*(5), 546–553. doi:10.1111/j.1365-2648.2005.03621.x

Wipf, A. (2017). *Professional Higher Education 4.0: A Change for Universities of Applied Sciences*. EURASHE. European Association of Institutions in Higher Education. Available at: https://www.eurashe.eu/library/mission-phe/EURASHE_AC_LeHavre_170330-31_portfolio.pdf

Wood, A. J., & Wollenberg, B. F. (2012). *Power Generation, Operation, and Control*. John Wiley & Sons.

Wooldridge, M., & Jennings, N. (1995). Intelligent Agents: Theory and Practice. *The Knowledge Engineering Review*, *10*(2), 115–152. doi:10.1017/S0269888900008122

WSDL-S. (2005). Retrieved from http://www.w3.org/Submission/WSDL-S/

Wu, Z. L., Wu, Q. H., Zhou, X. X., & Li, M. S. (2015, November). Hybrid quadratic programming and compact formulation method for economic dispatch with prohibited operating zones and network losses. In Innovative Smart Grid Technologies-Asia (ISGT ASIA), 2015 IEEE (pp. 1-6). IEEE.doi:10.1109/ISGT-Asia.2015.7386963

Wu, X., Ma, T., & Tian, Y. (2009). The Empirical Impact of Internal Knowledge Structure on the Endogenous Industrial Clusters Innovation - With Wuqueqiao industrial cluster in Suzhou city as example. *IEEE Conference Publications*, 721 – 725.

Wu, Z., Ding, J., Wu, Q. H., Jing, Z., & Zheng, J. (2017). Reserve constrained dynamic economic dispatch with valve-point effect: A two-stage mixed integer linear programming approach. *CSEE Journal of Power and Energy Systems*, *3*(2), 203–211. doi:10.17775/CSEEJPES.2017.0025

Xia, H., Li, Q., Xu, R., Chen, T., Wang, J., Hassan, M. A. S., & Chen, M. (2018). Distributed Control Method for Economic Dispatch in Islanded Microgrids With Renewable Energy Sources. *IEEE Access: Practical Innovations, Open Solutions*, *6*, 21802–21811. doi:10.1109/ACCESS.2018.2827366

Xiao, R., Wu, Z., & Sugiura, K. (2013). Slideware2.0: A prototype of presentation system by integrating web2.0 and second screen to promote education communication. *IEEE Conference Publications*, 181 - 185. 10.1109/ICICM.2013.38

Xiong, H., Fan, H., Jiang, G., & Li, G. (2017). A simulation-based study of dispatching rules in a dynamic job shop scheduling problem with batch release and extended technical precedence constraints. *European Journal of Operational Research, 257*(1), 13–24. doi:10.1016/j.ejor.2016.07.030

Xue, D. (2017). *Fractional-order Control Systems: Fundamentals and Numerical Implementations* (Vol. 1). Walter de Gruyter GmbH & Co KG. doi:10.1515/9783110497977

Yadav, G., Seth, D., & Desai, T. N. (2017). Analysis of research trends and constructs in context to lean six sigma frameworks. *Journal of Manufacturing Technology Management, 28*(6), 794–821. doi:10.1108/JMTM-03-2017-0043

Yan, T., & Azadegan, A. (2017). Comparing inter-organizational new product development strategies: Buy or ally; Supply-chain or non-supply-chain partners? *International Journal of Production Economics, 183*(Part A), 21-38. doi:10.1016/j.ijpe.2016.09.023

Yang, P., Zhu, C., Zhao, L., Wang, M., Ning, X., & Liu, Y. (2016, November). Solving dynamic economic dispatch with valve point effect by a two-step method. In Region 10 Conference (TENCON), 2016 IEEE (pp. 546-550). IEEE. doi:10.1109/TENCON.2016.7848060

Yazdani, M., Zandieh, M., Tavakkoli-Moghaddam, R., & Jolai, F. (2015). Two meta-heuristic algorithms for the dual-resource constrained flexible job-shop scheduling problem. *Scientia Iranica. Transaction E. Industrial Engineering (American Institute of Industrial Engineers), 22*(3), 1242.

Yousfi, N., Melchior, P., Rekik, C., Derbel, N., & Oustaloup, A. (2012). Design of Centralized CRONE Controller Combined with MIMOQFT Approach Applied to Non Square Multivariable Systems. *International Journal of Computer Applications, 45*(16).

Yuan, L.-S., & Dong, R.-P. (2015). A Novel Risk Assessment Algorithm for Accounting Information System Using Analytic Hierarchy Process. *IEEE Conference Publications*, 61 – 64.

Yuzuki, R., Hata, H., & Matsumoto, K. (2015). How We Resolve Conflict: An Empirical Study of Method-Level Conflict Resolution. *IEEE Conference Publications*, 21 – 24. 10.1109/SWAN.2015.7070484

Zanakis, S. H., Solomon, A., Wishart, N., & Dublish, S. (1998). Multi-attribute decision making: A simulation comparison of select methods. *European Journal of Operational Research, 107*(3), 507–529. doi:10.1016/S0377-2217(97)00147-1

Zhang, G., Gao, L., & Shi, Y. (2011). An effective genetic algorithm for the flexible job-shop scheduling problem. *Expert Systems with Applications, 38*(4), 3563–3573. doi:10.1016/j.eswa.2010.08.145

Zhang, L., & Swirski, M. (2002). Managing the Specification Process in Complex Projects. *IEEE Conference Publications*, 350 – 355. 10.1109/IEMC.2002.1038456

Zhang, L., Yuan, Y., Chen, B., & Su, D. (2017, May). A robust interval economic dispatch model accommodating large-scale wind power generation with consideration of price-based demand response. In *Control And Decision Conference (CCDC), 2017 29th Chinese* (pp. 6679-6684). IEEE.10.1109/CCDC.2017.7978379

Zhang, W. Y., Cai, M., Qiu, J., & Yin, J. W. (2009). Managing distributed manufacturing knowledge through multi-perspective modelling for Semantic Web applications. *International Journal of Production Research, 47*(23), 6525–6542. doi:10.1080/00207540802311114

Zhang, Y., Hajiesmaili, M. H., Cai, S., Chen, M., & Zhu, Q. (2018). Peak-aware online economic dispatching for microgrids. *IEEE Transactions on Smart Grid, 9*(1), 323–335. doi:10.1109/TSG.2016.2551282

Zhan, J. P., Wu, Q. H., Guo, C. X., & Zhou, X. X. (2014). Fast lambda-Iteration Method for Economic Dispatch With Prohibited Operating Zones. *IEEE Transactions on Power Systems*, *29*(2), 990–991. doi:10.1109/TPWRS.2013.2287995

Zhao, C. L., Ming, X. G., Wang, X. H., & Li, D. (2009). A Framework of Supplier Involved Collaborative Project Management. *IEEE Conference Publications*, 4130 - 4135. 10.1109/ICISE.2009.36

Zhu, Q., Geng, Y., Fujita, T., & Hashimoto, S. (2010). Green supply chain management in leading manufacturer's case studies in Japanese large companies. *Management Research Review*, *4*, 380–392. doi:10.1108/01409171011030471

Zitzler, E., & Kunzli, S. (2004). *Indicator-based selection in multiobjective search. In PPSN VIII Proceedings* (pp. 832–842). Springer.

Zitzler, E., Thiele, L., Laumanns, M., Fonseca, C. M., & Grunert da Fonseca, V. (2003). Performance assessment of multiobjective optimizers: An analysis and review. *IEEE Transactions on Evolutionary Computation*, *7*(2), 117–132. doi:10.1109/TEVC.2003.810758

Zsidisin, G. A., & Siferd, S. P. (2001). Environmental purchasing: A framework for theory development. *European Journal of Purchasing and Supply Management*, *7*(1), 61–73. doi:10.1016/S0969-7012(00)00007-1

About the Contributors

Prasanta Sahoo received a Ph.D. degree in Mechanical Engineering and M.Tech. degree in Mechanical Engineering (Machine Design) both from Indian Institute of Technology, Kharagpur, Bachelor of Mechanical Engineering degree, from Jadavpur University, Kolkata. Currently, he is Professor at the Department of Mechanical Engineering of Jadavpur University, Kolkata, India. He has more than 25 years of teaching and research experience in Manufacturing, Materials and Mechanical Engineering, with special emphasis in Tribology of Materials. He has also interest in Structural Mechanics, Contact Mechanics, Fractal Modeling, Composites and Coatings. He has guided large numbers of Ph.D. and master's students as well as coordinated & participated in several financed research projects. He has worked as evaluator of projects for international research agencies as well as examiner of Ph.D. thesis for many universities and Institutes. He is on the Editorial Board of several international journals and conferences. Presently, he acts as reviewer for more than 60 prestigious Web of Science journals. In addition, he has also authored 2 books, over 20 book chapters and 400 articles in peer-reviewed journals and conference proceedings (h-index 25+/3000+ citations).

* * *

Ahmad Al-Subhi is an assistant professor at electrical power technology department at Yanbu Industrial College in Royal Commission of Yanbu. He got his BS, MS and PhD in electrical engineering from King Fahd University of Petroleum and Minerals at 2005, 2012 and 2018, respectively. His interests include power system modeling and forecasting, control of renewable energy sources and DC grid analysis and investigation.

Hesham K. Alfares is Chairman and Professor in the Systems Engineering Department at King Fahd University of Petroleum & Minerals, Dhahran, Saudi Arabia. He has a PhD in Industrial Engineering from Arizona State University. He has more than 100 publications, including 50 journal papers and a U.S. patent. Dr. Alfares has been as a visiting scholar at MIT and the University of North Carolina in the USA, in addition to four universities in the UK. He won one research grant from the US Fulbright Foundation, and four grants from the British Council. He won the Almarai Prize for Scientific Innovation in Industrial Engineering. He has been a member in the editorial boards of three international journals and in the program committees of 27 international conferences.

Liliana Avelar-Sosa has a bachelor degree in Electronic Engineering, a master degree in Industrial Sciences, a PhD in Engineering Sciences. She is a National Researcher level 1 recognized by National Council of Science and Technology in Mexico, expert in logistics and supply chain aspects applied to manufacturing companies.

Sumit Bhowmik is working as an Assistant Professor in Department of Mechanical Engineering, National Institute of Technology Silchar, Assam, India. He received his PhD in Mechanical Engineering with specialization in Design from Jadavpur University in 2014. Composite material, application of optimization techniques and mechanical behaviour of material under different constraint are the current research area of interest. Dr. Bhowmik published no. of SCI paper and book chapters and guided PG and PhD scholar. He has over 10 years of teaching and research experience.

Patricia Cano-Olivos is a professor-researcher in the Department of Logistics and Supply Chain Management at Universidad Popular Autonoma del Estado de Puebla (UPAEP) in Mexico. She received a Ph.D. in Logistics and Supply Chain Management. Her research interests are risks in the Supply Chain and Logistics management models.

K. B. Chuah received degree in mechanical engineering from Loughborough University, Leicestershire, U.K. He received the Ph.D. degree as an award for his work on a U.S. Navy-funded project on roughness and hydrodynamic drag. He was a Senior Research Coordinator managing a multidisciplinary computer integrated manufacturing research team at the University of Teesside, Middlesbrough, U.K. before starting his lecturing career. He is currently an Associate Professor and Associate Head of the Department of Systems Engineering and Engineering Management, City University of Hong Kong, Hong Kong, China. Apart from being a core faculty member of various U.G. and P.G. programs, he has also supervised many M.Phil., Ph.D., and EngD candidates in mechanical engineering, engineering management, product/industrial strategy, and organizational learning.

Biplab Das is an Assistant Professor at the Department of Mechanical Engineering, National Institute of Technology Silchar Assam, India. He obtained Ph.D. in Mechanical Engineering with specialization in Thermal Engineering from NERIST, India. His areas of interests are Solar Thermal, Heat Transfer, CFD, Alternative fuel, Wind energy, and optimization. He has over 2 years of industrial experience and 8 years of teaching and research experience. He received an Institutional Award (Gold Medal) by institution of engineers India and best innovative award by springer for his outstanding research contribution and published more than 14 research papers and 3 book chapters. Currently, he is doing research in Solar Thermal, Heat Transfer, CFD, Alternative fuel, Wind energy optimization and decision-making tools.

K. G. Durga Prasad is a Professor of Mechanical Engineering, Gayatri Vidya Parishad School of Engineering, Technical Campus, GVP College for Degree & P.G.Courses (A), Visakhapatnam, Andhra Pradesh, India. He published 24 papers in various International and National Journals and 15 papers in International/ National Conferences. He also published 2 book chapter, one monograph and 2 text books for Diploma students. He is a Charted Engineer and a Life Member of Institution of Engineers (India). His research areas of interests are Product design and development, Supply Chain Management, Multi Criteria Decision Making, Composite materials.

Jorge Luis Garcia-Alcaraz is a full time researcher at Department of Industrial Engineering at the Universidad Autónoma de Ciudad Juárez. He received a MSc in Industrial Engineering from the Instituto Tecnológico de Colima (Mexico), a PhD in Industrial Engineering from Instituto Tecnológico de Ciudad Juárez (Mexico), a PhD in Innovation in Product Engineering and Industrial Process and a Postdoc in Manufacturing Process from University of La Rioja (Spain). His main research areas are Multicriteria decision making applied to lean manufacturing, production process modeling and statistical inference. He is founding member of the Mexican Society of Operation Research and active member in the Mexican Academy of Industrial Engineering. Currently, Dr. Garcia is a National Researcher recognized by the National Council of Science & Technology of Mexico (CONACYT) and is author/coauthor in around 50 papers published in journals indexed in the Journal Citation Reports, more than 150 international conferences and congress around the words. Dr. Garcia is author of several books published by international publishers as Springer and IGI Global, all then related to lean manufacturing and its tools and techniques. orcid.org/0000-0002-7092-6963, Scopus Author ID: 55616966800.

Suresh Gudala is a M.Tech scholar of Department of Mechanical Engineering, National Institute of Technology Silchar, Assam, India. He received his M.Tech (Design and Manufacturing) from National Institute of Technology Silchar Assam India. His areas of interests are rapid prototyping, optimization, decision making tools, etc.

Vivekananda Rajah Harindran received the M.Sc. degree in Technology Management, from Staffordshire University, UK, in 2010. He joined PETRONAS Group Technical Solutions in 1992, where he is currently the Sr. Custodian Engineer in the division of Instrumentation and Control. He has 27 years' experience in petrochemical, oil and gas industries with skills in project design, improving plant production, reliability and availability. His current research interests include design and implementation of instrumentation systems and control.

Sabo Miya Hassan is a graduate assistant at the department of electrical and electronic engineering, Universiti Teknologi PETRONAS, Malaysia. He received the B.Eng. (Hons.) degree in electrical and electronic engineering from Abubakar Tafawa Balewa University (ATBU), Bauchi, Nigeria, in 2008, and the M.Sc. Eng. (Hons.) degree in control systems from the University of Sheffield, U.K., in 2011. He is with the Department of Electrical and Electronic Engineering, ATBU. He is currently pursuing the Ph.D. degree in the electrical and electronic engineering department, Universiti Teknologi PETRONAS, Perak, Malaysia. His current research interests include wireless networked control systems, intelligent control, soft computing and optimization.

S. C. Ho received the Bachelor's Degree in Logistics and Supply Chain Management from University of Northumbria at Newcastle, Master's Degree in Global Supply Chain Management and Master's Degree in Integrated Engineering from The Hong Kong Polytechnic University. Before he started his Engineering Doctorate Degree studies, he was a Manager, Program & Projects Operations at Hong Kong R&D Centre for Logistics and Supply Chain Management Enabling Technologies (LSCM). His current research interests include innovation systems management, R&D technology commercialization, process and system re-engineering. He is a committee member in Education and Training Committee & Logistics Policy Committee and Chartered Member of The Chartered Institute of Logistics and Transport in Hong Kong (CILTHK).

Rosdiazli Ibrahim received the B.Eng. degree in electrical engineering from Universiti Putra Malaysia, Kembangan, Malaysia, in 1996, the M.Sc. degree in automation and control from Newcastle University, Newcastle upon Tyne, U.K., in 2000, and the Ph.D. degree in electrical and electronic engineering from the University of Glasgow, Glasgow, U.K., in 2008. He is an Associate Professor with the Department of Electrical and Electronics Engineering at Universiti Teknologi PETRONAS (UTP), Seri Iskandar, Perak, Malaysia. He is currently, The Dean, Centre for Graduate Studies, UTP. His current research interests include intelligent control and non-linear multi-variable process modeling for control application.

Jagadish is an Assistant Professor at the Department of Mechanical Engineering, National Institute of Technology Raipur Chhattisgarh, India. He obtained Ph.D. in Mechanical Engineering with specialization in Manufacturing Engineering from National Institute of Technology Silchar and Post graduation in Product Design and Development from National Institute of Technology Warangal, India. His areas of interests are advanced manufacturing process, polymer machining, green manufacturing, computer aided design, and applied soft computing techniques. He has over 3 years of industrial experience in the field of design and analysis and 5 years of teaching and research experience. He received an Institutional Award (Gold Medal) by institution of engineers India and best innovative award by springer for his outstanding research contribution and published more than 14 research papers and 3 book chapters. Currently, he is doing research in advanced machining process, green manufacturing, polymer machining, and decision-making tools.

Emilio Jiménez-Macías is a full Profesor at Department of Electrical Engineering in the University of La Rioja (Spain), where he is also currently head of the Modeling and Simulation in Science and Engineering official research group, Coordinator of the System Engineering and Automation Area, and Director of the Master on Industrial Engineering. He is also the President of the Spanish Modelling and Simulation Group, and President of EUROSIM (the Federation of European Simulation Societies), as well as the Spanish representative in the Discrete Event and Hybrid Systems Technical Committee (TC 1.3) of IFAC (International Federation of Automatic Control). His main research areas include Simulation, Systems Engineering and Automation, Industry 4.0. Prof. Emilio Jiménez has published 78 papers in Journals indexed in Journal Citation Report (JCR), and more than 150 works indexed in Scopus (with more than 1000 cites and h-index 17), has participated in more than 40 research projects, has advised 25 PhD Thesis, has elaborated more than 70 patents, has organized 5 international conferences, and has received several awards for his main research works. ORCID orcid.org/0000-0001-6749-4592, Scopus Author ID: 57200162568.

Marcell S. Kalman is Operations Director and Partner at ACOR furniture since January 2010. He is responsible for the manufacturing and planning operations of the company leading a team of 6 production and purchasing engineers and a manufacturing staff of 75 workers involving a hybrid techno-craft process. Marcell has previously held operations and production planning related positions as manufacturing manager of a packaging goods manufacturing company. Marcell earned a degree in industrial and systems engineering from Universidad Panamericana campus Guadalajara, graduating in 2005 and currently working on an Engineering Master's degree also at Universidad Panamericana.

Mohd Noh Karsiti received the B.Eng. degree in electrical engineering from California State University, Long Beach, CA, in 1985, the M.Sc. degree in electrical engineering (control) from California State University, Long Beach, CA, in 1987, and the Ph.D. degree in electrical and computer engineering (Hierarchical control) from University of California, Irvine, CA in 1991. He joined Universiti Teknologi PETRONAS, Perak, Malaysia, in 1999, where he is currently an Associate Professor. His current research interests include advanced process control, robotics systems manipulator design and control, control of highly nonlinear systems and automatic compensation of control systems.

Bingi Kishore received the B.Tech. (Hons.) degree in Electrical & Electronics Engineering from Bapatla Engineering College (BEC), Bapatla, Andhra pradesh, India, in 2012, and the M.Tech (Hons.) degree in Instrumentation and Control Systems from National Institute of Technology (NIT) Calicut, Calicut, Kerala, India, in 2014. He worked with TATA consultancy service as an Assistant systems Engineer from 2015 to 2016. He is currently pursuing the Ph.D. degree with the Electrical and Electronic Engineering Department, Universiti Teknologi Petronas (UTP), Perak, Malaysia. His current research interests include process modeling, control and optimization.

Aide Maldonado-Macías completed her Bachelor Degree in Industrial Engineering, the MSc in Industrial Engineering by the Technological Institute of Ciudad Juarez and Ph. D. in Sciences in Industrial Engineering by the same Institute. She is currently a full-time Professor-Investigator at the Autonomous University of Ciudad Juarez (UACJ). She has published in International Journal of Industrial Engineering and International Journal of Advanced Manufacturing Technology, Expert Systems with Applications, Work, Journal of Accounting and Management, Journal of Educational Research among others. She has also collaborated with book chapters for Springer, IGI Global.

B. L. Manasa is working as Assistant Professor in Mechanical Engineering Department, University College of Engineering, JNTUK, Kakinada, Andhra Pradesh, India. She has published 6 papers in International and National journals and 2 conference papers. Her teaching and research interests are Robotics, Machine design, Finite element analysis, Product design and development.

José-Luis Martínez-Flores is a researcher and Academic Director of the Postgraduate in Logistics and Supply Chain Management at Universidad Popular Autonoma del Estado de Puebla (UPAEP) in Mexico. He received a Ph.D. in Mathematics. His objectives in research are the optimization and implementation of models for problems in the field of logistics through the use of information technologies.

Pilar Novo Ibarra has a Bachelor's in Design Engineering from the Universidad Panamericana, Mexico, with specialization in operational research and optimization. She has interests in data analysis, optimization, and industrial design. She has work experience in Business Analytics at a large direct sales company and in Sales Support and Analysis at an important electronic component sourcing company.

Samuel Nucamendi-Guillén holds a Master of Science in Quality Systems and Productivity, as well as a Doctorate in Engineering Sciences, both awarded by the Tecnológico de Monterrey (with doctorate research in the intelligent computing group of the Universidad de La Laguna, in Tenerife, Spain). He is certified as Black Belt in Six Sigma by "Breakthrough Management Group International (BMGI)", an organization in which he also participated as an external consultant. He has also joined as an instructor for the "Faculty of Physical-Mathematical Sciences" of the UANL. He also has work experience in manufacturing and telecommunications companies. Since 2015, he works as a research professor at the Faculty of Engineering of the Universidad Panamericana, Campus Guadalajara. Additionally, since 2012, he is a member of the Mexican Operations Research Society (SMIO). He has several publications in peer-reviewed scientific journals, as well as participation in various international conferences.

Elias Olivares-Benitez is a Professor of Industrial Engineering at Universidad Panamericana, Mexico. He has a BSc degree on Metallurgical Engineering by Instituto Politecnico Nacional (Mexico). He has a MSE degree on Industrial Engineering by Arizona State University (USA). He has a MSc degree on Manufacturing Systems and a PhD in Engineering by Tecnologico de Monterrey (Mexico). He worked as Process and Quality Engineer in Steelmaking, Heat Treatment and Quality Labs at TAMSA (TECHINT-TENARIS). He was professor of Industrial Engineering at Tecnologico de Monterrey (Aguascalientes Campus) and UPAEP University. He has research interests in optimization, metaheuristics, supply chain management, logistics, and production planning. He has President of publications in many international journals. He is a member of the Mexican Society of Operations Research. https://orcid.org/0000-0001-7943-3869.

Pritam Pain has completed his B.Tech degree from WBUT in 2014 and then completed his M.Tech degree from Haldia Institute of Technology in 2016. He is currently working as assistant professor in Haldia Institute of Technology. He had several journal papers and book chapters regarding Non-Traditional machining. His main interest of research is in nature inspired modern optimization algorithms.

Kamalendu Pal is with the Department of Computer Science, School of Mathematics, Computer Science and Engineering, City University London. Kamalendu received his BSc (Hons) degree in Physics from Calcutta University, India, Postgraduate Diploma in Computer Science from Pune, India; MSc degree in Software Systems Technology from Sheffield University, Postgraduate Diploma in Artificial Intelligence from Kingston University, MPhil degree in Computer Science from University College London, and MBA degree from University of Hull, United Kingdom. He has published dozens of research papers in international journals and conferences. His research interests include knowledge-based systems, decision support systems, computer integrated design, software engineering, and service oriented computing. He is a member of the British Computer Society, the Institution of Engineering and Technology, and the IEEE Computer Society.

Arturo Realyvásquez-Vargas is a full time professor in the Department of Industrial Engineering at Instituto Tecnológico de Tijuana. He received a Master Degree in Industrial Engineering and a PhD in Engineering Sciences from Universidad Autónoma de Ciudad Juárez (Mexico). He has carried out research in Ergonomics since 2007. He has offered the classes of Manufacturing Systems, Projects Management, Operations Research, Fundamentals of Research, and Ergonomics. Some of his contributions in the Ergonomics field are the Macroergonomic Compatibility Questionnaire (MCQ), the Macroergonomic Compatibility Index (MCI). Moreover, he validated the construct of macroergonomic compatibility by means of structural equations models. Currently, he is a Nacional Researcher recognized by the Nacional Council of Science & Technology of Mexico (CONACYT).

Jagannath Reddy is a PhD scholar of Department of Mechanical Engineering, National Institute of Technology Silchar, Assam, India. He received his M.Tech (Thermal Engineering) from PDA College Gulbarga, Karnataka. His areas of interests are modeling of thermal systems using MCDM and soft computing techniques, green supply chain management, decision making tools etc.

Omar Rojas is a Professor of Decision Sciences at Universidad Panamericana, Mexico. He has a PhD in Mathematics by La Trobe University, Australia.

Sudeshna Roy obtained her Bachelor's Degree West Bengal University of Technology (now known as, Maulana Abul Kamal Azad University of Technology) in the year 2010. She completed her Master's Degree in Industrial Engineering and Management from West Bengal University of Technology, Kolkata in the year 2012 and was awarded Gold Medal as First Rank Holder in the said discipline from the University. She is the recipient of INSPIRE Fellowship by Department of Science & Technology (DST), Ministry of Science & Technology, India.

Supriyo Roy received Associate Membership in Mechanical Engineering from The Institution of Engineers, India, and Master of Mechanical Engineering from Jadavpur University, Kolkata, India. He obtained his Ph.D. from Jadavpur University. Presently he is working as Assistant professor in Mechanical Engineering Department of Hooghly Engineering & Technology College, Hooghly, India. His research domain is Tribology of Electroless Nickel Coatings. He has so far to his credit 15 publications in international journals and conference proceedings and also co-author of two book chapters.

Diana Sánchez-Partida is a Professor and Researcher of the Postgraduate in Logistics and Supply Chain Management at Universidad Popular Autonoma del Estado de Puebla (UPAEP) in Mexico. Since 2018 she has been a member of the National Council of Researchers (SNI 1) in Mexico. Her area of interest is in Logistics, Design and Optimization of Supply Chain. She has participated in applied projects like Humanitarian Logistics, timetabling process, optimization process in warehouses, levels and control of inventories, and Planning and Control of Production.

Michael Sony is working at the Faculty of Engineering, Department of Mechanical and Marine Engineering, Namibia. He holds a PhD and MEng degree. His research interest is Production & Operations Management, Industry 4.0, Discrete Event Simulations and Markov Models.

K. D. S. Sravani is a graduate student of Mechanical Engineering, A.U.College of Engineering for Women, Andhra University, India. She published two conference papers. Her areas of Research interest are Product development and Optimization.

Corneliu Octavian Turcu received the B.Sc. and Ph.D. degrees in automatic systems, from the University of Iasi, Romania, in 1991, and 1999, respectively. He also holds a degree in Informatics (M.Sc.) from the University of Suceava, Romania. Since 1991, he has been with the Faculty of Electrical Engineering and Computer Science, University of Suceava, where he is a full professor of System Theory and Intelligent Systems and also holds a joint appointment as head of Programmes Management Department. At University of Suceava he is also a supervisor for Ph.D. and M.S. theses. He has published over 90 research papers and 4 books. His research interests include intelligent systems, RFID, IoT and automatic control system design.

Cristina Elena Turcu is a Professor and currently vice-dean at the Faculty of Electrical Engineering and Computer Science, Stefan cel Mare University of Suceava, Romania. During the period 2004-2011, she was Head of Computer Department. Her research interests include software engineering, intelligent systems and Internet/Web of Things. She was an Editor for 6 books. She has served on various program committees of conferences in computing and Internet of Things. She also has served as a reviewer for numerous referred journals and conferences. She is an Associate Editor of the International Journal of Engineering Business Management. She has published over 80 publications in books or book chapters, refereed journals, technical reports, and refereed conference/workshop/seminar proceedings.

Index

L

M

O

P

Q

R

S

T

W

X

www.ingramcontent.com/pod-product-compliance
Lightning Source LLC
LaVergne TN
LVHW062118300926
849161LV00010B/2486
* 9 7 8 1 5 2 2 5 8 2 2 3 6 *